PLANTS THAT FIGHT CANCER

Also available from CRC Press:

Bioactive Compounds from Natural Sources
Edited by Corrado Tringali
ISBN 0-7484-0890-8 (hardback)

Biochemical Targets of Plant Bioactive Compounds
A pharmacological reference guide to sites of action and biological effects
Gideon Polya
ISBN 0-415-30829-I (hardback)

Phytochemical Dictionary (Second Edition)
A handbook of bioactive compounds from plants
Edited by Jefery B. Harborne, Herbert Baxter and Gerard Moss
ISBN 0-7484-0620-4 (hardback)

Taxus
The genus *Taxus*
Edited by Hideji Itokawa and Kuo-Hsiung Lee
ISBN 0-415-29837-7 (hardback)

Mistletoe
The genus *Viscum*
Edited by Arndt Bussing
ISBN 90-5823-092-9 (hardback)

Hypericum
The genus *Hypericum*
Edited by Edzard Ernst
ISBN 0-415-36954-I (hardback)

Oregano
The genera *Origanum* and *Lippia*
Edited by Spiridon E. Kintzios
ISBN 0-415-36943-6 (hardback)

Sage
The genius *Salvia*
Edited by Spiridon E. Kintzios
ISBN 90-5823-005-8 (hardback)

Sho-Saiko-To
Scientific evaluation and clinical applications
Edited by Yukio Ogihora and Mosaki Aburada
ISBN 0-415-30837-2 (hardback)

PLANTS THAT FIGHT CANCER

Edited by Spiridon E. Kintzios
and Maria G. Barberaki

CRC PRESS

Boca Raton London New York Washington, D.C.

Library of Congress Cataloging-in-Publication Data

Plants that fight cancer / edited by Spiridon E. Kintzios and Maria G. Barberaki.
p. ; cm.
Includes bibliographical references and index.
ISBN 0-415-29853-9 (hardback: alk. paper)
1. Herbs—Therapeutic use. 2. Cancer—Treatment. 3. Medicinal plants. 4. Materia medica, Vegetable. 5. Pharmacognosy.
[DNLM: I. Neoplasms—drug therapy. 2. Phytotherapy. 3. Plant Extracts—therapeutic use. QZ 267 P714 2003] I. Kintzios, Spiridon E. II. Barberaki, Maria G.

RC271.H47 P56 2003
616.99'4061—dc21 2003005700

International Standard Book Number 0-415-29853-9
Library of Congress Card Number 2003005700
Printed in the United States of America 1 2 3 4 5 6 7 8 9 0
Printed on acid-free paper

Eleni Grafakou, Vaggelis Kintzios and all people giving their personal fight against cancer

Contents

Contributors

Spiridon E. Kintzios
Agricultural University of Athens
Faculty of Agricultural Biotechnology
Laboratory of Plant Physiology
Iera Odos 75, Athens, Greece
Tel. +3-010-5294292
E-mail: skin@aua.gr

Maria G. Barberaki
Agricultural University of Athens
Faculty of Agricultural Biotechnology
Laboratory of Plant Physiology
Iera Odos 75, Athens, Greece
Tel. +3-010-5294292
E-mail: maria.barberaki@serono.com

Olga G. Makri
Agricultural University of Athens
Faculty of Agricultural Biotechnology
Laboratory of Plant Physiology
Iera Odos 75, Athens, Greece
Tel. +3-010-5294292
E-mail: olmak@hotmail.com

Theodoros Matakiadis
Agricultural University of Athens
Faculty of Agricultural Biotechnology
Laboratory of Plant Physiology
Iera Odos 75, Athens, Greece
Tel. +3-010-5294292
E-mail: teomatas@hotmail.com

Vassilios Roussis
University of Athens
School of Pharmacy
Department of Pharmacognosy and
Chemistry of Natural Products
Panepistimiopolis Zografou
Athens 157 71, Greece
Tel. +3-010-7274592
E-mail: roussis@pharm.uoa.gr

Costas Vagias
University of Athens
School of Pharmacy
Department of Pharmacognosy and
Chemistry of Natural Products
Panepistimiopolis Zografou
Athens 157 71, Greece

Leto – A. Tziveleka
University of Athens
School of Pharmacy
Department of Pharmacognosy and
Chemistry of Natural Products
Panepistimiopolis Zografou
Athens 157 71, Greece

Preface

This is a book about the most fearsome disease of modern times, which will strike every fourth citizen of a developed country sometime during his life: cancer. It is not a book about the prevention of cancer, but rather its treatment with plant-derived chemicals. It is an up-to-date and extensive review of plant genera and species with antitumor and antileukemic properties that have been documented in a strictly scientific sense. From the layman to the medical expert, the book is addressed to people seeking information on novel opportunities on disease therapy in order to make decisions about care programs. Purpose-wise, the book is written in colloquial style.

The volume comprises four chapters. In the first chapter, the current knowledge of the nature of cancer and the main types of the disease are briefly described. In the second chapter, the various approaches for treating cancer – including conventional, advanced and alternative methods – are presented, while a relative emphasis is given on the chemotherapy of cancer. The restrictions and risks of each approach are comparatively reviewed. The second chapter of the book is a general review of plant-derived groups of compounds with anticancer properties, including their chemistry, biosynthesis and mode of action. Evolutionary aspects of the anticancer properties of plants are presented and a separate chapter is devoted on the application of biotechnology in this field. The third, most extensive chapter of the book contains detailed information on each of more than 150 anticancer terrestrial plant genera and species. Topics include tradition and myth, distribution, botany, culture, active ingredients and product application (including an analysis of expected results and risks) along with photographs and illustrations of each species. In addition, further information can be found on plant species with equivocal or minor anticancer value. Although the traditional sources of secondary metabolites were terrestrial higher plants, animals and microorganisms, marine organisms have been the major targets for natural products research in the past decade. In the fourth chapter of the book, algal extracts and isolated metabolites having cytotoxic and antineoplastic activity and with the potential for pharmaceutical exploitation are reviewed, along with the phylogeny and physiology of the organisms. Emphasis is given to the chemical nature of these compounds, the novelty and complexity of which has no counterpart in the terrestrial world.

The chemical structures of the most important compounds derived from terrestrial higher plants are given in the Appendix of the book. An extensive list of publications provides an overview of published research for each species, to be used as extensive background information for the expert reader.

Finally, we feel compelled to state that this volume, as concise as it is, cannot include all existing plant species with anticancer properties; even during the stage of the final editing of the manuscript, many novel substances from other species have been identified as potential chemotherapeutic agents against various tumors. This fact is an evidence in itself for the rapidly growing interest of the international scientific and medical community in the utilization of plant-derived chemicals in cancer treatment.

The Editors
Athens, 2002

Chapter 1

What do we know about cancer and its therapy?

Spiridon E. Kintzios

1. A BRIEF OVERVIEW OF THE DISEASE AND ITS TREATMENT

1.1. Incidence and causes

Everybody thinks it cannot happen to them. And yet, six million people die of cancer every year. Approximately every fourth citizen of a developed country will be stricken sometime during his life and approximately 400 new incidents emerge per 100,000 people annually.

Once considered a mysterious disease, cancer has been eventually revealed to investigators (Trichopoulos and Hunter, 1996). Disease development begins from a genetic alteration (**mutation**) of a cell within a tissue. This mutation allows the cell to proliferate at a very high rate and to finally form a group of fast reproducing cells with an otherwise normal appearance (**hyperplasia**). Rarely, some hyperplastic cells will mutate again and produce abnormally looking descendants (**dysplasia**). Further mutations of dysplastic cells will eventually lead to the formation of a **tumor**, which can either remain localized at its place of origin, or invade neighboring tissues (**malignant tumor**) and establish new tumors (**metastases**).

Cancer cells have some unique properties that help them compete successfully against normal cells:

1 Under appropriate conditions cancer cells are capable of dividing almost infinitely. Normal cells have a limited life span. As an example, human epithelial cells cultured *in vitro* are commonly capable of sustaining division for no more than 50 times (the so-called **Hayflick number**) (Hayflick and Hayflick, 1961).

2 Normal cells adhere both to one another and to the **extracellular matrix**, the insoluble protein mesh that fills the space between cells. Cancer cells fail to adhere and, in addition, they possess the ability to migrate from the site where they began, invading nearby tissues and forming masses at distant sites in the body, via the bloodstream. This process is known as **metastasis** and examples include melanoma cells migrating to the lung, colorectal cancer cells to the liver and prostrate cancer cells to bone. Although metastatic cells are indeed a small percentage of the total of cancer cells (e.g. 10^{-4} or 0.0001%), tumors composed of such malignant cells become more and more aggressive over time.

In a general sense, cancer arises due to specific effects of environmental factors (such as smoking or diet) on a certain genetic background. In the hormonally related cancers like breast and prostate cancer, genetics seem to be a much more powerful factor than lifestyle.

Two gene classes play major roles in triggering cancer. **Proto-oncogenes** *encourage* such growth, whereas **tumor suppressor** genes *inhibit* it. The coordinated action of these two gene classes normally prevents cells from uncontrolled proliferation; however, when mutated, oncogenes promote excessive cell division, while inactivated tumor suppressor genes fail to block the division mechanism (Table 1.1). On a molecular level, control of cell division is maintained by the inhibitory action of various molecules, such as pRB, p15, p16, p21 and p53 on proteins promoting cell division, essentially the complex between cyclins and cyclin-dependent kinases (CDKs) (Meijer *et al.*, 1997). Under normal conditions, deregulation of the cell control mechanism leads to cellular suicide, the so-called **apoptosis** or **programmed cell death**. Cell death may also result from the gradual shortening of **telomeres**, the DNA segments at the ends of chromosomes. However, most tumor cells manage to preserve telomere length due to the presence of the enzyme telomerase, which is absent in normal cells.

Some oncogenes force cells to overproduce growth factors, such as the **platelet-derived growth factor** and the **transforming growth factor alpha** (sarcomas and gliomas). Alternatively, oncogenes such as the *ras* genes distort parts of the signal cascade within the cell (carcinoma of the colon, pancreas and lung) or alter the activity of transcription factors in the nucleus. In addition, suppressor factors may be disabled upon infection with viruses (e.g. a human papillomavirus). Tumor development is a **step-wise process** in that it requires an accumulation of mutations in a

Table 1.1 Examples of genes related to cancer incidence in humans

Type of gene	*Gene*	*Cancer type*
Oncogene	*PDGF*	Glioma
Oncogene	*Erb-B*	Glioblastoma, breast
Oncogene	*RET*	Thyroid
Oncogene	*Ki-ras*	Lung, ovarian, colon, pancreatic
Oncogene	*CDKN2*	Melanoma
Oncogene	*HPC1*	Prostate
Oncogene	*N-ras*	Leukemia
Oncogene	*c-myc*	Leukemia, breast, stomach, lung
Oncogene	*N-myc*	Neuroblastoma, glioblastoma
Oncogene	*Bcl-1*	Breast, head, neck
Oncogene	*MDM2*	Sarcomas
Oncogene	*BCR-ABL*	Leukemia
Tumor suppressor gene	*p53*	Various
Tumor suppressor gene	*RB*	Retinoblastoma, bone, bladder, small cell lung, breast
Tumor suppressor gene	*BRCA1*	Breast, ovarian
Tumor suppressor gene	*BRCA2*	Breast
Tumor suppressor gene	*APC*	Colon, stomach
Tumor suppressor gene	*MSH2, MSH6, MLH1*	Colon
Tumor suppressor gene	*DPC4*	Pancreas
Tumor suppressor gene	*CDK4*	Skin
Tumor suppressor gene	*VHL*	Kidney
Other	*Chromosome 3 (deletions)*	Lung

number of these genes. Altered forms of other classes of genes may also participate in the creation of a malignancy, particularly in enabling the emergence of metastatic cancer forms.

Environmental causes of cancer comprise an extremely diverse group of factors that may act as **carcinogens**, either by mutating genes or by promoting abnormal cell proliferation (Nagao *et al.*, 1985; Sugimura, 1986; Koehnlechner, 1987; Wakabayashi *et al.*, 1987; Greenwald, 1996). Most of these agents have been identified through epidemiological studies, although the exact nature of their activity on a biological level remains obscure. These factors include chemical substances (such as tobacco, asbestos, industrial waste and pesticides), diet (saturated fat, read meat, overweight), ionizing radiation, pathogens (such as the Epstein–Barr virus, the hepatitis B or C virus, papillomaviruses and *Helicobacter pylori*). However, in order for environmental factors to have a significant effect, one must be exposed to them for a relatively long time.

Cancer may also arise, or worsen, as a result of physiological stress. For example, a recent large-scale study in Israel demonstrated that survival rates declined for patients having lost at least one child in war (Anonymous, 2000).

1.2. Classification of cancer types

There are several ways to classify cancer. A general classification relates to the tissue type where a tumor emerges. For example, **sarcomas** are cancers of connective tissues, **gliomas** are cancers of the nonneuronal brain cells and **carcinomas** (the most common cancer forms) originate in epithelial cells. In the following box, a classification of major cancer diseases is given according to the currently estimated five-year survival rate of the affected patient.

Cancers with less than 20% five-year survival rate (at all stages)

1 *Lung cancer* is associated with exposure to environmental toxins like cigarette smoke and various chemicals and has an incidence higher than 17%. It can be distinguished in two types, small cell (rapidly spreading) and non-small cell disease. With a percentage of terminally affected patients more than 26%, it is one of the less curable cancer diseases.
2 *Pancreatic cancer* is associated with increasing age, smoking, consumption of fats, race and pancreatic diseases. Diagnosis usually lags behind metastasis.

Cancers with five-year survival rates (at all stages) between 40% and 60%

1 *Non-Hodgkin's lymphoma* is associated with dysfunctions of the immune system, including many different types of disease.
2 *Kidney cancer* is associated with sex (males), smoking and obesity.
3 *Ovarian cancer* is associated with increasing age and heredity, especially as far as mutations in the *BRCA1* or *BRCA2* genes are concerned.

Cancers with five-year survival rates (at all stages) between 60% and 80%

1 *Uterine (cervical and endometrial cancer) cancer* is associated with hormonal treatment (such as estrogen replacement therapy), race, sexual activity and pregnancy history. Can be efficiently predicted by the **Pap** test (named after its inventor, the physician G. *Pap*anikolaou).

2 *Leukemia* is distinguished in acute lymphocytic (common among children), acute myelogenous and chronic lymphocytic leukemia. The disease is associated with genetic abnormalities, viral infections and exposure to environmental toxins or radiation.
3 *Colorectal cancer* is associated with heredity, obesity, polyps and infections of the gastrointestinal tract. Prevention of metastases in the liver is crucial. The disease is presumably associated with elevated concentrations of the **cancer embryonic antigen** (CEA).
4 *Bladder cancer* is associated with race, smoking and exposure to environmental toxins.

Cancers with five-year survival rates (at all stages) higher than 80%

1 *Prostate cancer* is associated with increasing age, obesity and race. The incidence of the disease is high (>15%). The disease can be efficiently detected at an early stage by using the prostate-specific antigen (PSA) blood test.
2 *Breast cancer* is associated with increasing age, heredity (especially as far as mutations in the *BRCA1* or *BRCA2* genes are concerned), sexual activity, obesity and pregnancy history. Although the incidence of the disease is high (>24%), survival rates have been remarkably increased. The disease can be efficiently detected at an early stage by self-examination and mammography. In addition, the disease is presumably associated with elevated concentrations of CA15-3.
3 *Skin cancer (basal cell skin cancer, squamous cell skin cancer, melanoma)* is mainly associated with prolonged exposure to the sun and race. Detection at an early stage is extremely crucial.

Most cancers are currently increasing in incidence. However, growth in the major pharmacologically treated cancers, namely breast, colorectal, lung, ovarian and prostate cancer is driven by shifting demographics rather than any underlying increase in the risk of developing the disease. Breast cancer is the most prevalent cancer today, followed by cancer of the prostate, colon/rectum, lung and ovaries respectively. Unsurprisingly, given that cancer is a disease driven by imperfections in DNA replication, the risk of developing most cancers increases with increasing age. For some hormonally driven female cancers, the risk of developing the disease increases rapidly around the time of the menopause.

Diagnosis rates are consequently very high, at over 95% of the prevalent population diagnosed for prostate cancer, and over 99% for breast, colorectal, lung and ovarian cancers (Sidranski, 1996). The stage of the patient's cancer at diagnosis varies highly with each individual cancer with survival times associated with the disease falling rapidly with increasing stage of diagnosis.

1.3. Therapy

1.3.1. Conventional cancer treatments

Conventional cancer treatments include surgery, radiation and chemotherapy.

Surgery is used for the excision of a tumor. It is the earliest therapy established for cancer and the most widely used. Its disadvantages include the possible (and often unavoidable) damage of healthy tissues or organs (such as lymph nodes) and the inability to remove metastasized cancer cells or tumors not visible to the surgeons. In addition, surgery can activate further proliferation of "latent" small tumors, the so-called "pet-cancers" (Koehnlechner, 1987).

Radiation (X-rays, gamma rays) of a cancerous tumor, thus causing cancer cell death or apoptosis preserves the anatomical structures surrounding the tumor and also destroy non-visible cancer cells. However, they cannot kill metastasized cancer cells. Radiation treatment presents some side effects (such as neurotoxicity in children), but patients usually recover faster than from surgery. Additional side effects include weakening of the immune system and replacement of damaged healthy tissue by connecting tissue (Koehnlechner, 1987).

Chemotherapy is based on the systemic administration of anticancer drugs that travel throughout the body via the blood circulatory system. In essence, chemotherapy aims to wipe out all cancerous colonies within the patients body, including metastasized cancer cells. However, the majority of the most common cancers are not curable with chemotherapy alone. This kind of treatment also has many side effects, such as nausea, anemia, weakening of the immune system, diarrhea, vomiting and hair loss. Finally, cancer cells may develop resistance to chemotherapeutic drugs (Koehnlechner, 1987; Barbounaki-Konstantakou, 1989).

Drugs in adjunct therapy do not attack the tumor directly, but instead treat side effects and tolerance problems associated with the use of chemotherapy. For example, anti-emetics such as *ondansetron* or *granisetron* reduce levels of nausea associated with some chemotherapies. This improves compliance rates, and enables patients to tolerate higher doses of chemotherapy than would normally be the case. Similarly, some drugs such as *epoetin alpha* target deficiencies in red blood cell counts that often result from the use of chemotherapy and enable normal physical function to be restored to some degree.

Many different compounds are currently used (often in combination). Chemotherapy is the most rapidly developing field of cancer treatment, with new drugs being constantly tested and screened. These include also plant metabolites (the topic of this book) and regulators of the endocrine system (important in cases of hormone-dependent cancers, like breast and prostate cancer). Chemotherapeutic drugs are classified in ten general groups:

1 **Antimetabolites** act as nonfunctional analogues of essential metabolites in the cell, thus blocking physiological functions of the tumor.
2 **Alkylating agents** chemically bond with DNA through alkyl groups, thus disrupting gene structure and function, or with proteins, thus inhibiting enzymes.
3 **Topoisomerase inhibitors** inhibit DNA replication in rapidly dividing cells, as in the case of tumors.
4 **Plant alkaloids** also inhibit tumor cell division by blocking microtubule depolymerization, an essential step for chromosome detachment during mitosis. However, novel plant alkaloids act through other mechanisms as well, which will be analyzed further in this book.
5 **Antibiotics** are derived from diverse groups of microorganisms or synthesized and block DNA replication and protein synthesis.
6 **Anthracyclins** are a subgroup of antibiotics, associated with considerable toxic side effects on the heart and bone marrow.
7 **Enzymes**, in particular proteolytic and fibrinolytic ones, as well as tyrosinase inhibitors, such as *Gleevec*, a new cytotoxic drug used for treating chronic myeloid leukemia.
8 **Hormones** are substances interfering with other chemotherapeutic agents by regulating the endocrine system. They find specific application against carcinomas of breast, prostate and endometrium.

9 **Immunomodulators** act by inhibiting tumor proliferation through the stimulation of the host's immune system (see section on immunotherapy).
10 Various substances not falling in any of the above categories.

Some representative chemotherapeutic agents are listed in Table 1.2.

The success of chemotherapy depends on the type of cancer that is being treated. It can have curative effects on some less common cancers, like Burkitt-Lymphoma, Wilms-Tumor, teratomas and lymphoblastic leukemia. A less satisfactory, though life-prolonging effect is observed on myloblastic leukemia, multiple myeloma, ovarian, prostate, and cervical and breast cancer. Much poorer results must be expected against bronchial, lung, stomach, colorectal, pancreatic, kidney, bladder, brain, glandular and skin cancer, as well as against bone sarcomas.

Use of pharmacological therapy for cancer vary by both geographic area and tumor type. Lung cancer patients are most likely to be treated with drugs, with around 99% of them being treated with drugs at the first-line treatment stage. Prostate cancer patients are least likely to be treated with drugs, with only around 42% of them being treated with drugs at the first-line treatment stage.

For those cancers which manifest themselves as a solid tumor mass, the most efficient way to treat them is to surgically resect or remove the tumor mass, since this reduces both the tumor's ability to grow and metastasize to distant sites around the body. If a tumor can be wholly resected, there are theoretically no real advantages in administering drug treatment, since surgery has essentially removed the tumor's ability to grow and spread. For early stage I and II tumors, which are usually golf ball sized and wholly resectable, drug therapy is therefore infrequently used. At stages III and IV, the tumor has usually grown to such a size and/or has spread around the body to such an extent that it is not wholly resectable. For example, rectal tumors at stage III have usually impinged upon the pelvis, which reduces the ability of the surgeon to wholly remove the tumor. In these cases, drug therapy is used either to reduce the size of the tumor before resection, or else "mop up" stray cancer cells. Drug therapy therefore features prominently for tumors diagnosed at stage III and IV, together with those cancers that have recurred following initial first-line treatment and/or metastasized to distant areas around the body.

1.3.2. Advanced cancer treatments

Immunotherapy

Infectious agents entering the body are encountered by the immune system. They bear distinct molecules called **antigens**, which are the target of **antigen-presenting cells**, such as macrophages, that roam the body and fragment antigens into antigenic peptides. These, in turn, are joined to the **major histocompatibility complex** (MHC) molecules which are displayed on the cell surface. Macrophages bearing different MHC-peptide combinations activate specific **T-lymphocytes**, which divide and secrete **lymphokines**. Lymphokines activate **B-lymphocytes**, which can also recognize free-floating antigens in a molecule-specific manner. Activated B-cells divide and secrete **antibodies**, which can bind to antigens and neutralize them in various ways (Nossal, 1993).

Lymphocytes are produced in *primary lymphoid organs*: the thymous (T cells) and the bone marrow (B cells). They are further processed in the *secondary lymphoid organs*, such as the lymph nodes, spleen and tonsils before entering the bloodstream.

In an ideal situation, cancer cells would constitute a target of the patient host immune system. To single out cancer cells, an immunotherapy must be able to distinguish them from normal cells. During the last years, **monoclonal antibodies** have revealed a large array of antigens that exist

Table 1.2 Some of the compounds currently used in cancer chemotherapy

Class	*Compound*
Antimetabolites	Azathioprin Cytosine arabinoside 5-fluorouracile 6-mercaptopurine 6-thioguanine Methotrexate Hydroxyurea
Alkylating agents	Busulfan Chlorambucile Cyclophosphamide Ifosfamide Melphalan hydrochloride Thiotepa Mechlorethamine hydrochloride *Nitrosoureas*: Lomustine Carmustine Streptozocin
Topoisomerase inhibitors	Amsacrine
Plant alkaloids	Etoposide Teniposide Vinblastine Vincristine Vindesine
Antibiotics	Bleomycin Plicamycin Mitomycin Dactinomycin
Anthracyclines	Daunorubicin Doxorubicin hydrochloride Rubidazone Idarubicine Epirubicin (investigational drug) Aclarubicin chlorhydrate
Enzymes	L-aspariginase Tyrosine kinase inhibitors
Hormones	Adrenocorticoids Estrogens Anti-androgens Luteinizing hormone release hormone (LHRH) analogues Progestogens Antiestrogens (investigational) Aromatase inhibitors
Immunomodulators	Interferons Interleukins
Various	Cisplatin Dacarbazine Procarbazine Mitoxantrone

on human cancer cells. Many of them are related to abnormal proteins resulting from genetic mutations which turn normal cells into cancer ones. However, cancer cells can elude attack by lymphocytes even if they bear distinctive antigens, due to the absence of proper co-stimulatory molecules, such as B7 or the employment of immunosuppression mechanisms. The ultimate goal of cancer immunotherapy research is the production of **an effective vaccine**. This may include whole cancer cells, tumor peptides or DNA molecules, other proteins or viruses (Koehnlechner, 1987; Old, 1996). The idea of a vaccine is an old one, indeed. In 1892, William B. Coley at the Memorial Hospital in New York treated cancer patients with killed bacteria in order to elicit a tumor-killing immunoresponse.

The immunotherapy of cancer can be roughly classified in four categories:

1 *Non-specific:* involves the general stimulation of the immune system and the production of **cytokines**, such as interferons, tumor necrosis factor (TNF), interleukins (IL-2, IL-12) and GM-CSF.
2 *Passive:* involves the use of "humanized" mice-derived monoclonal antibodies bearing a toxic agent (such as a radioactive isotope or a chemotherapeutic drug).
3 *Active:* vaccines are made on the basis of human antitumor antibodies.
4 *Adoptive:* involves lymphocytes from the patient himself.

Table 1.3 Substances that stimulate the immune system

Substances	*They activate*
Bordetella pertussis Bacillus–Calmette–Guerin (BCG) (tuberculosis bacterium a.d. Rind.)[1] *Escherichia coli* Vitamin A	Macrophages
Corynobacterium parvum[2] *C. granulosum* *Bordetella pertussi* *Escherichia coli* Vitamin A[3]	B-lymphocytes
Bordetella pertussis BCG (tuberculosis bacterium a.d. Rind.)[1] *Escherichia coli* Vitamin A[3] Poly-adenosin-poly-urakil Saponine Levamisol[4] Lentinan Diptheriotoxin Thymus factors	T-lymphocytes

Notes
1 In combination with radiotherapy can cause a 40% reduction of leukemia incidence in mice. Has been reported to prolong life expectancy in cancer and leukemia patients who received conventional treatment.
2 Has been used for the treatment of melanomas, lung and breast cancer.
3 Has been used for the treatment of various skin cancers.
4 A former anti-worm veterinarian drug, levamisol has displayed slight post-operative immunostimulatory and survival-increasing properties in patients suffering from bronchial, lung and intestinal cancer.

Apart from plant-derived compounds, several other agents can stimulate the immune system in a more or less antitumor specific manner. Some of the most prominent substances and/or organisms are presented in Table 1.3 (adapted from Koehnlechner, 1987). Other compounds include trace elements (selenium, zinc, lithium), hemocyanin, propionibacteria.

Angiogenesis inhibition

A promising therapeutic strategy focuses on blocking tumor **angiogenesis**, that is, the inhibition of the growth of new blood vessels in tumors. Such drugs have not only performed impressively in experimental animal models but also offer an alternative means of tackling multidrug-resistant tumors that have proved intractable to conventional chemotherapy. The link between angiogenesis and tumor progression was first established by Judah Folkman of Boston Children's Hospital (Folkman, 1996; Brower, 1999). His observations led to the notion of an "angiogenic switch", a complex process by which a tumor mass expands and overtakes the rate of internal apoptosis by developing blood vessels, thereby changing into an angiogenic phenotype. Drugs that target blood vessel growth should have minimal side effects, even after prolonged treatments. The ready accessibility of the vasculature to drugs and the reliance of potentially hundreds of tumor cells on one capillary add to the benefits of such therapies, which however are limited to the subfraction of tumor capillaries expressing the immature angiogenic phenotype. Another problem is the heterogeneity of the vasculature within tumors. Many approaches for inhibiting angiogenesis are still very early in development, approximately 30 antiangiogenic drugs are in clinical trial. Among them, endogenous angiogenic inhibitors such as *angiostatin, troponin-I* and *endostatin* are in Phase I trials, while synthetic inhibitors, such as TNP-470, various proteolysis inhibitors and signaling antagonists are in Phase II and III trials. At this point it is worth mentioning that the angiogenesis-inhibitor *squalamine* is based on dogfish shark liver. Shark cartilage has been sold as an alternative treatment for cancer since the early 1990s when a book entitled "Sharks Don't Get Cancer" by William Lance was published. It suggested that a protein in shark's cartilage kept the fish from getting cancer by blocking the development of small blood vessels that cancer cells need to survive and grow. The idea spawned a market for shark cartilage supplements that is estimated to be worth $50 million a year. Researchers have since discovered that sharks do get cancer but they have a lower rate of the disease than other fish and humans. Danish researchers tested the treatment on 17 women with advanced breast cancer that had not responded to other treatments. The patients took 24 shark cartilage capsules a day for three months, but the disease still progressed in 15 and one developed cancer of the brain. The Danish results support earlier research that found powdered shark cartilage did not prevent tumor growth in 60 patients with an advanced cancer.

1.3.3. Other advanced therapies

Advanced cancer therapies also include the use of tissue-specific cytotoxic agents. For example, novel mutagenic cytotoxins (*interleukin* 13 – IL13) have been developed against brain tumors, which do not interact with receptors of the normal tissue but only with brain gliomas (Beljanski and Beljanski, 1982; Beljanski *et al.*, 1993).

1.3.4. Alternative cancer treatments

These include diverse, mostly controversial methods for treating cancer while avoiding the debilitating effects of conventional methods. The alternative treatment of cancer will probably gain in significance in the future, since it has been estimated that roughly half of all cancer patients currently turn to alternative medicine. The most prominent alternative cancer treatments include:

1 The delivery of **antineoplastons**, peptides considered to inhibit tumor growth and first identified by Stanislaw Burzynski in blood and urine. According to the Food and Drug Administration (FDA) the drug can be applied only in experimental trials monitored by the agency and only on patients who have exhausted conventional therapies. However, the therapy has found a significant amount of political support, while attracting wide publicity (Keiser, 2000).
2 *Hydrazine sulfate*, a compound reversing cachexia of cancer patients, thus improving survival.
3 Various **herbal extracts**, some of which are dealt with in this book.

1.4. From source to patient: testing the efficiency of a candidate anticancer drug

Drug development is a very expensive and risky business. On average, a new drug takes 15 years from discovery to reach the market, costing some $802 m. Considerable efforts have been made by public organizations and private companies to expedite the processes of drug discovery and development, by expanding on promising results from preliminary *in vitro* screening tests. The United States National Cancer Institute (NCI) has set forward exemplary strategies for the discovery and development of novel natural anticancer agents. Over the past 40 years, the NCI has been involved with the preclinical and/or clinical evaluation of the overwhelming majority of compounds under consideration for the treatment of cancer. During this period, more than 4,00,000 chemicals, both synthetic and natural, have been screened for antitumor activity (Dimitriou, 2001).

Plant materials under consideration for efficacy testing are usually composed of complex mixtures of different compounds with different solubility in aqueous culture media. Furthermore, inert additives may also be included. These properties render it necessary to search for appropriate testing conditions. In the past, model systems with either high complexity (animals, organ cultures) or low molecular organization (subcellular fractions, organ and cell homogenates) were used for evaluating the mechanism of action of phytopharmaceuticals. The last decade, however, has seen an enormous trend towards isolated cellular systems, primary cells in cultures and cell lines (Gebhardt, 2000). In particular, the combination of different *in vitro* assay systems may not only enhance the capacity to screen for active compounds, but may also lead to better conclusions about possible mechanisms and therapeutic effects.

1.4.1. Preclinical tests

Preclinical tests usually comprise evaluating the cytotoxicity of a candidate antitumor agent *in vitro*, that is, on cells cultured on a specific nutrient medium under controlled conditions. Certain neoplastic animal cell lines have been repeatedly used for this purpose. Alternatively, animal systems bearing certain types of cancer have been used. For example, materials entering the NCI drug discovery program from 1960 to 1982 were first tested using the L1210 and P-388 mouse leukemia models. Most of the drugs discovered during that period, and currently available for cancer therapy, are effective predominantly against rapidly proliferating tumors, such as leukemias and lymphomas, but with some notable exceptions such as paclitaxel, show little useful activity against the slow-growing adult solid tumors, such as lung, colon, prostatic, pancreatic and brain tumors.

A more efficient, disease-oriented screening strategy should employ multiple disease-specific (e.g. tumor-type specific) models and should permit the detection of either broad-spectrum or disease-specific activity. The use of multiple *in vivo* animal models for such a screen is not practical, given the scope of requirements for adequate screening capacity and specific tumor-type representation. The availability of a wide variety of human tumor cell lines representing many different forms of human cancer, however, offered a suitable basis for development of a disease-oriented *in vitro* primary screen during 1985 to 1990. The screen developed by NCI currently comprises 60 cell lines derived from nine cancer types, and organized into subpanels representing leukemia, lung, colon, central nervous system, melanoma, ovarian, renal, prostate and breast cancer. A protein-staining procedure using sulforhodamine B (SRB) is used as the method of choice for determining cellular growth and viability in the screen. Other, more sophisticated methods are referred to in the literature. In addition, cell lines used in the *in vitro* screen can be analyzed for their content of molecular targets, such as p-glycoprotein, p53, Ras and BCL2. Each successful test of a compound in the full screen generates 60 dose–response curves, which are printed in the NCI screening data report as a series of composites comprising the tumor-type subpanels, plus a composite comprising the entire panel. Data for any cell lines failing quality control criteria are eliminated from further analysis and are deleted from the screening report. The *in vitro* human cancer line screen has found widespread application in the classification of compounds according to their chemical structure and/or their mechanism of action. Valuable information can be obtained by determining the degree of similarity of profiles generated on the same or different compounds.

Some of the most commonly used animal and cell culture lines used for primary screening are listed in following: (for more detailed information on each method see Miyairi *et al.*, 1991; Mockel *et al.*, 1997; Gebhardt, 2000, and cited references in Chapter 3)

- Ehrlich Ascites tumor bearing mice
- P-388 lymphocytic leukemia bearing mice
- The 9KB carcinoma of the nasopharynx cell culture assay
- The human erythroleukemia K562 cell line
- The MOLT-4 leukemic cell line
- The RPMI, and TE671 tumor cells
- *ras*-expressing cells
- Alexander cell line (a human hepatocellular carcinoma cell line secreting HbsAg)
- The human larynx (HEp-2) and lung (PC-13) carcinoma cells
- The mouse B16 melanoma, leukemia P-388, and L5178Y cells
- The liver-metastatic variant (L5)
- 7,12-dimethyl benzanthracene (DMBA) induced rat mammary tumors
- Ehrlich ascites carcinoma (EAC), Dalton's lymphonia ascites (DLA) and Sarcoma-180 (S-180) cells
- MCA-induced soft tissue sarcomas in albino mice.

Sophisticated methods for determining cellular growth and viability in primary screens include:

- Suppression of 12-O-tetradecanoylphorbol-13-acetate (TPA)-stimulated ^{32}Pi-incorporation into phospholipids of cultured cells.
- Epstein–Barr virus activation.
- Suppression of the tumor-promoting activity induced by 7,12-dimethylbenz[a]anthracene (DMBA) plus TPA, (calmodulin involved systems).
- Production of TNF, possibly through stimulation of the reticuloendothelial system (RES).
- Stimulation of the uptake of tritiated thymidine into murine and human spleen cells.
- Inhibition of RNA, DNA and protein synthesis in tumoric cells.
- Analysis of endogenous cyclic GMP: cyclic GMP is thought to be involved in lymphocytic cell proliferation and leukemogenesis. In general, the nucleotide is elevated in leukemic vs. normal lymphocytes and changes have been reported to occur during remission and relapse of this disease.
- Determination of DNA damage in Ehrlich ascites tumor cells by the use of an alkaline DNA unwinding method, followed by hydroxylapatite column chromatography of degraded DNA.
- The brine shrimp lethality assay for activity-directed fractionation.
- Suppression of the activities of thymidylate synthetase and thymidine kinase involved in de novo and salvage pathways for pyrimidine nucleotide synthesis.
- Suppression of the induction of the colonic cancer in rats treated with a chemical carcinogen 1,2-dimethylhydrazine (DMH).
- Inhibition of Epstein–Barr virus early antigen (EBV-EA) activation induced by 12-*0*-tetradecanoylphorbol-13-acetate (TPA).
- Inhibition of calmodulin-dependent protein kinases(CaM kinase III). These enzymes phosphorylate certain substrates that have been implicated in regulating cellular proliferation, usually via phosphorylation of elongation factor 2. The activity of CaM kinase III is increased in glioma cells following exposure to mitogens and is diminished or absent in nonproliferating glial tissue.
- Inhibition of the promoting effect of 12-*0*-tetradecanoylphorbol-13-acetate on skin tumor formation in mice initiated with 7,12-dimethylbenz-[a]anthracene.
- Inhibition of two-stage initiation/promotion [dimethylbenz[a]anthracene (DMBA)/croton oil] skin carcinogenesis in mice.
- The MTT [3-(4,5-dimethylthiazol-2-yl)-2,5-diphenyl tetrazolium bromide] colorimetric assay.

Clinical trials are studies that evaluate the effectiveness of new interventions. There are different types of cancer clinical trials. They include:

- prevention trials designed to keep cancer from developing in people who have not previously had cancer;
- prevention trials designed to prevent a new type of cancer from developing in people who have had cancer;

- early detection trials to find cancer, especially in its early stages;
- treatment trials to test new therapies in people who have cancer;
- quality of life studies to improve comfort and quality of life for people who have cancer;
- studies to evaluate ways of modifying cancer-causing behaviors, such as tobacco use.

1.4.2. Phases of clinical trials

Most clinical research that involves the testing of a new drug progresses in an orderly series of steps (Dimitriou, 2001; NCI, 2001). This allows researchers to ask and answer questions in a way that expands information about the drug and its effects on people. Based on what has been learned in laboratory experiments or previous trials, researchers formulate hypotheses or questions that need to be answered. Then they carefully design a clinical trial to test the hypothesis and answer the research question. It is customary to separate different kinds of trials into phases that follow one another in an orderly sequence. Generally, a particular cancer clinical trial falls into one of three phases.

Phase I trials

These first studies in people evaluate how a new drug should be administered (orally, intravenously, by injection), how often, and in what dosage. A Phase I trial usually enrols only a small number of patients, as well as about 20 to 80 normal, healthy volunteers. The tests study a drug's safety profile, including the safe dosage range. The studies also determine how a drug is absorbed, distributed, metabolized and excreted, and the duration of its action. This phase lasts about a year.

Phase II trials

A Phase II trial provides preliminary information about how well the new drug works and generates more information about safety and benefit. Each Phase II study usually focuses on a particular type of cancer. Controlled studies of approximately 100 to 300 volunteer patients assess the drug's effectiveness and take about two years.

Phase III trials

These trials compare a promising new drug, combination of drugs, or procedure with the current standard. Phase III trials typically involve large numbers of people in doctors' offices, clinics, and cancer centers nationwide. This phase lasts about three years and usually involves 1,000 to 3,000 patients in clinics and hospitals. Physicians monitor patients closely to determine efficacy and identify adverse reactions.

Some use the term Phase IV to include the continuing evaluation that takes place after FDA approval, when the drug is already on the market and available for general use (post-marketing surveillance).

1.4.3. Clinical trial protocols

Clinical trials follow strict scientific guidelines. These guidelines deal with many areas, including the study's design, who can be in the study, and the kind of information people must be given when they are deciding whether to participate. Every trial has a chief investigator, who is usually a doctor. The investigator prepares a study action plan, called a protocol. This plan explains what the trial will do, how and why. For example, it states:

- How many people will be in the study.
- Who is eligible to participate in the study.
- What study drugs participants will take.
- What medical tests they will have and how often.
- What information will be gathered.

Chapter 2

Plants and cancer

Spiridon E. Kintzios and Maria G. Barberaki

2. THE PLANT KINGDOM: NATURE'S PHARMACY FOR CANCER TREATMENT

2.1. Brief overview of the general organization of the plant cell (see also Figure 2.1)

Although plant cells exhibit considerable diversity in their structure and function, their basic morphology is relatively unique. A typical plant cell consists of a *cell wall* (primary and secondary) surrounding a *protoplast*, which is delineated by the plasma membrane (or plasmalemma). The *protoplasm* (the protoplast without the plasmalemma) contains bodies bounded by membranes, known as organelles, as well as membrane structures, which do not enclose a body. The *cytoplasm* is the part of the protoplasm including various membrane structures, filaments and various particles, but not organelles. The *cytosol* is the aqueous phase of the cytoplasm, devoid of all particulate material. All membranes (including plasmalemma) chemically consist of a phospholipid bilayer carrying various proteins. Thanks to the existence of these internal compartments, specific functions can be executed in different parts or organelles of the plant cell (Anderson and Beardall, 1991). For example, the cell membrane permits the controlled entry and exit of compounds into and out of the cell while preventing excessive gain or loss of water and metabolic products. The *nucleus* is a large organelle containing chromatin, a complex of DNA and protein. It is the main center for the control of gene expression and replication. Chlorophyll-containing *chloroplasts* are the site for photosynthesis. *Mitochondria* contain enzymes important for the process of oxidative phosphorylation, that is, the phosphorylation of ADP to ATP with the parallel consumption of oxygen. *Vacuoles* are large organelles (usually only one vacuole is found in mature cells, representing up to 90% of the cell volume). They store water, salts, various organic metabolites, toxic substances or waste products and water-soluble pigments. Generally, the vacuole content (the cell sap) is considered to represent, together with the cytosol, the hydrophilic part of the plant cell. *Ribosomes* are small spheroid particles (attached to the cytoplasmic side of the *endoplasmic reticulum*, mitochondria and chloroplasts), which serve as sites for protein synthesis. *Golgi bodies* (or *dictyosomes*) consist of a stack of about five flattened sacs (cisternae) and are the sites for the synthesis of most of the matrix polysaccharides of cell walls, glycoproteins and some enzymes. *Microbodies* are small organelles containing various oxidases. Finally, *microtubules* are tubular inclusions within the cytoplasm, consisting of filamentous polymers of the protein tubulin, which can polymerize and depolymerize in a reversible manner. They direct the physical orientation of various components within the cytoplasm.

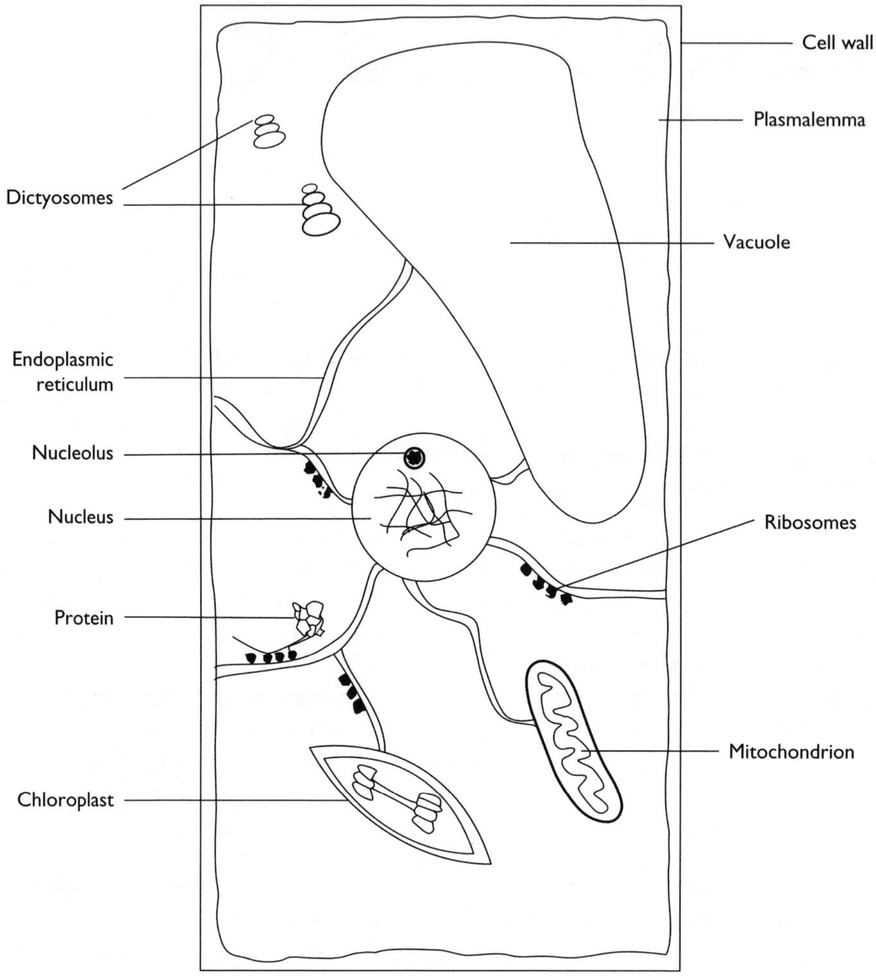

Figure 2.1 General outline of the structure of a plant cell.

2.2. The chemical constituents of the plant cell

Throughout human history, plants have been an indispensable source of natural products for medicine. The chemical constituents of the plant cell that exert biological activities on human and animal cells fall into two distinct groups, depending on their relative concentration in the plant body, as well as their major function: **primary metabolites**, the accumulation of which satisfies nutritional and structural needs, and **secondary metabolites**, which act as hormones, pharmaceuticals and toxins.

2.2.1. *Primary metabolites*

By definition, primary metabolism is the *total of processes* leading to the production of sugars (carbohydrates) (structural and nutritional elements), amino acids (structural elements and enzymes), lipids (constituents of membranes, nutritional elements) and nucleotides (constituents of genes). These account for about 90% of the biological matter and are required for the growth of plant cells (Payne *et al.*, 1991). These compounds occur principally as components of macromolecules, such as cellulose or amylose (from sugars), proteins (from amino acids) and nucleic acids, such as DNA (from nucleotides). Primary metabolites mainly contain carbon, nitrogen and phosphorous, which are assimilated into the plant cell by three main catabolic pathways: glycolysis, the pentose phosphate pathway and the tricarboxylic (TCA) cycle. Primary metabolism in plants is distinct from its animal counterpart, since it is a light-dependent process, known as *photosynthesis*. In other words, carbon assimilation in plant biological matter is mediated by chlorophyll and other photosynthetic pigments, which are found in chloroplasts of mesophyll cells.

2.2.2. Secondary metabolites

Secondary metabolites are compounds belonging to extremely varied chemical groups, such as organic acids, aromatic compounds, terpenoids, steroids, flavonoids, alkaloids, carbonyles, etc., which are described in detail in Section 2.4. Their function in plants is usually related to metabolic and/or growth regulation, lignification, coloring of plant parts and protection against pathogen attack (Payne *et al.*, 1991). Even though secondary metabolism generally accounts for less than 10% of the total plant metabolism, its products are the main plant constituents with pharmaceutical properties.

Despite the diversity of secondary metabolites, a few key intermediates in primary metabolism supply the precursors for most secondary products. These are mainly sugars, acetyl-CoA, nucleotides and amino acids (Robinson, 1964; Jakubke and Jeschkeit, 1975; Payne *et al.*, 1991).

- Cyanogenic glycosides and glucosinolates are derived from sugars.
- Terpenes and steroids are produced from isoprene units which are derived from acetyl-CoA.
- Nucleotide bases are precursors to purine and pyrimidine alkaloids.
- Many different types of aromatic compounds are derived from shikimic acid pathway intermediates.
- The non-aromatic amino acid arginine is the precursor to plyamines and the tropane alkaloids.

In addition, many natural products are derived from pathways involving more than one of these intermediates:

- Phenylpropanoids are derived from the amino acid phenylalanine, with acetyl-CoA and sugar units being added later in the biosynthetic pathway.
- The indole and the quinoline alkaloids are derived from the amino acid tryptophan and from monoterpenes.
- The aglycon moieties of cyanogenic glycosides and glucosinolates are derived from amino acids.

Primary and secondary metabolic pathways in plants are summarized in Figure 2.2.

2.3. Why do plant compounds have an anticancer activity?

Some secondary metabolites are considered as metabolic waste products, for example, alkaloids may function as nitrogen waste products. However, a significant portion of the products derived

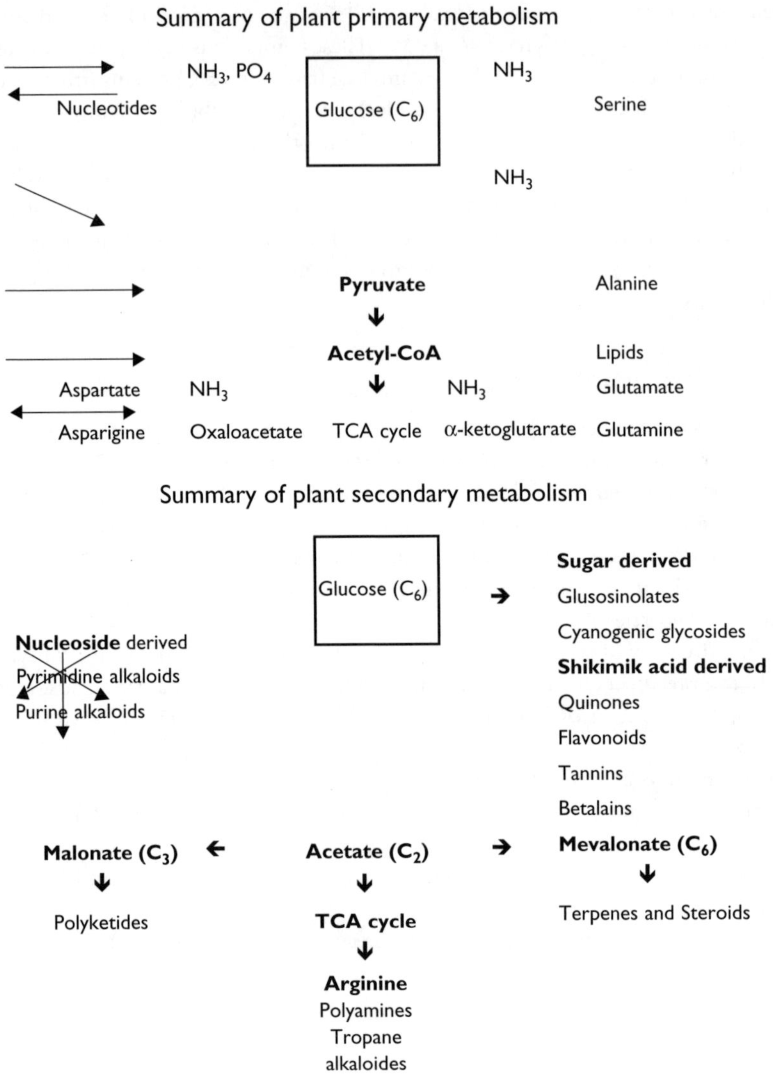

Figure 2.2 Summary of primary and secondary metabolic pathways in plants (adapted from Payne *et al.*, 1991).

form secondary pathways serve either as *protective agents against various pathogens* (e.g. insects, fungi or bacteria) or *growth regulatory molecules* (e.g. hormone-like substances that stimulate or inhibit cell division and morphogenesis). Due to these physiological functions, secondary metabolites are potential anticancer drugs, since either direct cytotoxicity is effected on cancer cells or the course of tumor development is modulated, and eventually inhibited. Administration of these compounds at low concentrations may be lethal for microorganisms and small animals, such as herbivorous insects, but in larger organisms, including humans, they may specifically affect the fastest growing tissues such as tumors.

2.4. Chemical groups of natural products with anticancer properties

Plant-derived natural products with documented anticancer/antitumor properties can be classified into the following 14 chemical groups:

1 Aldehydes
2 Alkaloids
3 Annonaceous acetogenins
4 Flavonoids
5 Glycosides
6 Lignans
7 Lipids
8 Lipids (unsaponified)
9 Nucleic acids
10 Phenols and derivatives
11 Polysaccharides
12 Proteins
13 Terpenoids
14 Unidentified compounds.

Aldehydes are volatile substances found (along with alcohols, ketones and esters) in minute amounts and contributing to the formation of odor and flavor of plant parts.

Structure and properties: They are aliphatic, usually unbranched molecules, with up to twelve carbon atoms (C_{12}). They can be extracted from plants by distillation, solvent extraction or aeration.

Biosynthesis in plant cells: It is suggested that the biosynthesis of aldehydes is related to fatty acids.

Basis of anticancer/antitumor activity: Some aldehydes are cytotoxic against certain cancer types *in vitro*, mainly due to inhibition of tyrosinase. Immunomodulatory properties have been also ascribed to this group of secondary metabolites.

Some plants containing aldehydes with anticancer properties are indicated in Table 2.1 (for more details on each plant, please consult Chaper 3 of this book).

Table 2.1 Plants containing aldehydes with anticancer properties

Species	*Target disease or cell line (if known)*	*Mode of action (if known)*	*Page*
Cinnamomum cassia	Human cancer lines, SW-620 xenograft	Cytotoxic, immunomodulatory	68
Mondia whitei	Under investigation in various cell lines	Tyrosinase inhibitor	183
Rhus vulgaris	Under investigation in various cell lines	Tyrosinase inhibitor	183
Sclerocarya caffra	Under investigation in various cell lines	Tyrosinase inhibitor	183

Alkaloids are widely distributed throughout the plant kingdom and constitute a very large group of chemically different compounds with diversified pharmaceutical properties.

Many alkaloids are famous for their psychotropic properties, as very potent narcotics and tranquilizers. Examples are *morphine, cocaine, reserpine and nicotine*. Several alkaloids are also very toxic.

Structure and properties: They are principally nitrogen-containing substances with a ring structure that allows their general classification in the groups described in Table 2.2 (Jakubke and Jeschkeit, 1975). Most alkaloids with anticancer activity are either *indole, pyridine, piperidine* or *aminoalkaloids*.

Table 2.2 General structural classification of alkaloids

Group name	*Base structure*
Pyrrolidine	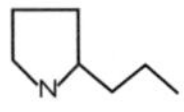
Pyrrolizidine	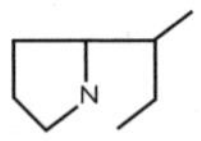
Tropane	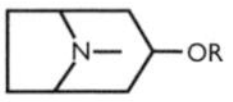
Piperidine	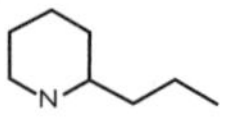
Punica, Sedum and *Lobelia* alkaloids	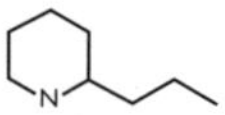
Quinolizidine	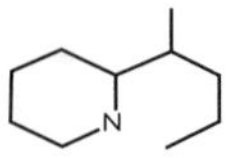
Isoquinolizidine	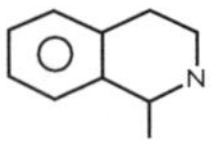
Indole	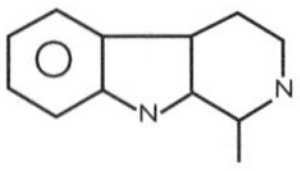
Rutaceae alkaloids	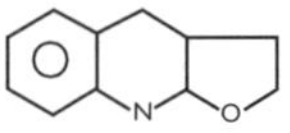
Terpene alkaloids	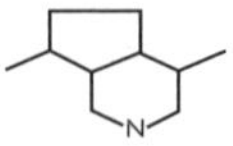

Alkaloids are weak bases, capable of forming salts, which are commonly extracted from tissues with an acidic, aqueous solvent. Alternatively, free bases can be extracted with organic solvents.

Distribution: Quite abundant in higher plants, less in gymnosperms, ferns, fungi and other microorganisms. Particularly rich in alkaloids are plants of the families Apocynaceae, Papaveraceae and Fabaceae.

Biosynthesis in plant cells: Rather complicated, with various amino acids (phenylalanine, tryptophan, ornithine, lysine and glutamic acid) serving as precursor substances.

Basis of anticancer/antitumor activity: Alkaloids are mainly cytotoxic against various types of cancer and leukemia. They also demonstrate antiviral properties. More rarely, they demonstrate immuno-modulatory properties.

Some plants containing alkaloids with anticancer properties are indicated in Table 2.3 (for more details on each plant, please consult Chapter 3 of this book).

Table 2.3 Plants containing alkaloids with anticancer properties

Species	*Target disease or cell line (if known)*	*Mode of action (if known)*	*Page*
Aconitum napellus	Under investigation in various cell lines	Poisonous	160
Acronychia baueri, A. haplophylla	Possible against KB cells	Cytotoxic	74
Annona purpurea	Under investigation in various cell lines	Cytotoxic	81
Brucea antidysenterica	Leukemia	Cytotoxic	81
Calycodendron milnei	Antiviral	Cytotoxic	182
Cassia leptophylla	Under investigation in various cell lines	DNA-damaging (piperidine)	86
Chamaecyparis sp.	P-388	Cytotoxic: inhibition of cyclic GMP formation	169
Chelidonium majus	Various cancers, lung	Immunomodulator (clinical)	86
Colchicum autumnale	P-388, esophageal	Tubulin inhibitor	93
Ervatamia microphylla	k-ras-NRK (mice) cells	Growth inhibition	101
Eurycoma longifolia	Various human cell lines	Cytotoxic *in vitro*	172
Fagara macrophylla	P-388	Cytotoxic	101
Nauclea orientalis	Antitumor, *in vitro* human bladder carcinoma	Antiproliferative	178
Psychotria sp.		Antiviral	181
Strychnos usabarensis	Various *in vitro* (liver damage)	Cytotoxic	162

Annonaceous acetogenins are antitumor and pesticidal agents of the Annonaceae family.

Structure and properties: They are a series of C-35/C-37 natural products derived from C-32/C-34 fatty acids that are combined with a 2-propanol unit. They are usually characterized by a long aliphatic chain bearing a terminal methyl-substituted α,β-unsaturated γ-lactone ring with 1-3 tetrahydrofuran (THF) rings located among the hydrocarbon chain and a number of oxygenated moieties and/or double bonds. Annonaceous acetogenins are classified according to their relative stereostructures across the THF rings (Alali *et al.*, 1999).

Annonaceous acetogenins are readily soluble in most organic solvents. Ethanol extraction of the dried plant material is followed by solvent partitions to concentrate the compounds.

Distribution: Exclusively in the Annonaceae family.

Biosynthesis in plant cells: Derived from the polyketide pathway, while the tetrahydrofuran and epoxide rings are suggested to arise from isolated double bonds through epoxidation and cyclization.

Basis of anticancer/antitumor activity: Annonaceous acetogenins are cytotoxic against certain cancer species and leukemia. They are the most powerful inhibitors of complex I in mammalian and insect mitochondrial electron transport system, as well as of NADH oxidase of the plasma membranes of cancer cells. Therefore they decrease cellular ATP production, causing apoptotic cell death.

Some plants containing Annonaceous acetogenins are indicated in Table 2.4 (for more details on each plant, please consult Chapter 3 of this book).

Table 2.4 Plants containing Annonaceous acetogenins

Species	*Target disease or cell line (if known)*	*Mode of action (if known)*	*Pages*
Annona muricata, A. squamosa	Prostate adenocarcinoma, pancreatic carcinoma	Cytotoxic	81
A. bullata	Human solid tumors *in vitro* (colon cancer)	Cytotoxic	80
Eupatorium cannabinum, E. semiserratum, E. cuneifolium	Leukemia	Cytotoxic	99
Glyptopetalum sclerocarpum	*In vitro* various human cancers	Non-specific cytotoxic	173
Goniothalamus sp.	Breast cancer, *in vitro* various human cancers	Cytotoxic	109
Helenium microcephalum	Leukemia	Cytotoxic	112
Passiflora tetrandra	P-388	Cytotoxic	179
Rabdosia ternifolia	Various human cancer cells	Cytotoxic	141

Flavonoids are widely distributed colored phenolic derivatives. Related compounds include *flavones, flavonols, flavanonols, xanthones, flavanones, chalcones, aurones, anthocyanins* and *catechins*.

Structure and properties: Flavonoids may be described as a series of C_6–C_3–C_6 compounds, that is, they consist of two C_6 groups (substituted benzene rings) connected by a three-carbon-aliphatic chain. The majority of flavonoids contain a pyran ring linking the three-carbon chain with one of the benzene rings. Different classes within the group are distinguished by additional oxygen-heterocyclic rings and by hydroxyl groups distributed in different patterns. Flavonoids frequently occur as glycosides and are mostly water-soluble or at least sufficiently polar to be well extracted by methanol, ethanol or acetone; however they are less polar than carbohydrates and can be separated from them in an aqueous solution.

Distribution: They are widely distributed in the plant kingdom, since they include some of the most common pigments, often fluorescent after UV-irradiation. They also act as metabolic regulators and protect cells from UV-radiation. Finally, flavonoids have a key function in the mechanism of biochemical recognition and signal transduction, similar to growth regulators.

Biosynthesis in plant cells: Flavonoids are derived from shikimic acid via the phenylpropanoid pathway. Related compounds are produced through a complex network of reactions: isoflavones, aurones, flavanones and flavanonols are produced from chalcones, leucoanthocyanidins, flavones and flavonols from flavanonols and anthocyanidins from leucoanthocyanidins.

Basis of anticancer/antitumor activity: Flavonoids are cytotoxic against cancer cells, mostly *in vitro*.

Some plants containing flavonoids with anticancer properties are indicated in Table 2.5 (for more details on each plant, please consult Chapter 3 this book).

Table 2.5 Plants containing flavonoids with anticancer properties

Species	*Target disease or cell line (if known)*	*Mode of action (if known)*	*Pages*
Acrougehia porteri	KB	Cytotoxic	72
Angelica keiskei	Antitumor promoting activity (mice)	Calmodulin inhibitor	78
Annona densicoma, A. reticulata	Various mammalian cell cultures	Cytotoxic	80
Claopodium crispifolium	Potential anticarcinogenic agent	Cytotoxic	80
Eupatorium altissimum	P-338, KB	Cytotoxic	99
Glycyrrhiza inflata	HeLa cells (mice)	Cytotoxic	68
Gossypium indicum	B16 melanoma	Cytotoxic	111
Polytrichum obioense	Hela, leukemia (mice)	Cytotoxic	80
Psorospermum febrifigum	KB	Cytotoxic	80
Rhus succedanea	Under investigation in various cell lines	Cytotoxic	182
Zieridium pseudobtusifolium	KB	Cytotoxic	74

Glycosides are carbohydrate ethers that are readily hydrolyzable in hot water or weak acids. Most frequently, they contain glucose and are named by designating the attached alkyl group first and replacing the *–ose* ending of the sugar with *–oside*.

Basis of anticancer/antitumor activity: Glycosides are mainly cytotoxic against certain types of cancer and also demonstrate antiviral and antileukemic properties.

Some plants containing glycosides with anticancer properties are indicated in Table 2.6 (for more details on each plant, please consult Chapter 3 of this book).

Lignans are colorless, crystalline solid substances widespread in the plant kingdom (mostly as metabolic intermediaries) and having antioxidant, insecticidal and medicinal properties.

Structure and properties: They consist of two phenylpropanes joint at their aliphatic chains and having their aromatic rings oxygenated. Additional ring closures may also be present. Occasionally they are found as glycosides.

Lignans may be extracted with acetone or ethanol and are often precipitated as slightly soluble potassium salts by adding concentrated potassium hydroxide to an alcoholic solution.

Distribution: Wide.

Biosynthesis in plant cells: Lignans are originally derived from shikimic acid via the phenylpropanoid pathway, with *p*-hydroxycinnamyl alohol and coniferyl alcohol being key intermediates of their biosynthesis.

Basis of anticancer/antitumor activity: Some lignans are cytotoxic against certain cancer types, such as mouse skin cancer, or tumor and leukemic lines *in vitro*.

Some plants containing lignans with anticancer properties are indicated in Table 2.7 (for more details on each plant, please consult Chapter 3 of this book).

Table 2.6 Plants containing glycosides with anticancer properties

Species	*Target disease or cell line (if known)*	*Mode of action (if known)*	*Page*
Phlomis armeniaca	Liver cancer, Dalton's lymphoma (mice), Leukemia	Antiviral, cytotoxic, chemopreventive (human)	154
Phyllanthus sp.	Liver cancer, Dalton's lymphoma (mice), P-388	Cytotoxic	136
Plumeria rubra (*iridoids*)	P-388, KB	Cytotoxic	138
Scutellaria salviifolia	Various cancer cell lines	Cytotoxic	154
Wikstroemia indica	Leukemia, Ehrlich ascites carcinoma (mice), P-388	Antitumor	159

Table 2.7 Plants containing lignans with anticancer properties

Species	*Target disease or cell line (if known)*	*Mode of action (if known)*	*Page*
Brucea sp.	KB, P-388	Cytotoxic	81
Juniperus virginiana	Liver cancer (mice)	Tumor inhibitor	117
Magnolia officinalis	Skin (mice)	Tumor inhibitor	178
Plumeria sp.	P-388, KB	Cytotoxic	138
Wikstroemia foetida	P-388	Cytotoxic	160

Table 2.8 Plants containing lipids with anticancer properties

Species	*Target disease or cell line (if known)*	*Mode of action (if known)*	*Page*
Nigella sativa	Ehrilch ascites carcinoma	Cytotoxic *in vitro*	96
Sho-saiko-to, Juzen-taiho-to (*extract*)	Dalton's lymphoma, sarcoma-180 (clinical)	Immunomodulator, antitumor	68

Lipids (saponifiable) include fatty acids (aliphatic carboxylic acids), fatty acid esters, phospholipids and glycolipids.

Structure and properties: By definition, lipids are soluble only in organic solvents. On heating with alkali, they form water-soluble salts (therefore the designation *saponifiable lipids*). Fatty acids are usually found in their ester form, mostly having an unbranched carbon chain and differ from one another in chain length and degree of unsaturation.

Distribution: Lipids are widely distributed in the plant kingdom. They both serve as nutritional reserves (particularly in seeds) and structural elements (i.e.phospholipids of the cell membrane, fatty acid esters in the epidermis of leaves, stems, fruits etc.).

Biosynthesis in plant cells: They are derived by condensation of several molecules of acetate (more specifically malonyl-coenzyme A), thus being related to long-chain fatty acids.

Basis of anticancer/antitumor activity: Saponifiable lipids are cytotoxic against a limited number of cancer types.

Some plants containing lipids with anticancer properties are indicated in Table 2.8 (for more details on each plant, please consult Chapter 3 of this book).

Unsaponifiable lipids (in particular **quinones**) are a diverse group of substances generally soluble in organic solvents and not saponified by alkali. They are yellow to red pigments, often constituents of wood tissues and have toxic and antimicrobial properties.

Structure and properties: *Naphthoquinones* are yellow-red plant pigments, extractable with non-polar solvents, such as benzene. They can be separated from lipids by stem distillation with weak alkali treatment. *Anthraquinones* represent the largest group of natural quinines, are usually hydroxylated at C-1 and C-2 and commonly occur as glycosides (water-soluble). Thus, their isolation is carried out according to the degree of glycosidation. Hydrolysis of glycosides (after extraction in water or ethanol) takes place by heating with acetic acid or dilute alcoholic HCl. Phenanthraquinones have a rather more complex structure and can be extracted in methanolic solutions.

Distribution: Anthraquinones are particularly found in the plant families Rubiaceae, Rhamnaceae and Polygonaceae. Phenanthraquinones are rare compounds having important medicinal properties (e.g. *hypericin* from *Hypericum perforatum*, *tanshinone* from *Salvia miltiorrhiza*).

Biosynthesis in plant cells: They are derived by condensation of several molecules of acetate (more specifically malonyl-coenzyme A), thus being related to long-chain fatty acids.

Basis of anticancer/antitumor activity: Several quinines are cytotoxic against certain cancer types, such as melanoma, or tumor lines *in vitro*.

Some plants containing quinones with anticancer properties are indicated in Table 2.9 (for more details on each plant, please consult Chapter 3 of this book).

Table 2.9 Plants containing quinones with anticancer properties

Species	*Target disease or cell line (if known)*	*Mode of action (if known)*	*Page*
Kigelia pinnata	*In vitro* melanoma, renal cell carcinoma	Tumor inhibitor	174
Koelreuteria henryi	Src-Her-2/neu, ras oncogenes	Tumor inhibitor	174
Landsburgia quercifolia	P-388	Cytotoxic	176
Mallotus japonicus	*In vitro*: human lung carcinoma, B16 melanoma, P-388, KB	Cytotoxic	120
Nigella sativa	MDR human tumor	Cytotoxic *in vitro*	96
Rubia cordifolia	*In vitro* human cancer lines	Antitumor	142
Sargassum tortile	P-388	Cytotoxic	150
Wikstroemia indica	Ehrlich ascites carcinoma, MK, P-388	Antitumor	159

Nucleic acids, *deoxyribonucleic acid* (DNA) and *ribonucleic acid* (RNA) are known as the "genetic molecules," the building blocks of genes in each cell or virus.

Structure and properties: Each nucleic acid contains four different nitrogen bases (*purine* and *pyrimidine bases*), phosphate and either deoxyribose or ribose. DNA contains the bases *adenine, quanine, cytosine, thymine* and *5-methylcytosine*. The macromolecular structure of DNA is a two-stranded helix with the strands bound together by hydrogen bonds.

Like proteins and polysaccharides, nucleic acids are water-soluble and non-dialyzable. They can be separated from a water extract by denaturating proteins in chloroform-octyl alcohol and then precipitate polysaccharides in a weakly basic solution.

Distribution: Wide.

Biosynthesis in plant cells: Bases are derived originally from ribose-5-phosphate, purines from inosinic acid and pyrimidines from uridine-5-phosphate. Nucleic acids are formed after nucleotide transformation and condensation.

Basis of anticancer/antitumor activity: Some nucleotides, like cyclopentenyl cytosine (derived from *Viola odorata*), present cytotoxicity against certain cancer species *in vitro*.

Phenols and derivatives are the main aromatic compounds of plants, whose structural formulas contain at least one benzene ring. They serve as odors, fungicidals or germination inhibitors. *Coumarins* are especially common in grasses, orchids, citrus fruits and legumes.

Structure and properties: Simple phenols are colorless solids, which are oxidized by air. Water solubility increases with the number of hydroxyl groups present, but solubility in organic solvents is generally high. Natural aromatic acids are usually characterized by having at least one aliphatic chain attached to the aromatic ring.

Coumarins are lactones of *o*-hydroxycinnamic acid. Almost all natural coumarins have oxygen (hydroxyl or alkoxyl) at C-7. Other positions may also be oxygenated and alkyl side-chains are frequently present. *Furano-* and *pyranocoumarins* have a pyran or furan ring fused with the benzene ring of a coumarin.

Phenolic acids may be extracted from plant tissues or their ether extract in 2% sodium bicarbonate. Upon acidification, acids often precipitate or may be extracted with ether. After removal of carboxylic acids, phenols may be extracted with 5% sodium hydroxide solution. Phenols are usually not steam-distillable, but their ethers or esters can be. Coumarins can be purified from a crude extract by treatment with warm dilute alkali which will open the lactone ring and form a water-soluble coumarinate salt. After removal of organic impurities with ether, coumarins can be reconstituted by acidification.

Distribution: Wide, abundant in herbs of the families Lamiaceae and Boraginaceae.

Biosynthesis in plant cells: Phenolic compounds generally are derived from shikimic acid via the phenylpropanoid pathway.

Basis of anticancer/antitumor activity: Phenolic compounds are cytotoxic against certain cancer types *in vitro*. They usually interfere with the integrity of the cell membrane or inhibit various protein kinases. Coumarins, in particular furanocoumarins, are highly toxic.

Some plants containing phenols with anticancer properties are indicated in Table 2.10 (for more details on each plant, please consult Chapter 3 of this book).

Table 2.10 Plants containing phenols with anticancer properties

Species	*Target disease or cell line (if known)*	*Mode of action (if known)*	*Page*
Acronychia laurifolia	Under investigation in various cell lines	Under investigation	73
Angelica gigas, A. decursiva, A. keiskei	No data	Cytotoxic	77
Gossypium indicum	Murine B16 melanoma, L1210 lymphoma	Cytotoxic	111

Polysaccharides and generally carbohydrates represent the main carbon sink in the plant cell. Polysaccharides commonly serve nutritional (e.g. *starch*) and structural (e.g. *cellulose*) functions in plants.

Structure and properties: They are polymers of monosaccharides (and their derivatives) containing 10 or more units, usually several thousand. Despite the vast number of possible polysaccharides, only few of the structural possibilities actually exist. Generally, *structural polysaccharides* are strait-chained (not very soluble in water), while *nutritional* (*reserve food*) *polysaccharides* tend to be branched, therefore forming viscous hydrophilic colloid systems. *Plant gums* and *mucilages* are hydrophilic heteropolysaccharides (i.e. they contain more than one type of monosaccharide), with the common presence of uronic acid in their molecule.

Depending on their degree of solubility in water, polysaccharides can be extracted from plant tissues either with hot water (pectic substances, nutritional polysaccharides, mucilages, fructans) or alkali solutions (hemicelluloses).

Distribution: They are universally distributed in the plant kingdom. Structural polysaccharides are the main constituents of the plant cell wall (cellulose, hemicelluloses, xylans, pectins, galactans). Nutritional polysaccharides include starch, fructans, mannans and galactomannans. Mucilages abound in xerophytes and seeds. Polysaccharides also have a key function in the mechanism of biochemical recognition and signal transduction, similar to growth regulators.

Biosynthesis in plant cells: There exists a complex network of interrelated biosynthetic pathways, with various monosaccharides (glucose, fructose, mannose, mannitol, ribose and erythrose) serving as precursor substances. Phosphorylated intermediates are found in subsequent biosynthetic steps and branching points. The glycolytic, pentose and UDP-glucose pathways have been defined in extend.

Basis of anticancer/antitumor activity: Some polysaccharides are cytotoxic against certain types of cancer, such as mouse skin cancer, or tumor lines *in vitro* (e.g. mouse Sarcoma-180). However, most polysaccharides exert their action through stimulation of the immune system (cancer immunotherapy).

Some plants containing polysaccharides with anticancer properties are indicated in Table 2.11 (for more details on each plant, please consult Chapter 3 of this book).

Table 2.11 Plants containing polysaccharides with anticancer properties

Species	*Target disease or cell line (if known)*	*Mode of action (if known)*	*Page*
Angelica acutiloba	Epstein–Barr, skin(mice)	Cytotoxic, immunological	78
Angelica sinensis	Ehrlich Ascites (mice)	Cytotoxic, immunological	77
Brucea javanica	Leukemia, lung, colon, CNS, melanoma, brain	Cytotoxic	83
Cassia angustifolia	Solid Sarcoma-180 (mice)	Cytotoxic	86
Sargassum thunbergii	Ehrlich Ascites (mice)	Immunostim activates the reticuloenthothelial system	150
S. fulvellum	Sarcoma-180 (mice)	Immunomodulator	150
Tamarindus indica	Potential activity in various cell lines	Immunomodulator	184

Proteins, like carbohydrates, belong to the most essential constituents of the plant body, since they are the building molecules of structural parts and the enzymes.

Structure and properties: Proteins are made up from amino acids, the particular combination of which defines the physical property of the protein. Thus, protein sequences differing in only one amino acid will correspond to entirely different molecules, both structurally (tertiary structure) and functionally. *Peptides* are small proteins, amino acid oligomers with a molecular weight below 6000. In nature, 24 different amino acids are widely distributed. Sixteen to twenty different amino acids are usually found on hydrolysis of a given protein, all having the L-configuration. Conjugate proteins comprise other substances along with amino acids. Particularly important are *glycoproteins*, partially composed of carbohydrates. Proteins may be soluble in water and dilute salt solutions (albumins), in dilute salt solutions (globulins), in very dilute acids and bases (glutelins) or in ethanolic solutions (prolamines).

Peptides and proteins can be isolated from a plant tissue by aqueous extraction or in less polar solvents (depending on the water solubility of a particular protein). Fractionation of the proteins can frequently be achieved by controlling the ionic strength of the medium through the use of salts. However, one must always take precautions against protein denaturation (due to high temperature).

Distribution: Extremely wide.

Biosynthesis in plant cells: Proteins are synthesized in ribosomes from free amino acids under the strict, coordinated control of genomic DNA, mRNA and tRNA (gene transcription and translation).

Basis of anticancer/antitumor activity: Proteins are indirectly cytotoxic against certain cancer types, acting mainly through the inhibition of various enzymes or by inducing apoptotic cell death.

Some plants containing proteins with anticancer properties are indicated in Table 2.12 (for more details on each plant, please consult Chapter 3 of this book).

Table 2.12 Plants containing proteins with anticancer properties

Species	*Target disease or cell line (if known)*	*Mode of action (if known)*	*Pages*
Acacia confusa	Sarcoma 180/HeLa cells	Trypsin inhibitor	167
Ficus cunia	Under investigation in various cell lines	White cell aglutination	103
Glycyrrhiza uralensis	Under investigation	SDR-enzymes (antimutagenic)	68
Momordica charantia	Leukemia	Inhibits DNA synthesis *in vitro*, immunostimulant	124
Momordica indica	Leukemia	Antiviral, cytotoxic	124
Rubia sp. *R. cordifolia*	P338	Under investigation	142

Terpenoids are diverse, widely distributed compounds commonly found under groups such as essential oils, sterols, pigments and alkaloids. They exert significant ecological functions in plants. *Mono-* and *sesquiterpenoids* are found as constituents of steam-distillable essential oils. *Di-* and *triterpenoids* are found in resins.

Structure and properties: They are built up of isoprene or isopentane units linked together in various ways and with different types of ring closures, degrees of unsaturation and functional groups. Depending on the number of isoprene molecules in their structure, terpenoids are basically classified as *monoterpenoids* (2), *sesquiterpenoids* (3), *diterpenoids* (4) and *triterpenoids* (6). *Sterols* share the core structure of lanosterol and other tetracyclic triterpenoids, but with only two methyl groups at positions 10 and 13 of their ring system. *Steroids* occur throughout the plant kingdom as free sterols and their lipid esters.

There exists no general method for isolating terpenoids from plants, however many of them are non-polar and can be extracted in organic solvents. After saponification in alcoholic alkali and extraction with ether, most terpenoids will accumulate into the ether fraction.

Distribution: Wide.

Biosynthesis in plant cells: Terpenoids are all derived from mevalonic acid or a closely related precursor. The pyrophosphate of alcohol farnesol is a key intermediate in terpenoid biosynthesis, particularly leading to the formation of diterpenoids, triterpenoids and sterols. Monoterpenoids are derived from geranyl pyrophosphate.

Basis of anticancer/antitumor activity: Terpenoids and sterols often possess alkaloidal properties, thus being cytotoxic *in vivo* and *in vitro* against various cancer types, such as human prostate cancer, pancreatic cancer, lung cancer and leukemia.

Some plants containing terpenoids and sterols with anticancer properties are indicated in Table 2.13 (for more details on each plant, please consult Chapter 3 of this book).

Unidentified compounds usually refer to complex mixtures or plant extracts the composition of which has not been elucidated in detail or the bioactive properties of which can not be assigned to a particular substance only. Ironically, unidentified extracts are usually more potent against various types of cancer than single, well-studied molecules.

Some plants containing unidentified compounds with anticancer properties are indicated in Table 2.14 (for more details on each plant, please consult Chapter 3 of this book).

Table 2.13 Plants containing terpenoids and sterols with anticancer properties

Species	*Target disease or cell line (if known)*	*Mode of action (if known)*	*Pages*
Aristolochia versicolar	Under investigation	Under investigation	169
Brucea antidysenterica	Under investigation	Cytotoxic	81
Casearia sylvestris	Under investigation	Cytotoxic *in vitro*, apoptotic	172
Crocus sativus	KB, P-388, human prostate, pancreatic, *in vitro*	Cytotoxic	93
Glycyrrhiza sp.	P-388	No data	68
Mallotus anomalus	P-388	No data	120
Melia sp.	Carcinoma, sarcoma, leukemia, AS49, VA13	Apoptotic/inhibits DNA synthesis	122
Maytenus sp.	Leukemia	Cytotoxic	121
Neurolaena lobata	Human carcinoma *in vitro*	Cytotoxic	178
Polyalthia barnesii	Human carcinoma *in vitro*	Cytotoxic	180
Rabdosia trichocarpa	HeLa cells, P-388	Cytotoxic	141
Seseli mairei	KB, P-388, L1210		183
Stellera chamaejasme	Human leukemia, stem, lung, P-388, L1210	Proteinokinase C activator	154

Table 2.14 Plants containing unidentified compounds with anticancer properties

Species	*Target disease or cell line (if known)*	*Mode of action (if known)*	*Page*
Chelidonium majus	Esophageal squamous cell carcinoma clinical	Immunostimulant	86
Menispernum dehuricum	Intestinal metaplasia, atypical hyperplasia of the gastric	Anti-estrogen, LH-RH antagonist (mice)	90
Paeonia sp.	Esophageal squamous cell carcinoma clinical	Immunostimulant	131
Phyllanthus amarus	Antiviral (HBV)		136
Phyllanthus emblica	NK cells	Immunostimulant	136
Trifolium pratense	Various human cell lines	Chemopreventive	155

2.5. Biotechnology and the supply issue

In spite of the plethora of plant metabolites with tumor cytotoxic or immuno-stimulating properties, plants may not be an ideal source of natural anticancer drugs. There are a number of reasons that make the recovery of plant-derived products less attractive than the alternative methods of biotechnology and chemical synthesis.

First of all, there is the **supply issue**. In several cases, compounds of interest come from slow growing plant species (e.g. woody species), or species that are endangered. In addition, *in vivo* productivity can be considerably low, thus necessitating the use of an overwhelming amount of plant biomass in order to obtain a satisfactory portion of the natural product (especially in the case of secondary metabolites). For example, taxol concentration in needles and dried bark of *Taxus brevifolia* is approx. 0.01–0.1%. Supply can also be hindered due to inadequate plant production, which, in turn, may be caused by a number of problems related to disease, drought or socioeconomic factors. As demonstrated previously for taxol the issue of supply can be partly resolved by breeding for high-yielding varieties, introduction of wild species or derivation (in abundant amounts) of precursors for the hemisynthesis of the desired product (Cragg *et al.*, 1993; Gragg, 1998).

Second, plant-derived pharmaceutical extracts frequently lack the necessary **standardization** that could render them reliable for large-scale, clinical use. This problem is related to a number of factors, such as the dependence of the production on environmental factors and the plant developmental stage (e.g. flowering), as well as the heterogeneity of the extracts, which makes further isolation and purification of the product an indispensable, though costly step. A representative example is mistletoe lectin extract, which presents a remarkable seasonal variation in the levels of ML isolectins (ML I, ML II and ML III). Furthermore, the specific bioactivity of the extract fluctuates over a prolonged (e.g. two-year) storage time (Lorch and Troger, 2000).

Biotechnology could offer an alternative method for the production of considerable plant biomass or natural products in a relatively short time (Payne *et al.*, 1991; Kintzios and Barberaki, 2000). *In vitro* techniques are a major component of plant biotechnology, since they permit artificial control of several of the parameters affecting the growth and metabolism of cultured tissues. Plainly put, plant tissue culture works on the principle of inoculating an explant (that is a piece of plant tissue, such as a leaf or stem segment) from a donor plant on a medium containing nutrients and growth regulators and causing thereof the formation of a more or less dedifferentiated, rapidly growing **callus tissue**. Production of plant-derived anticancer agents could be advantageous over derivation from plants *in vivo* since:

1. By altering the culture parameters, it might be possible to control the quantity, composition and timing of production of mistletoe extracts. In this way, problems associated with the standardization of plant extracts could be overcome.
2. By feeding cultures with precursor substances for the biosynthesis of certain metabolites, a higher productivity can be achieved from cultured cells (*in vitro*) than from whole plants.
3. Potentially, entirely novel substances can be synthesized through *biotransformation* or by taking advantage of *somaclonal* variation, that is, a transient or heritable variability of metabolic procedures induced by the procedure of *in vitro* culture.
4. The establishment of a callus culture is the first step required in order to obtain genetically modified cells or plants, for example, crop plants able to specifically produce a desired product in excessive amount.

5 *Protoplasts* are plant cells having their cell wall artificially removed. In this way, they can be used in gene transfer experiments and for the creation of hybrid cells that result from the direct fusion of two protoplast cells that might have been derived from entirely different species.
6 Plant species that are difficult to propagate (such as mistletoe, which is exclusively accomplished with the aid of birds, carrying distantly mistletoe seeds) could be *clonally micropropagated*, thus obtaining thousands of seedlings from a very limited mass of donor tissue (essentially from one donor plant only). This can be achieved by plant regeneration via *organogenesis* (induction of shoots and roots from callus cultures) or somatic *embryogenesis* (the process of embryo formation from somatic (sporophytic) tissues without fertilization).

Promising as the perspectives of plant cell culture may be, established plant-derived commercial anticancer drugs (such as *vinblastine* and *vincristine* from *Catharanthus roseus*) are still produced by isolation from growing plants; eventually drugs are semisynthetically produced from natural precursors also isolated from plant sources *in vivo*. Currently, there are only a few plant-derived natural compounds with antineoplastic properties that are being produced biotechnologically, mostly on the laboratory level:

Periwinkle (*Vinca rosea* or *Catharanthus roseus*): Numerous studies have been conducted on the scale-up indole alkaloid production from cell suspension cultures of *C. roseus*. Several factors affecting production have been evaluated, including medium nutrient and growth regulator composition, elicitors, osmotic stress and precursor (tryptophan) feeding. *Vinblatine*, an antileukemic dimeric indole alkaloid dimmer cannot be directly produced from *C. roseus in vitro*, due to under-expression of the enzyme acetyl CoA:deacetylvindoline *O*-acetyl transferase, which catalyzes the formation of *vindoline*, one of the substrates leading to *anhydrovinblastine*. Yield values of *catharanthine* (the second substrate for vinblastine synthesis) up to $17\,\mu g\,l^{-1}$ after fungal induction have been reported (Bhadra *et al.*, 1993).

Pacific Yew (*Taxus brevifolia*): In 1977, NCI awarded contracts for the investigation of plant tissue culture as a source of anticancer drugs, and two of these studies related to *taxol* production. Unfortunately, these contracts were terminated in 1980 before any positive results had been obtained. Considerable research effort has once more been focused on the application of this technology to taxol production. Ketchum *et al.* (1999) reported the production of up to 1.17% of paclitaxel within five days of elicitation with methyl jasmonate, along with other taxoids, such as *13-acetyl-9-dihydrobaccatin, 9-dihydrobaccatin III* and *baccatin VI*. Two companies (ESCAgenetics Corporation and Phyton Catalytic) reported on their plans for a scale-up production of taxol in the near future.

American mandrake (*Podophyllum hexandrum*): Cell suspensions of *P. hexandrum* have been established which accumulate up to 0.1% *podophyllotoxin*, a cytotoxic lignan used for the hemisynthesis of *etoposide* and *teniposide*. Accumulation of podophyllotoxin has been increased twelve-fold after precursor feeding with coniferin, a glucosylated intermediate of the phenylpropanoid pathway (Smollny *et al.*, 1998).

Mistletoe (*Viscum album* L.): Becker and Schwarz (1971) were the first to mention the possible use of mistletoe callus cultures as a source of bioactive products. In 1990, Fukui *et al.* reported on the induction of callus from leaves of *V. album* var. *lutescens*: they were able to identify in the callus two *galactose-binding lectins* which were originally observed in mistletoe leaves. Kintzios and Barberaki (2000) succeeded in inducing callus and protoplast cultures from mistletoe leaves and stems in a large number of different growth regulator and media treatments.

They have also studied the effects of different plant parts (stems and leaves), harvest time (winter or summer), explant disinfection methods, growth regulators, culture medium composition and cell wall digestion treatments. Finally, they observed a relatively low (8%) somaclonal variation, in the aspect of both the quantitative and the qualitative mistletoe protein production *in vitro* (Kintzios and Barberaki, 2000; Kintzios *et al.*, 2002). Langer *et al.* (1997) cloned different fragments of the ML gene from mistletoe genomic DNA, constructed expression vectors (A- and B-chain coding region) and the single chains were expressed in *E. coli* separately. Experimental investigations on the activity of recombinant mistletoe lectin (rML) were promising.

Chapter 3

Terrestrial plant species with anticancer activity

A presentation

Spiridon E. Kintzios, Maria G. Barberaki**
and Olga G. Makri

3.1. Introduction: general botanical issues

In this chapter a detailed analysis will be given on a number of species with documented anticancer properties either *in vitro* or in clinical use. Before we proceed with analysis, however, and for the purpose of a better understanding of the description of each species, a brief overview of botanical terms is given in following:

Life cycle

Plants can be distinguished according to their life cycle (germination, growth, flowering and seed production) as annuals, biennials and perennials. **Annuals** complete their life-cycle within a year. **Biennials** grow without flowering in the first year, coming into flowering in the second. Both these groups are herbs, which flower only once, produce seeds, and then die. **Perennials** flower for several or many years in succession.

Plant anatomy

The stem is made up of **internodes**, separated by *nodes*. The leaves arise at these nodes. The stem is either unbranched, or has *side branches* emerging from buds in the leaf axils. The side branches may themselves branch. Shoots continue to grow at the tip, and develop new leaves, with buds in the axils, which can grow into branches. The shoot can either be hairless, or it may carry hairs of various kinds, often glandular.

Roots serve *to* anchor a plant (in the soil or another host plant) and to facilitate the uptake of water and mineral salts. The *main* or *tap root* is normally vertical. From this grow *lateral roots*, which may themselves branch, and in this way the full root system develops. Many plants have swollen roots which contain stores of food.

A fully developed **leaf** consists of the *blade*, the *leafstalk* (petiole) and *leaf base*. Sometimes there is no leafstalk, in which case the leaf is termed *sessile*, or unstalked; otherwise it is known as *petiolate*, or stalked. The leaf base is often inconspicuous, but sometimes has a *leaf sheath*. The leaf base may have blunt or pointed extensions at either side of the stem (*amplexicaul*), or even completely encircle and fuse with the stem (*perfoliate*). In *decurrent* leaves, the leaf blade extends some distance down the sides of the stem.

Leaves can have different shapes, which often serve as taxonomic characters. They are distinguished in **simple** leaves with undivided blade, and **compound** leaves, consisting of several separate leaflets. Some have **parallel** or curved veins, without a central midrib; others have **pinnate** veins, with an obvious midrib and lateral veins. A leaf can have anyone of a number of shapes, including *linear, lanceolate, elliptic, ovate, hastate* (spear-shaped), *reniform* (kidney-shaped), *cordate* (heart-shaped), *rhombic, spatulate* or *spathulate* (spoon-shaped) or *sagittate* (arrow-shaped). There are also differences in leaf *margins* including *entire, crenate* (bluntly toothed *margins*), *serrate*

* These authors contributed equally to this chapter.

(serrated), *dentate* (*toothed*), *sinuate/undulate* (wavy margins), *pinnately lobed*, or *palmately lobed* leaves. Accordingly, compound leaves can be found as **pinnate** (**imparipinnate** if there is a terminal leaflet and **paripinnate** if not). Leaves grow as lateral appendages of the stem, from nodes. In the case of **alternate** leaves, there is a single leaf at each node, and successive leaves are not directly above each other. **Opposite** leaves are placed as a pair, one at each side of the node. When there are three or more leaves at each node, they are described as **whorls** (in whorls) (Podlech, 1996).

The *inflorescence* is the part of the stem which carries the flowers. A *spike* is a flowerhead in which the individual flowers are stalkless. It can be short and dense, or long and loose. A *raceme* is similar, but consists of stalked flowers. A *panicle* is an inflorescence whose main branches are themselves branched. In an *umbel*, the flower stalks are of equal length and arise from the same point on the stem (Podlech, 1996). A *head* consists of many unstalked or short-stalked flowers growing close together at the end of a stem. The particularly densely clustered head of composites is known as a *capitulum*.

The **flower** is a thickened shoot which carries the reproductive parts of the plant. Its individual parts can be interpreted as modified leaves. The **perianth** consists either of *perianth segments*, or of **sepals** and **petals**. More commonly, these are differentiated into an outer ring of usually green sepals (the **calyx**), and an inner ring of usually coloured petals (the *corolla*). The male part of the flower (*androecium*) consists of the *stamens*; the female part (*gynoecium*) consists of the *ovary, style* and *stigma*, together known as the **pistil**. Each stamen consists of a thin *filament* and an *anther*, the latter containing the *pollen*. In the center of the flower is the **pistil** (gynoecium). This consists of at least one *carpel*, often more, either free or fused. The pistil is divided into *ovary, style* and *stigma*.

The **fruit** develops from the *ovary*, after pollination. It protects the seeds until they are ripe and often also has particular adaptations for seed dispersal. *Dehiscent* fruits open to release the seeds, while *indehiscent* do not (Podlech, 1996).

3.2. Species-specific information

3.2.1. The guardian angels: plant species used in contemporary clinical cancer treatment

Camptotheca acuminata (Camptotheca) (Nyssaceae) — **Antitumor, Tumor inhibitor**

Synonyms: It is well known as the Chinese happy tree – xi shu, or Cancer Tree.

Location: Most provinces south of the Yangtze River. Origin: Asia, specially in Southern China and Tibet. Degree of rarity: low as it is commonly cultivated, mainly on roadsides. It is also cultivated for the production of camptothecins (CPTs).

Appearance:
Stem: trees, deciduous, to 20 m high; the bark is light gray.
Leaves: simple, alternate, exstipulate; blade oblong-ovate or oblong-elliptic.
Flowers: calyx cup-shaped, shallowly 5-lobed; petals 5, light green.
In bloom: May–July

Tradition: It has been used in medications prepared for centuries, in China, to treat different kind of cancers, especially cancers of the stomach, liver and leukemia.

Part used: Bark, wood and lately young leaves.

The status of mistletoe application in cancer therapy: During a screening program conducted by the National Cancer Institute in late 50s, it was confirmed that a compound from *Camptotheca acuminata* had anticancer properties (in 1958 by Dr Monroe E. Wall of the USDA and Jonathon Hartwell of the NCI). Later, in 1966, a quinoline alkaloid *camptothecin* (CPT) was isolated from bark (and wood), by Wall and other researchers of the Research Triangle Institute ('Description and Natural History of Camptotheca', Duke and Ayensu, 1985). Although animal studies confirmed anti-cancer properties, clinical trials were suspended because of high toxicity and severe side effects. Only in 1985 was interest renewed, when it was discovered that CPT inhibited *topoisomerase I* and therefore inhibited DNA replication; and CPT was developed as an anticancer drug. Because of the high toxicity of CPT itself, researchers developed several semisynthetic derivatives that had fewer side effects. After that CPTs became the second most important source of anti-cancer drugs.
Three semi-synthetic drugs from CPT have been approved by the FDA:

1 *topotecan*, as a treatment for advanced ovarian cancers (approved in May 1996). It is manufactured by Smith Kline Beecham Pharmeceuticals and sold under the trade name *Hycamtin*.
2 injectable *irinotecan HCl*, as a treatment for metastatic cancer of the colon or rectum (approved in June 1996). It is usually prescribed in cases that have not responded to standard chemotherapy treatment. It is marketed by *Pharmacia & Upjohn* under the trade name *Camptosar*. It helps fight cancer but also has more tolerable side effects than the original plant extract. (Information for Patients from Pharmacia & Upjohn Company.)
3 *9-nitro camptothecin*, as a treatment for pancreatic cancer. It is marketed as *Rubitecan*.

There are more CPTs used in clinical trials for the treatment of breast cancer, colon cancers, malignant melanoma, small-cell lung cancer and leukemia, and also testing for antiviral (anti-HIV) uses. Although, many attempts for the chemical synthesis of CPT have been made with success, because of their high cost, natural supplies are still the main source of production (Cyperbotanica: Plants used in cancer treatment). Because of the production of CPTs, *C. acuminata* is a protected species and export of its seeds is prohibited. In US the cultivation of this tree has been successfully carried out, but the yields of CPT levels seem to be lower than that growing in China.

Active ingredients: *quinoline alkaloid camptothecin (CPT): topotecan, irinotecan HCl, 9-nitro camptothecin.*

Particular value: It is known and used in medicine as a chemotherapeutic drug. It is used against tumors of the esophagus, stomach, rectum, liver, urinary bladder and ovary, chronic granulocytic leukemia, acute lymphatic leukemia and lymphosarcoma. Also, against psoriasis (20 percent of ointment of the fruit is used for external application; injection of seed for intramuscular injection). (Ovarian Cancer Research Notebook: *Fructus Camptothecae.*)

Precautions: It must be used carefully, as it is poisonous. The major potential side effects of camptothecin drugs are severe diarrhea, nausea and lowered leukocyte counts. It can also damage bone marrow.

Indicative dosage and application (against ovarian cancer)

- 1.25 mg m^{-2}/day^{-1} (topotecan as a 30 min infusion for 5 days, every 3 weeks) (Goldwasser *et al.*, 1999)

- Intravenous (i.v.) dose of 1.5 mg m^{-2} was administered as a 30 min continuous infusion on day 2.

Further doses are under investigation for ovarian cancer, lung cancer and other types of cancer.

Documented target cancers

- Ovarian cancer
- Lung cancer
- Pancreatic cancer.

Further details

Related compounds

- *Topotecan* has shown relative effectiveness when compared to *taxol*, in several clinical trials, although when combined with *taxol*, it may not have the desired dose intensity because of toxicity. Response rates between 13% and 25% are comparable to *paclitaxel* (Heron, 1998).
- Topotecan has demonstrated value as a second-line therapy in recurrent / refractory ovarian cancer, although the hematologic toxicity of topotecan is significant. In a clinical trial decreased topotecan platelet toxicity with successive topotecan treatment cycles in advanced ovarian cancer patients:patients were treated with 1.25 mg m^{-2}/day^{-1} topotecan as a 30 min infusion for 5 days, every 3 weeks (Goldwasser, 1999). Mean platelet nadir values were significantly less after the second and subsequent treatment cycles, suggesting that current treatment schedules are feasible without G-CSF support and that treatment should be able to continue without dose reduction. Other clinical trials have shown that twenty-one-day infusion is a well-tolerated method of administering topotecan. The objective response rate of 35–38% in this small multicenter study is at the upper level for topotecan therapy in previously treated ovarian cancer. Prolonged topotecan administration therefore warrants further investigation in larger, randomized studies comparing this 21-day schedule with the once-daily-for-5-days schedule.
- CPT-11 (Irinotecan) is a drug similar in activity to topotecan. CPT-11 combined with platinum has demonstrated significant response in ovarian cancer trials. CPT-11 combined with *mitomycin-c* is active for clear cell ovarian cancer. CPT-11 are derivatives of camptothecin, derived from the bark of the Chinese tree *C. accuminata*. Both topotecan and CPT-11 are unique in their ability to inhibit *topoismerase I* (*topoisomerases* are responsible for the winding and unwinding of the supercoiled DNA composing the chromosomes. If the chromosomes cannot be unwound, transcription of the DNA message cannot occur and the protein cannot be synthesized.). Both of these drugs have shown significant activity in advanced malignancies (Ovarian Cancer Research Book: *Camptothecin*). DNA *Topoisomerase I (Topo I)* is the unique target for both *topotecan* and CPT-11. *Topo I* transiently breaks a single strand of DNA, thereby reducing the torsional strain (supercoiling) and unwinding the DNA ahead of the replication fork. Although eukaryotic cell lines lacking *Topo I* can survive

in culture, the enzyme has an important role in chromatic organization, in mitosis, and in DNA replication, transcription and recombination. *Topo I* binds to the nucleic acid substrate (DNA) noncovalently. The bound enzyme then creates a transient break in one DNA strand and concomitantly binds covalently to the *3′-phosphoryl* end of the broken DNA strand. *Topo I* then allows the passage of the unbroken DNA strand through the break site and religates the cleaved DNA. The intermediate, covalently bound enzyme–DNA complex is called a "cleavable complex," because protein-linked single DNA breaks can be detected when the reaction is aborted with a strong protein denaturant. The cleavable compound is in equilibrium with the noncovalently bound complex (the "noncleaveable complex"), which does not result in single-strand DNA breaks when exposed to denaturing conditions (Ovarian Cancer Research Book: *Camptothecin*).

- In recents clinical trials oral forms of topotecan have been tested. Results of the pharmacokinetic analyses showed that orally administered topotecan has a lower peak plasma concentration (C_{max}) and longer mean residence time than intravenously administered drug. Preliminary data suggest that the oral formulation has efficacy similar to that of the i.v. formulation in patients with recurrent or refractory ovarian and small-cell lung cancer. The type and degree of toxicity appeared to be related to the dosing schedule (number of days of consecutive treatment), but overall, oral topotecan appeared to be associated with less hematologic toxicity than the IV formulation (Burris 3rd, 1999). In another clinical trial from the Department of Medical Oncology, Rotterdam Cancer Institute, Netherlands by Schellens JH, and other researchers (1996), the results of preclinical and clinical studies indicate enhanced antineoplastic activity of topotecan (SKF 104864-A) when administered as a chronic treatment. We determined the apparent bioavailability and pharmacokinetics of topotecan administered orally to 12 patients with solid tumors in a two-part crossover study. The oral dose of 1.5 $mg\,m^{-2}$ was administered as a drinking solution of 200 ml on day 1. The i.v. dose of 1.5 $mg\,m^{-2}$ was administered as a 30 min continuous infusion on day 2. The bioavailability was calculated as the ratio of the oral to i.v. area under the curve (AUC) calculated up to the last measured time point. The oral drinking solution was well tolerated. The bioavailability revealed moderate inter-patient variation and was 30% $\pm$ 7.7% (range 21–45%). The time to maximum plasma concentration after oral administration (T_{max}) was 0.78 h (median, range 0.33–2.5). Total i.v. plasma clearance of topotecan was 824 $\pm$ 154 $ml\,min^{-1}$ (range 535–1068 $ml\,min^{-1}$). The AUC ratio of topotecan and the lactone ring-opened hydrolysis product (hydroxy acid) was of the same order after oral (0.34–1.13) and i.v. (0.47–0.98) administration. The bioavailability of topotecan after oral administration illustrates significant systemic exposure to the drug which may enable chronic oral treatment.

Antineoplastic activity

- A significant clinical trial is ongoing for *Rubitens* – Phase III – targeted at treating pancreatic cancer. The Phase II clinical data that has been presented on Rubitecan for pancreatic cancer has been nothing short of astounding. *Rubitecan* showed a 63%

response or stable disease in pancreatic cancer patients. Median survival was 16.2 months among responders, which is the longest survival rate ever reported among pancreatic cancer patients. Among stable patients, median survival was 9.7 months and among nonresponders, 5.9 months. Data shows that among 61 patients, 33% were responders, 30% were stable, and 37% were nonresponders following treatment with Rubitecan.

- Pancreatic cancer kills approximately 29,000 Americans annually, and is the fourth leading cause of cancer deaths. Duplicating the Phase II results in much larger Phase III trials. Currently, three separate Phase III trials (a total of 1,800 patients) for Rubitecan are going on. The largest of the three trials is the *Rubitecan* versus *Gemzar* comparison in patients who have not undergone chemotherapy. Rubitecan's once-daily oral formulation, which the patient takes his or her medication five days on followed by two days off, mild side effect profile, and antitumor activity could propel Rubitecan above the competition. Gemzar is a once-weekly, 30 min i.v. administration that requires at least one trip per week to a medical facility (doctor's office, hospital, clinic, etc.) (Tzavlakis, 2000).

Antitumor activity:

- Antitumor effects of CPT-11 as a single drug was examined in 52 patients with prior chemotherapy including *cisplatin*-containing regimens who were enrolled in a Phase II study. These patients were randomly divided into two groups, and CPT-11 was administered once weekly at a dose of $100\,mg\,m^{-2}$ (Method A, 27 cases) or once biweekly at a dose of $150\,mg\,m^{-2}$ (Method B, 25 cases). Dose intensity was $72\,mg\,m^{-2}/week^{-1}$ in Method A and $61\,mg\,m^{-2}/week^{-1}$ in Method B. Method A was more effective than Method B, that is, response rates of Method A and B were 29.6% and 16.0%, respectively. The duration with 50% response was 94 days, and the 50% survival time was 233 days. It was remarkable that cases of serous adenocarcinoma as well as those of mucinous carcinoma and clear-cell carcinoma which were considered to be less sensitive to cisplatin responded to CPT-11 (Sugiyama *et al.*, 1997). At the end, it was considered that CPT-11 will be a useful drug for salvage chemotherapy for ovarian cancer.
- The cytotoxicity of CPT-11 on human ovarian epithelial malignancies was tested *in vitro* utilizing the ATP chemosensitivity assay. Flow cytometry was also performed on the fresh carcinoma specimens.

Methods: Fresh tumor samples were obtained at laparotomy from 20 patients with primary adenocarcinoma of the ovary and 1 patient with heavily pretreated recurrent ovarian carcinoma. Tumors were plated in an *in vitro* system and treated with varying doses of both CPT-11 and its active metabolite SN-38 (7-ethyl-10-hydroxycamptothecin), in addition to a panel of standard chemotherapeutic agents used in treating ovarian cancer. The results showed that it is a promising agent for further use in ovarian cancer (O'meara and Sevin, 1999)

- Another study by Noriyuki Katsumata, and other researchers of the National Cancer Center Hospital, Tokyo, Japan, in 1999 was focused on the advantage of using CPT

11 against ovarian cancer. CPT-11 and CBDCA are active agents in the treatment of ovarian cancer. They conducted phase I trial of the CPT-11 and CBDCA in advanced ovarian cancer. The objective of the study was to determine the maximum tolerated dose (MTD) in escalating doses of CPT-11 and CBDCA. Eligible patients had ovarian cancer failing to first-line chemotherapy, adequate organ functions, and PS 0 and 1, dose limiting toxicity (DLT) defined as grade 4 (G4) neutropenia or thrombocytopenia lasting >3 days or non-hematologic toxicity $\geq$G3. CPT-11 and CBDCA were administered as i.v. infusion on d1, d8 and d15, respectively. CBDCA dosage was estimated by CBDCA clearance (CL) $\times$ target AUC, and CL was calculated by Chatelut's formula. The initial dose of CPT-11 was $50\,\mathrm{mg\,m^{-2}}$, and the dose was escalated to 50 and 60. Treatment was repeated at 28-day interval. Twelve patients were registered and evaluated for toxicity. Median age was 55 (range 40–63) and median number of previous treatment regimens were 2 (range 1–4). Symptoms of toxicity (G1–4) in 12 patients have been diarrhea (5/12, 42%) and nausea/vomiting (8/12, 67%). Grade 3/4 toxicities of diarrhea have not been observed. As of 12–98, MTD was not yet reached. No DLT or grade ≥ 3 non-hematologic toxicity was observed up to now. Further dose escalation is under evaluation.

References

Armstrong, D. and O'Reilly, S. (1998) Clinical guidelines for managing topotecan-related hematologic toxicity. *Oncologist* 3(1), 4–10.

Berger, N.A. (2001) TOPO I Inhibitors in Breast, Ovary, and Lung Cancers. Project Number: 5 U01 CA63200-03. Case Western Reserve University, 10900 Euclid Avenue, Cleveland, OH 44106–4937.

Bokkel, ten H.W., Carmichael, J., Armstrong, D. and Gordon, A. (1997) Malfetano J Efficacy and safety of topotecan in the treatment of advanced ovarian carcinoma. *Semin Oncol.* 24(1 Suppl 5), S5-19–25.

Bookman, M.A. (1999) Extending the platinum-free interval in recurrent ovarian cancer: the role of topotecan in second-line chemotherapy. *Oncologist* 4(2), 87–94.

Budavari, S. (ed.) (1989) *The Merck Index: An Encyclopedia of Chemicals, Drugs and Biologicals. Merck & Co.*, Rahway, NJ.

Burris, H.A. 3rd (1999) The evolving role of oral topotecan. *Semin. Hematol.* 36(4 Suppl 8), 26–32.

Chan, S., Carmichael, J., Ross, G., Wheatley, A.L. and Julian D., (1999) Sequential Hycamtin Following Paclitaxel and Carboplatin in Advanced Ovarian Cancer. American Society of Clinical Oncology Annual Meeting. Abstract: 1436.

Creemers, G.J., Bolis, G., Gore, M., Scarfone, G., Lacave, A.J., Guastalla, J.P., Despax, R., Favalli, G., Kreinberg, R., Van Belle, S., Hudson, I., Verweij, J. and ten Bokkel Huinink, W.W. (1996) Topotecan, an active drug in the second-line treatment of epithelial ovarian cancer: results of a large European phase II study. *J. Clin. Oncol.* 14, 3056–61.

Dobelis, I.N. (ed.) (1989) *Magic and Medicine of Plants.* Reader's Digest Books, Pleasantville, NY.

Duke, J.A. and Ayensu, E.S. (1985) *Medicinal Plants of China.* Reference Publications, Inc., Algonac, MI.

Duke, J.A. and Foster, S. (1990) *A Field Guide to Medicinal Plants of Eastern and Central North America.* Houghton Mifflin Co., New York, NY.

Ferrari, S., Danova, M., Porta, C., Comolli, G., Brugnatelli, S., Pugliese, P. and Riccardi, A. (1999) Ascari E Circulating progenitor cell release and functional characterization after topotecan plus G-CSF and erythropoietin in small cell lung cancer patients. *Int. J. Oncol.* 15(4), 811–5.

Flora of North America Editorial Committee (1993) *Flora of North America*. Oxford University Press, Oxford, UK.
Goldwasser, F., Buthaud, X., Gross, M., Bleuzen, P., Cvitkovic, E., Voinea, A., Jasmin, C., Romain, D. and Misset, J.L. (1999) Decreased topotecan platelet toxicity with successive topotecan treatment cycles in advanced ovarian cancer patients. *Anticancer Drugs* 10(3), 263–5.
Goodman, L., Sanford, A., Goodman G.A. (eds) (1990) *The Pharmacological Basis of Therapeutics*, 8th edition. Pergamon Press, Elmsford, NY.
Heron, J.F. (1998) Topotecan: an oncologist's view. *Oncologist* 3(6), 390–402.
Heywood, V.H. (ed.) (1993) *Flowering Plants of the World*. Oxford University Press, New York, NY.
Hochster, H., Wadler, S., Runowicz, C., Liebes, L., Cohen, H., Wallach, R., Sorich, J., Taubes, B. and Speyer, J. (1999) Activity and pharmacodynamics of 21-Day topotecan infusion in patients with ovarian cancer previously treated with platinum-based chemotherapy. New York Gynecologic Oncology Group. *J. Clin. Oncol.*, 17(8), 2553–61.
Katsumata, N., Tsunematsu, R., Hida, K., Kasamatsu, T., Yamada, T., Tanemura, K. and Ohmi, K. (1999) Phase I Trial of Irinotecan (CPT-11) and Carboplatin (CBDCA) in Advanced Ovarian Cancer. American Society of Clinical Oncology Annual Meeting. Abstract: 1406
Koshiyama, M., Fujii, H., Kinezaki, M., Ohgi, S., Konishi, M., Hidetaka, N, Hayashi, M. and Yoshida, M. (2000) Chemosensitivity testing of irinotecan (CPT-11) in ovarian and endometrial carcinomas: a comparison with cisplatin. *Anticancer Res* 20(3A), 1353–8.
Lane, S.R., Cesano, A., Fitts, D. and Fields, S.Z. (1999) SmithKline Relationship Between Tumor Response and Survival in SCLC and Ovarian Cancer Patients Treated with IV Topotecan as Second-Line Therapy. American Society of Clinical Oncology *Annual Meeting*. Abstract: 1643.
McGuire, W.P., Blessing, J.A., Bookman, M.A., Lentz, S.S. and Dunton, C.J. (2000) Topotecan has substantial antitumor activity as first-line salvage therapy in platinum-sensitive epithelial ovarian carcinoma: a gynecologic oncology group study. *J. Clin. Oncol.* 18(5), 1062–7.
Mutschler, E. and Derendorf, H. (1995) *Drug Actions: Basic Principles and Therapeutic Aspects*. Medpharm Scientific Publishers, Stuttgart, Germany.
O'meara, A.T. and Sevin, B.U. (1999) *In vitro* sensitivity of fresh ovarian carcinoma specimens to CPT-11 (Irinotecan). *Gynecol Oncol* 72(2), 143–7.
Pettit, G.R., Pierson, F.H. and Herald, C.L. (1994) *Anticancer Drugs From Animals, Plants and Microorganisms*. John Wiley & Sons, Inc., New York, NY.
Physician's Desk Reference © (1995) Medical Economics Data Production Company, Montvale, NJ.
Saltz, L.B., Spriggs, D., Schaaf, L.J., Schwartz, G.K., Ilson, D., Kemeny, N., Kanowitz, J., Steger, C., Eng, M., Albanese, P., Semple, D., Hanover, C.K., Elfring, G.L., Miller, L.L, Kelsen, D. (1998) Phase I clinical and pharmacologic study of weekly cisplatin combined with weekly irinotecan in patients with advanced solid tumors. *J. Clin. Oncol.* 16, 3858–65.
Schellens, J.H., Creemers, G.J., Beijnen, J.H., Rosing, H., de Boer-Dennert, M., McDonald, M., Davies, B. and Verweij, J. (1996) Bioavailability and pharmacokinetics of oral topotecan: a new topoisomerase I inhibitor. *Br. J. Cancer* 73, 1268–71.
Simpson, Beryl Brintnall and Molly Conner-Ogorzaly. (1986) *Economic Botany: Plants in Our World*. McGraw-Hill Publishing Co., New York, NY.
Sugiyama, T., Yakushiji, M., Nishida, T., Ushijima, K., Okura, N., Kigawa, J. and Terakawa, N. (1998) Irinotecan (CPT-11) combined with cisplatin in patients with refractory or recurrent ovarian cancer. *Cancer Lett.* 128(2), 211–8.
Sugiyama, T., Nishida, T., Ookura, N., Yakushiji, M., lkeda, M., Noda, K., Kigawa, J., Itamochi, H. and Takeuchi, S. (1997) Is CPT-11 useful as a salvage chemotherapy for recurrent ovarian cancer? (Meeting abstract). *Proc Annu. Meet. Am. Soc. Clin. Oncol.* 16A, 1347.
Tyler, Varro E. (1993) *The Honest Herbal: A Sensible Guide to The Use of Herbs and Related Remedies*.Pharmaceutical Products Press, New York.
Tzavlakis, M., (2000) Super Potential. *Annual Meeting of the American Society of Clinical Oncology* May 20–23 N. Orleans, Louisiana, USA.

Catharanthus

See in *Vinca*.

Cephalotaxus

See in *Taxus* under Further details.

Podophyllum peltatum (Mandrake, American) (Berberidaceae) — Antitumor

Synonyms: Wild lemon, Ground lemon, May Apple, Racoonberry.

Location: Of North America origin. Common in the eastern United States and Canada, North America, growing there profusely in wet meadows and in damp, open woods.

Appearance (Figure 3.1)
Stem: solitary, mostly unbranched, 0.3–0.5 m high.
Root: is composed of many thick tubers, fastened together by fleshy fibres, which spread greatly underground.

Figure 3.1 Podophyllum peltatum.

Leaves: smooth, stalked, peltate in the middle like an umbrella, of the size of the hand, composed of 5–7 wedge-shaped divisions.
Flowers: solitary, drooping white, about 2 cm across, with nauseous odour.
Fruit: size and shape of a common rosehip, being 3–6 cm long. Yellow in colour, sweet in taste.
In bloom: May.

Tradition: North American Indians used it as an emetic and vermifuge.

Biology: The rhizome develops underground for several years before a flowering stem emerges (only one shoot per root). The plant can be propagated either by runners or by seed. For cultivation, adequate fertilization is recommended.

Part used: root, resin

Active ingredients: *podophyllotoxin* (a neutral crystalline substance), *podophylloresin* (amorphous resin), *diphyllin* and *aryltetralin* (podophyllum lignan), *etoposide* (VP–16), *teniposide* (semisynthetic derivative of 4′-demethylepipodophyllotoxin, naturally occurring compounds).

Particular value: It was included in the British Pharmacopoeia in 1864. It is considered as one of the medicine with the most extensive service: it is used for all hepatic complaints, as antibilious, cathartic, hydragogue, purgative.

Precautions: Leaves and roots are poisonous, *podophyllotoxins* are classical spindle poisons causing inhibition of mitosis by blocking mitrotubular assembly, and should be avoided during pregnancy.

Indicative dosage and application

- *Etoposide* is used as etoposide phosphate (Etopophos; Bristol-Myers Squibb Company, Princeton, NJ) and because it is water soluble can be made up to a concentration of 20 mg ml^{-1}, however, it can be given as a 5 min bolus, in high doses in small volumes, and as a continuous infusion.
- Penile warts in selected cases can be safely treated with 0.5–2.0% *podophyllin* self applied by the patient at a fraction of the cost of commercially available podophyllotoxin (White *et al.*, 1997).

Documented target cancers

- Antiproliferative effects on human peripheral blood mononuclear cells and inhibition of *in vitro* immunoglobulin synthesis.
- *Etoposide* appears to be one of the most active drugs for small cell lung cancer, testicular carcinoma (the Food and Drug Administration approved indication), ANLL and malignant lymphoma. Etoposide also has demonstrated activity in refractory pediatric neoplasms, hepatocellular, esophageal, gastric and prostatic carcinoma, ovarian cancer, chronic and acute leukemias and non-small-cell lung cancer, although additional single and combination drug studies are needed to substantiate these data (Schacter, 1996).
- *Proresid* (a mixture of natural extracts from *Podophyllum* sp.) has been used to a triple drug therapy with high doses of *Endoxan* and *Methotrexate* instead of the earlier long-term Endoxan treatment in addition of surgery (Vahrson *et al.*, 1977).

Further details

Related compounds

- *Podophyllotoxin* is a natural product isolated from *Podophyllum peltatum* and *Podophyllum emodi* and has long been known to possess medicinal properties. Etoposide (VP-16), a podophyllotoxin derivative, is currently in clinical use in the treatment of many cancers, particularly small-cell lung carcinoma and testicular cancer. This compound arrests cell growth by inhibiting DNA topoisomerase II, which causes double strand breaks in DNA. VP-16 does not inhibit tubulin polymerization, however, its parent compound, podophyllotoxin, which has no inhibitory activity against DNA topoisomerase II, is a potent inhibitor of microtubule assembly. In addition to these two mechanisms of action, an unknown third mechanism of action has also been proposed for some of the recent modifications of podophyllotoxins. Some of the congeners exhibited potent antitumor activity, of which etoposide and teniposide are in clinical use, NK 611 is in phase II clinical trials and many compounds are in the same line. Recent developments on podophyllotoxins have led structure–activity correlations which have assisted in the design and synthesis of new podophyllotoxin derivatives of potential antitumor activity. Modification of the A-ring gave compounds having significant activity but less than that of etoposide, whereas modification of the B-ring resulted in the loss of activity. One of the modifications in the D-ring produced GP-11 which is almost equipotent with etoposide. E-ring oxygenation did not affect the DNA cleavage which led to the postulation of the third mechanism of action. It has also been observed that free rotation of E-ring is necessary for the antitumor activity. The C4-substituted aglycones have a significant place in these recent developments. Epipodophyllotoxin conjugates with DNA cleaving agents such as distamycin increased the number of sites of cleavage. The substitution of a glycosidic moiety with arylamines produced enhanced activity. Modification in the sugar ring resulted in the development of the agent, NK 611 which is in clinical trial at present (Nudelman *et al.*, 1997).
- Aryltetralin lignan is a constituent of the resins and roots/rhizomes of *P. hexandrum* and *P. peltatum*. A method confirms that *P. hexandrum* resins and roots/rhizomes contain approximately four times the quantity of lignans as do those of *P. peltatum* and also that there is a significant variation in the lignan content of *P. hexandrum* resins (But *et al.*, 1997).

Related species

- Podophyllotoxin is also, one of the main Related compounds of the *Bajiaolian* root. Bajiaolian (*Dysosma pleianthum*), one species in the Mayapple family, has been widely used as a general remedy and for the treatment of snake bite, weakness, condyloma accuminata, lymphadenopathy and tumors in China for thousands of years. The herb was recommended by either traditional Chinese medical doctors or herbal pharmacies for postpartum recovery and treatment of a neck mass, hepatoma, lumbago and dysmenorrhea (Sarin *et al.*, 1997).

- *Podophyllum emodi:* Indian *Podophyllum*, a native of Northern India. The roots are much stouter, more knotty, and twice as strong as the American. It contains twice as much podophyllotoxin. It is official in India and in close countries and it is used in place of ordinary *Podophyllum* (Grieve, 1994).

Other medical activity

- A mixture of natural and semisynthetic (modified) glycosides from *Podophyllum emodi* has been used for many years in the treatment of rheumatoid arthritis, but its use is hampered by gastrointestinal side effects. Highly purified podophyllotoxin (CPH86) and a preparation containing two semisynthetic podophyllotoxin glycosides (CPH82) are currently being tested in clinical trials. In this study these drugs were shown to inhibit *in vitro* [3H]-thymidine uptake of human peripheral blood mononuclear cells stimulated by the mitogens concanavalin A, phytohemagglutinin and pokeweed mitogen. Complete inhibition was observed with CPH86 in concentrations $\geq 20\,\mathrm{ng\,ml^{-1}}$ and with CPH82 in concentrations $\geq 1\,\mu\mathrm{g\,ml^{-1}}$ (Truedsson, *et al.*, 1993).
- In conclusion, both CPH86 and CPH82 inhibit mitogen-induced lymphocyte proliferation and immunoglobulin synthesis and the results may be of help in determining optimal dose levels if related to treatment effects (Truedsson *et al.*, 1993).

References

Bhattacharya, P.K., Pappelis, A.J., Lee, S.C., BeMiller, J.N. and Karagiannis, C.S. (1996) Nuclear (DNA, RNA, histone and non-histone protein) and nucleolar changes during growth and senescence of may apple leaves. *Mech. Ageing Dev.* 92(2–3), 83–99.

But, P.P., Cheng, L. and Kwok, I.M. (1997) Instant methods to spot-check poisonous podophyllum root in herb samples of clematis root. *Vet. Hum. Toxicol.* 39(6), 366.

But, P.P., Tomlinson, B., Cheung, K.O., Yong, S.P., Szeto, M.L. and Lee, C.K. (1996) Adulterants of herbal products can cause poisoning. *BMJ*, 313(7049), 117.

Damayanthi, Y. and Lown, J.W. (1998) Podophyllotoxins: current status and recent developments. *Curr. Med. Chem.* 5(3), 205–52.

Dorsey, K.E., Gallagher, R.L., Davis, R. and Rodman, O.G. (1987) Histopathologic changes in condylomata acuminata after application of *Podophyllum. J. Natl Med. Assoc.* 79(12), 1285–8.

Frasca, T., Brett, A.S. and Yoo, S.D. (1997) Mandrake toxicity. A case of mistaken identity. *Arch. Intern. Med.* 157(17), 2007–9.

Goel, H.C., Prasad, J., Sharma, A. and Singh, B. (1998) Antitumour and radioprotective action of *Podophyllum hexandrum. Indian J. Exp. Biol.* 36(6), 583–7.

Kadkade, P.G. (1981) Formation of podophyllotoxins by *Podophyllum peltatum* tissue cultures. *Naturwissenschaften* 68(9), 481–2.

McDow, R.A. (1996) Cryosurgery and podophyllum in combination for condylomata. *Am. Fam. Physician* 53(6), 1987–8.

Mack, R.B. (1992) Living mortals run mad. Mandrake (podophyllum) poisoning. *NC Med. J.*, 53(2), 98–9.

Nudelman, A., Ruse, M., Gottlieb, H.E. and Fairchild, C. (1997) Studies in sugar chemistry. VII. Glucuronides of podophyllum derivatives. *Arch. Pharm.* (*Weinheim*), 30(9–10), 285–9.

Perkins, J.A., Inglis, A.F. Jr and Richardson, M.A. (1998) Latrogenic airway stenosis with recurrent respiratory papillomatosis. *Arch. Otolaryngol. Head Neck Surg.* 124(3), 281–7.

Sarin, Y.K., Kadyan, R.S. and Simon, R. (1997) Condyloma accuminata. *Indian Pediatr.* 34(8), 741–2.

Schacter, L. (1996) Etoposide phosphate: what, why, where, and how? *Semin. Oncol.* 23(6 Suppl 13), 1–7.

Takeya, T. and Tobinaga, S. (1997) Weitz' aminium salt initiated electron transfer reactions and application to the synthesis of natural products. *Yakugaku Zasshi.* 117(6), 353–67.

Truedsson, L., Geborek, P. and Sturfelt, G. (1993) Antiproliferative effects on human peripheral blood mononuclear cells and inhibition of in vitro immunoglobulin synthesis by *Podophyllotoxin* (CPH86) and by semisynthetic lignan glycosides (CPH82). *Clin. Exp. Rheumatol.* 11(2), 179–82.

Vahrson, H., Wolf, A. and Jankowski, R. (1977) Combined therapy of ovarian cancer. *Geburtshilfe Frauenheilkd* 37(2), 131–8.

White, D.J., Billingham, C., Chapman, S., Drake, S., Jayaweera, D., Jones, S., Opaneye, A. and Temple, C. (1997) Podophyllin 0.5% or 2.0% v podophyllotoxin 0.5% for the self treatment of penile warts: a double blind randomised study. *Genitourin Med.* 73(3), 184–7.

Vinca rosea Linn. (Periwinkle) (Apocynaceae)

Immunomodulator
Cytotoxic
Antitumor

Madagascar periwinkle is a modern day success story in the search for naturally occurring anticancer drugs.

Synonyms: *Catharanthus roseus* (G. Don), *Lochnera rosea* (Reichb.), Madagascar periwinkle, and rose periwinkle.

Location: Of Madagascar, tropical Africa and generally Tropics origin. It can be found in East Indies, Madagascar and America. It has escaped cultivation and naturalized in most of the tropical world where it often becomes a rampant weed. Over the past hundreds of years, the periwinkle has been widely cultivated and can now be found growing wild in most warm regions of the world, including several areas of the southern United States. Madagascar periwinkle is grown commercially for its medicinal uses in Australia, Africa, India and southern Europe.

Appearance (Figure 3.2)
Stem: small under-shrub up to 40–80 cm high in its native habitat. the broken stem of Madagascar periwinkle exudes a milky latex sap.
Leaves: retains its glossy leaves throughout the winter; are always placed in pairs on the stem.
Flowers: springing from their axils, five-petaled flowers are typically rose pink, but among the many cultivars are those with pink, red, purple and white flowers. The flowers are tubular, with a slender corolla tube about an inch long that expands to about 25 mm and a half across. They are borne singly throughout most of the summer.
In bloom: Summer.

Biology: It propagates itself by long, trailing and rooting stems, and by their means not only extends itself in every direction, but succeeds in obtaining an almost exclusive possession of the soil. Because of the dense mass of stems, the periwinkle deprives the weaker plants of light and air.

Tradition: It was one of the plants believed to have power to exorcise evil spirits. Apuleius, in his Herbarium (printed 1480), writes: "this wort is of good advantage for many purposes, first against devil sinks and demoniacal possessions and against snakes and wild beasts and against poisons and for various wishes and for envy and for terror...." "The periwinkle is a great binder," said an old herbalist, and both Dioscorides and Galen commended it against fluxes. It was

Figure 3.2 Vinca rosea.

consider a good remedy for cramp. An ointment prepared from the bruised leaves with lard has been largely used in domestic medicine and is reputed to be both soothing and healing in all inflammatory ailments of the skin and an excellent remedy for bleeding piles. In India, juice from the leaves was used to treat wasp stings. In Hawaii, the plant was boiled to make a poultice to stop bleeding and throughout the Caribbean, an extract from the flowers was used to make a solution to treat eye irritation and infections. In France, it is considered an emblem of friendship. Parts used: leaves, stems, flower buds.

Active ingredients

- *Ajmalicine, vindoline, catharanthine.*
- *Vindoline* is enzymatically coupled with *catharanthine* to produce the powerful cytotoxic dimeric alkaloids: *vinblastine* (*VBL*), *vincristine* (*VCR*) and *leurosidine.*

Particular value: Cure for diabetes, anticancer drug. The plant has been used for centuries to treat diabetes, high blood pressure, asthma, constipation and menstrual problems. In the 1950s researchers learned of a tea that Jamaicans had been drinking to cure diabetes. A native who had been drinking the tea sent a small envelope full of leaves to researchers explaining that the leaves came from a plant known as the Madagascar periwinkle. The native explained that the tea was used in the absence of insulin treatment and apparently already had a worldwide reputation and was being sold as a remedy under the name *Vinculin*.

The status of vinca application in cancer therapy: The plant was used in traditional medicine. When tested in scientific studies it was demonstrated that it could be used in diabetes and anticancer research with great advantages. In the 1950s, a Dr Johnston who had been practicing in the Jamaica area, was quite convinced that his diabetic patients had received some benefit from drinking extracts of the periwinkle leaves. Therefore, it was decided among the researchers to send these leaves to a Dr Collip at the University of Western Ontario. The doctor had already been working with another group on insulin derived from a hormone, so it seemed logical to send the leaves of the periwinkle to him. Dr Collip decided to make a water extract to determine if, when given orally, they would lower blood sugar levels. These extracts were given to animals, but were not found to have any effect on the blood sugar or on the disease. One of Dr Collip's colleagues, a Dr McAlpine decided to give the water extract to a few of his diabetic patients, who had volunteered to try it. There was no effect except in one mildly diabetic woman. In the absence of oral activity and as a final resort, Dr Collip decided to give the most concentrated dose to a few rats by intraperitoneal injection. The rats survived for about five days, but then died rather unexpectedly from diffuse multiple abscesses. This intrigued the doctor because the extracts that had been given had been sterilized. Dr Collip became very excited because another colleague of his had published that overdoses of cortisone in rats also led to their death from multiple abscesses. Dr Collip wondered if perhaps the periwinkle plant might be a source of cortisone. Unfortunately it was found that the two had very different mechanisms involved. In Cortisone, lymphocytes are destroyed resulting in its well-known immunosuppressive effect; in the case of the periwinkle extracts, it was found that after a single injection there was a rapid but transient depression of the WBC count, which was traced to the destruction of the bone marrow. In view of the dramatic effect on the bone marrow, it looked like there might be one or more compounds present in the periwinkle that might be useful in the treatment of cancers of the hematopoietic system such as lymphomas and leukemias. Therefore, it was decided to try and identify and isolate the component in the extracts responsible for the effects on the WBC counts and bone marrow.

In 1954 Dr Charles T. Beer came to work in Dr Collip's laboratory on a one-year fellowship. He looked at the problem of isolating active compounds from the periwinkle plant. When he started working on the project, the supply of periwinkle leaves was a problem. Dr Johnston in Jamaica was still convinced that the research was headed in the wrong direction. He felt like researchers should look for a cure for diabetes instead of a cure for cancer. So he decided to continue to supply Dr Beer with dried periwinkle leaves. Unfortunately it took so many leaves to make the extract that he decided to grow the periwinkle himself in Ontario. After working on the project for a year, Dr Beer finally isolated a small amount of unknown alkaloid. In rats, this alkaloid was highly active and there was a dramatic decrease in the WBC counts and a marked depletion of the bone marrow. He decided to name the alkaloid *vincaleukoblastine* (the name was shortened later to *vinblastine*). He found upon further observation of the plant that the periwinkle contained tons of useful alkaloids (70 in all at last count). Some of the alkaloids isolated contained properties that lowered blood sugar levels, others lowered blood pressure, some acted as hemostatics. Upon further investigation of VBL, Dr Beer also noted some activity but in smaller amounts. The related alkaloid was VCR, but was present in an amount insufficient for isolation in the laboratory. VCR was later isolated in crystalline form by chemists at the Eli Lilly Co.

Later, there were isolated about 100 alkaloids, but there were not all suitable for clinical use. The most of the part of this investigation was done by the American pharmaceutical company: Eli Lilly and the responsible professor was: Dr Gordon H. Svoboda. The *C. roseus* bisindoles, VBL and VCR, were the first plant products to be approved by the FDA for cancer treatment in the early of 1970, and are still currently used. The needs of production of final product for medical use are high, without satisfactory cover, because of the low concentration of *vinblastine* and *vincristine* in *C. roseus*, although it is cultivated in several tropical countries (Samuelsson, 1992).

Precautions: Madagascar periwinkle is poisonous if ingested or smoked. It has caused poisoning in grazing animals. Even under a doctor's supervision for cancer treatment, products from Madagascar periwinkle produce undesirable side effects.

The principal *DLT* of *VCR* is peripheral neurotoxicity. In the beginning, only symmetrical sensory impairment and parasthesias may be encountered. However, neuritic pain and motor dysfunction may occur with continued treatment. Loss of deep tendon reflexes, foot and wrist drop, ataxia and paralysis may also be observed with continued use. These effects are almost always symmetrical and may persist for weeks to months after discontinuing the drug. These effects usually begin in adults who have received a cumulative dose of 5–6 mg and the toxicity may occasionally be profound after a cumulative dose of 15–20 mg. Children generally tolerate this toxicity better than adults do, and the elderly are particularly susceptible. Other toxicities involving VCR are gastrointestinal with symptoms such as constipation, abdominal cramps, diarrhea, etc. Cardiovascular symptoms include hypertension and hypotension and a few reports of massive myocardial infarction. The principle toxicity of VBL is myelosuppression or in particular neutropenia. Neurotoxicity occurs much less commonly with VBL than VCR and is generally observed in patients who have received protracted therapy. Hypertension is the most common cardiovascular toxicity of VBL. Sudden and massive myocardial infarctions and cerebrovascular events have also been associated with the use of single agent VBL and multiagent regimens. Pulmonary toxicities include acute pulmonary edema and acute bronchospasm. Pregnant women and people with neuromuscular disorders should steer clear of these drugs. With pregnant women, VCR and VBL have been found to cause severe birth defects.

Indicative dosage and application

- VCR is routinely given to children as a bolus intravenous injection at doses of $2.0\,mg\,m^{-2}$ weekly.
- For adults, the conventional weekly dose is $1.4\,mg\,m^{-2}$.
- A restriction of the absolute single dose of VCR to $2.0\,mg\,m^{-2}$ has been adopted by many clinicians over the last several decades, mainly because of reports that show an increasing neurotoxicity at higher doses.
- VBR has been given by several schedules. The most common schedule involves weekly bolus doses of $6\,mg\,m^{-2}$ incorporated into combination chemotherapy regimens such as ABVD (adriamycin, bleomycin, VBL, dacarbazine) and the MOPP–AVB hybrid regimen (nitrogen mustard, VCR, prednisone, procarbazine, adriamycin, bleomycin, VBL) (Canellos, 1992).

Documented target cancers: extracts from Madagascar periwinkle have been shown to be effective in the treatment of various kinds of leukemia, skin cancer, lymph cancer, breast cancer and Hodgkin's disease.

VCR is used against childhood's leukemia, Hodgkin's disease and other lymphomas. VBL is mainly used for the treatment of Hodgkin's disease, testicular cancer, breast cancer, Kaposi's sarcoma and other lymphomas (Canellos, 1992; Samuelsson, 1992).

Further details

Related species

- *Vinca major* (Apocynaceae family), with common names: large periwinkle, big periwinkle; it is a fast growing herbaceous perennial groundcover with evergreen foliage and pretty blue flowers. It is native to France and Italy, and eastward through the Balkans to northern Asia Minor and the western Caucasus. *V. major* and *V. minor* are the most commonly cultivated. Herbalists for curing diabetes have long used it, because it can prove an efficient substitute for insulin. It is used for in herbal practice for its astringent and tonic properties in menorrhagia and in hemorrhages generally. For obstructions of mucus in the intestines and lungs, diarrhea, congestions, hemorrhages, etc., periwinkle tea is a good remedy. In cases of scurvy and for relaxed sore throat and inflamed tonsils, it may also be used as a gargle. For bleeding piles, it may be applied externally. Apparently all the *vincas* are poisonous if ingested. Numerous alkaloids, some useful to man, have been isolated from big and common periwinkle (Grieve, 1994).
- Common periwinkle (*V. minor*) is similar but has smaller leaves (less than 5 cm long) and smaller flowers (2.5 cm or less across) than *V. major*, and is more cold hardy and more tolerant of shade. It is used for producing *Catharanthus* alkaloids. Also, a homoeopathic tincture is prepared from the fresh leaves of it and is given medicinally for the milk-crust of infants as well as for internal hemorrhages. Its flowers are gently purgative, but lose their effect on drying. If gathered in the spring and made into a syrup, they will impart thereto all their virtues and this is excellent as a gentle laxative for children and also for overcoming chronic constipation in grown-ups (Grieve, 1994).

Related compounds

- *VBR* and *VCR* are dimeric *Catharanthus* alkaloids isolated from *Vinca* plants. Both VBR and VCR are large, dimeric compounds with similar but complex structures. They are composed of an indole nucleus and a dihydroindole nucleus. They are both structurally identical with the exception of the substituent attached to the nitrogen of the *vindoline* nucleus where VCR possesses a formyl group and VBL has a methyl group. However, VCR and VBL differ dramatically in their antitumor spectrum and clinical toxicities. Both alkaloids are therapeutically proven to be effective in the treatment of various neoplastic diseases. Consequently, the determinations of these compounds in plant samples, as well as biological fluids, are of interest to many scientists. Many gas and high-performance liquid chromatographic (HPLC) and mass spectrometric methods have been developed for the determination of VCR and VBL in either plant samples or biological systems. The potential use of information-rich detectors such as mass spectrometry with capillary zone electrophoresis (CZE) has made this a more attractive separation method (Chu *et al.*, 1996).
- VBL and VCR, which belong to the group of *Vinca* alkaloids, induce cytotoxicity by direct contact with tubulin, which is the basic protein subunit of microtubules.

Other biochemical effects that have been associated with VBL and VCR include: competition for transport of amino acids into cells; inhibition of purine biosynthesis; inhibition of RNA, DNA and protein synthesis; inhibition of glycolysis; inhibition of release of histamine by mast cells and enhanced release of epinephrine; and disruption in the integrity of the cell membrane and membrane functions. Microtubules are present in eukaryotic cells and are vital to the performance of many critical functions including maintenance of cell shape, mitosis, meiosis, secretion and intracellular transport. VBL and VCR exert their antimicrotubule effects by binding to a site on tubulin that is distinctly different from the binding sites of others. They have a binding constant of 5.6×10^{-5} M and initiate a sequence of events that lead to disruption of microtubules. The binding of VBL and VCR to tubulin, in turn, prevents the polymerization of these subunits into microtubules. The net effects of these processes include the blockage of the polymerization of tubulin into microtubules, which may eventually lead to the inhibition of vital cellular processes and cell death. Although most evidence suggests that mitotic arrest is the principal cytotoxic effect of the alkaloids, there is also evidence that suggests that the lethal effects of these agents may be attributed in part to effects on other phases of the cell cycle. The alkaloids also appear to be cytotoxic to nonproliferating cells *in vitro* and *in vivo* in both G1 and S cell cycle phases. In other words, VBL and VCR work by inhibiting mitosis in metaphase (Danieli, 1998; Garnier *et al.*, 1996).

- Studies with germinating seedlings have suggested that alkaloid biosynthesis and accumulation are associated with seedling development. Studies with mature plants also reveal this type of developmental control. Furthermore, alkaloid biosynthesis in cell suspension cultures appears to be coordinated with cytodifferentiation. Vindoline biosynthesis in *Catharanthus roseus* also appears to be under this type of developmental control (Noble, 1990). Vindoline as well as the dimeric alkaloids are restricted to leaves and stems, whereas *catharanthine* is distributed equally throughout the aboveground and underground tissues. The developmental regulation of vindoline biosynthesis has been well documented in *C. roseus* seedlings, in which it is light inducible (Kutney *et al.*, 1988). This is in contrast to catharanthine, which also accumulates in etiolated seedlings. Furthermore, cell cultures that accumulate catharanthine but not vindoline recover this ability upon redifferentiation of shoots. These observations suggest that the biosynthesis of catharanthine and vindoline is differentially regulated and that vindoline biosynthesis is under more rigid tissue–development and environment-specific control than is that of catharanthine. The early stages of alkaloid biosynthesis in *C. roseus* involve the formation of tryptamine from tryptophan and its condensation with *secologanin* to produce the central intermediate *strictosidine*, the common precursor for the monoterpenoid *indole* alkaloids. The enzymes catalyzing these two reactions are *tryptophan decarboxylase* (TDC) and *strictosidine* synthase (STR1), respectively. *Strictosidine* is the precursor for both the *Iboga* (catharanthine) and *Aspidosperma* (tabersonine and vindoline) types of alkaloids. The condensation of vindoline and catharanthine leads to the biosynthesis of the bisindole alkaloid vinblastine (St-Pierre *et al.*, 1999).
- A successful attempt of production of Indole alkaloids by selected hairy root lines of *C. roseus* has been done. Approximately 150 hairy root clones from four varieties

were screened for their biosynthetic potential. Two key factors affecting productivity, growth rate and specific alkaloid yield. The detection of vindoline in these clones may potentially present a new source for the *in vitro* production of VBL. Production of vindoline and catharanthine by plant tissue culture and subsequent catalytic coupling *in vitro* is a possible alternative to using tissue culture alone to produce VBL and VCR. Recently, enzyme catalyzed techniques have been developed for the conversion of vindoline and catharanthine to bisindole alkaloids. Catharanthine is readily produced in cell suspension and hairy root cultures in amounts equal to or above that found in intact plant (Rajiv *et al.*, 1993).

References

Bhadra, R., Vani, S., Jacqueline, V. and Shanks (1993) Production of indole alkaloids by selected hairy root lines of *Catharanthus roseus. Biotech. Bioeng.* 41, 581–92.

Canellos, George P. (1992) Chemotherapy of Advanced Hodgkin's Disease with MOPP, BVD, or MOPP alternating with ABVD. *N Eng J. Med.* 327, 1478–84.

Chu, I., Bodnar, J.A., White, E.L. and Bowman, R.N. (1996) Quantification of vincristine and vinblastine in *Catharanthus roseus* plants by capillary zone electrophoresis. *J. Chromat.* A. 755, 281–8.

Danieli, B. (1998) *Vinblastine*-type antitumor alkaloids: a method for creating new C17 modified analogues. *J. Org. Chem.*, 63, 8586–8.

Garnier, F., Label, Ph., Hallard, D., Chenieux, J.C., Rideau, M. and Hamdi, S. (1996) Transgenic periwinkle tissues overproducing cytokinins do not accumulate enhanced levels of indole alkaloids. *Plant Cell, Tissue Organ Culture* 45, 223–30.

Gurr, Sarah J. (1996) The Hidden Power of Plants. *The Garden* 121, 262–4.

Jageti, G.C., Krishnamurthy, H. and Jyothi, P. (1996) Evaluation of cytotoxic effects of different doses of *vinblastine* on mouse spermatogenesis by flow cytometry. *Toxicology* 112, 227–36.

Jordan, M.A., Thrower, D. and Wilson, L. (1991) Mechanism of Inhibition of cell proliferation by *Vinca* alkaloids. *Cancer Res.* 51, 2212–22.

Jordan, M.A., Thrower, D. and Wilson, L. (1992) Effects of *vinblastine, podophyllotoxin* and *nocodzole* on mitotic spindles. *J. Cell Sci.* 102, 401–16.

Joyce, C. (1992) What past plants hunts produced. *BioScience*, 42, 402.

Kallio, M., Sjoblom, T. and Lahdetie, J. (1995) Effects of *vinblastine* and *colchicine* on male rat meiosis *in vivo*: disturbances in spindle dynamics causing micronuclei and metaphase arrest. *Environ Mol Mutagen.* 25, 106–17.

Kutney, J.P., Choi, L.S.L., Nakano, J., Tsukamoto, H., McHugh, M. and Boulet, A. (1988) A highly efficient and commercially important synthesis of the antitumor *catharanthus* alkaloids *vinblastine* and *leurosidine* from *catharanthine* and *vindoline. Heterocycles*, 27(8), 1845–53.

Madoc-Jones, H. and Mauro, F. (1968) Interphase action of *vinblastine* and *vincristine*: differences in their lethal actions through the mitotic cycle of cultured mammalian cells. *J. Cell Physiol* 72, 185–96.

Noble, R.L. (1990) The discovery of the vinca alkaloids – chemotherapeutic agents. *Biochem Cell Biol.*, 68, 1344–51.

Pollner, F. (1990) Chemo edging up on four so-far intractable tumors: U.S. and European teams report the first clinical successes – some dramatic – from novel attacks. *Medical World News* 31, 13–16.

Powell, J. (1991) Senior Seminar Presentation: Fall, BIOL 4900.

Rowinsky, E.K. and Donehower, R.C. (1991) The clinical pharmacology and use of antimicrotuble agents in cancer therapeutics. *Pharmacol. Therapeutics* 52, 35–84.

Samuelsson, G. (1992) *Drugs of Natural Origin – A textbook of Pharmacognosy*. Third revised, enlarged and translated edition. Swedish Pharmaceutical Press.

St-Pierre, B., Vazquez-Flota, F.A. and De Luca, V. (1999) Multicellular compartmentation of *Catharanthus roseus* alkaloid biosynthesis predicts intercellular translocation of a pathway intermediate. *Plant Cell* 11, 887–900.

Viscum album (Mistletoe) (Loranthaceae) — Immunomodulator, Cytotoxic

Location: Throughout Europe, Asia, N. Africa. It can be easily found, though not in abundant numbers.

Appearance
Stem: yellowish-green, branched, forming bushes 0.6–2 m in diameter.
Root: Nonexistent. The plant is a semiparasitic evergreen shrub growing on branches of various tree hosts, mostly apple, poplar, ash, hawthorn and lime, more rarely on oak and pear.
Leaves: opposite, tongue-shaped, yellowish-green.
Flowers: small, inconspicuous, clustered in groups of three.
Fruit: globular, pea-sized white berry, ripening in December.
In bloom: March–May.

Biology: Mistletoe is propagated exclusively by seed, which is carried distantly with the aid of birds (mostly the thrush). According to host specifity three different races can be distinguished. The plant is dioecious with very reduced male and female flowers. The life cycle of *V. album* is described starting from seed germination to the development of the leaves. The parasitism affords special adaptation to mineral nutrition.

Tradition: Following their visions, the Druids used to cut mistletoe from trees with a golden knife at the beginning of the year. They held that the plant protected its possessor from all evil. According to a Scandinavian legend, Balder, the god of Peace, was slain with an arrow made of mistletoe. Later, however, mistletoe was rendered an emblem of love rather than hate. Its poisonous nature has been further exploited for the construction of knifes as a defensive weapon.

Parts used: Leaves and young twigs.

Active ingredients: viscotoxin, mistletoe alkaloids and three lectins (lactose-specific lectin, galactose-specific lectin, *N*-acetylgalactosamine-specific lectin).

Particular value: Mistletoe preparations are well-tolerated with no significant toxicities observed so far.

The status of mistletoe application in cancer therapy: Mistletoe was introduced in the treatment of cancer in 1917. Rudolf Steiner (1861–1925), founder of the Society for Cancer Research, in Arlesheim (Switzerland) was the first to mention the immunoenhancing properties of mistletoe, suggesting its use as an adjutant therapy in cancer treatment.

Therapy of cancer with a *Viscum* extract has been carried out in Europe for over six decades in thousands of patients. Extracts from the plant are used mainly as injections.

Currently, there is a number of mistletoe preparations used in many countries against different kinds of cancer:

- *Iscador* and *Helixor* are licensed medications made from plants growing on different host trees, like oak, apple, pine and fir, and administered in different kinds of cancer therapy. Some

Iscador preparations also include metal, for example silver, mercury and copper. Iscador is usually given by injection. However, it can also be taken orally. The injection treatment typically lasts 14 days with one injection each day. It has been approved for use in Austria, Switzerland and West Germany; it apparently is also being used in France, Holland, Eastern Europe, Britain and Scandinavia. Proponents of the treatment claim that in 1978 almost 2,000,000 ampules were sold in countries where Iscador is prescribed and that about 30,000 patients are treated with it each year. Iscador is manufactured by the Verein fuer Krebsforschung (Cancer Research Association), a nonprofit organization in Arlesheim, Switzerland.

- *Iscusin-Viscum* preparations contain mistletoe from eight different host-trees and are produced according to a particular "rhythmic" procedure and additionally "potentialized." Sterilization is achieved by the addition of oligodynamic silver. The indications given are: precancerous conditions, postoperative tumor prevention, operable tumors, and inoperable tumors. Each of the eight preparations (according to host-tree) has its own list of indications. Iscucin is supposed to be injected close to the tumor between 5 and 7 p.m.; the dosage and the frequency depend on body temperature. However, no preclinical studies have been published on *iscucin*. In the clinical field, only individual case histories are available, four of which have minimal documentation, and results that can be explained without *iscucin*. *Iscucin* is produced and distributed by Wala-Heilmittel GmbH, Eckwalden.
- *Isorel* is an aqueous extract from whole shoots of mistletoe, the subspecies fir (Isorel A), apple (Isorel M) and pine (Isorel P) in each case. The preparation is injected hypodermically. It is usually applied for the medicative treatment of malignant tumors, postoperative and recidivation and prophylaxis of metastases, malignant illness of the hemopoietic system and defined precancerous stages. Isorel A is used principally for the treatment of male patients, while Isorel M is the respective preparation for female patients. Isorel is produced and distributed by Novipharm, Austria.

However, mistletoe preparations are not approved by the US Food and Drug Administration.

Precautions: It is generally recommended that treatment be stopped during menstrual period and pregnancy. According to a report of the Swiss Cancer League, fermented Iscador products contain large numbers of both dead and live bacteria and some yeast.

Home-made mistletoe preparations can be very poisonous. Reported minor side-effects (for Isorel) include a small increase in temperature of 1–1.5 °C which disappear after 1–2 days. For Helixor, if the dosage is increased too rapidly, temperature rises of 1–1.5 °C and headache may occur. Several clinical studies of the fermented form of Iscador have noted that patients experience moderate fever (a rise of 2.3–2.4 °C) on the day of the injections. Local reactions around the injection site, temporary headaches and chills are also associated with the fever. It is recommended to wait for the normalization of the temperature before a new injection is administered. In the case of hyperthyroidism, it is recommended to start with low doses and increase gradually.

Indicative dosage and application:

- In all 11 melanoma cell lines tested: lectins isolated from *V. album* showed an antiproliferative effect at concentrations of $1–10\,ng\,ml^{-1}$, *viscotoxin*'s antiproliferative effect rises at concentrations of $0.5–1\,\mu g\,ml^{-1}$ and alkaloids' antiproliferative effect begin at $10\,\mu g\,ml^{-1}$ (Yoon *et al.*, 1998).
- *Lectins ML I, ML II* and *ML III*, at concentrations from 0.02 to $20\,pg\,ml^{-1}$, were able to enhance the secretion of the cytokines tumor necrosis factor (TNF) α, interleukin (IL)-1 α, IL-1 β and IL-6 by human monocytes (Ziska, 1978).

Documented target cancers:

- *Viscumin*, a galactoside-binding lectin, is a powerful inflammatory mediator able to stimulate the immune system (Heiny and Benth, 1994).
- A purified lectin (MLI) from *V. album* has immunomodulating effects in activating monocytes/macrophages for inflammatory responses (Metzner *et al.*, 1987).
- *Viscum album* L. extracts have been shown to provide a DNA stabilizing effect (Woynarowski *et al.*, 1980).
- Since Iscador stimulates the production of the natural killer cells, it can be applied in order to stabilize the number of T4 cells and thus the clinical condition of HIV positive persons. Laboratory tests suggested that the progress of the HIV infection was inhibited (Rentea *et al.*, 1981; Schink *et al.*, 1992).
- Iscador has an increased action against breast cancer cells and colon cancer cells (Heiny *et al.*, 1994).
- In most patients (but healthy individuals, as well) the quality of life increased remarkably.
- Water-soluble polysaccharides of *V. album* exert a radioprotective effect, which could be a valuable complement to radiotherapy of cancer.
- Iscador therapy proved to be clinically and immunologically effective and well tolerated in immuno-compromised children with recurrent upper respiratory infections, due to the Chernobyl accident (Lukyanova *et al.*, 1992).
- When whole mistletoe preparations are employed, the effect is host tree-specific.

Further details

Related species

- The Chinese herb *V. alniformosanae* is the source of a conditioned medium (CM), designated as 572-CMF-, which is capable of stimulating mononuclear cells. This CM has the capacity to induce the promyelocytic cell line HL-60 to differentiate into morphologically and functionally mature monocytoid cells. Investigations have shown that 572-CM did not contain IFN-r, TNF, IL-1 and IL-2 (Chen *et al.*, 1992).
- Hexanoic acid extracts of *Viscum cruciatum* Sieber parasitic on *Crataegus monogyna* Jacq. (I), *C. monogyna* Jacq. parasitized with *V. cruciatum* Sieber (II), and *C. monogyna* Jacq. Non-parasitized (III), and of a triterpenes enriched fractions isolated from I, II and III (CFI, CFII, CFIII, respectively) demonstrated significant cytotoxic activity against cultured larynx cancer cells (HEp-2 cells) (Gomez *et al.*, 1997).

Related compounds

- A galactose-specific lectin from *Viscum album* (VAA) was found to induce the aggregation of human platelets in a dose- and sugar-dependent manner. Small non-aggregating concentrations of VAA primed the response of platelets to known aggregants (ADP, arachidonic acid, thrombin, ristocetin and A23187). VAA-induced platelet aggregation was completely reversible by the addition of the sugar inhibitor lactose and the platelets from disrupted aggregates maintained the response to other aggregants. The lectin-induced aggregation of washed platelets was more resistant to metabolic inhibitors than thrombin- or arachidonic acid-dependent cell interaction (Büssing and Schietzel, 1999).

- Partially and highly purified lectins from *V. album* cause a dose-dependent decrease of viability of human leukemia cell cultures, MOLT-4, after 72 h treatment. The LC50 of the partially purified lectin was 27.8 ng ml^{-1}, of the highly purified lectin 1.3 ng ml^{-1}. Compared to the highly purified lectin a 140-fold higher protein concentration of an aqueous mistletoe drug was required to obtain similar cytotoxic effects on MOLT-4 cells. The cytotoxicity of the highly purified lectin was preferentially inhibited by D-galactose and lactose, cytotoxicity of the mistletoe drug and the partially purified lectin were preferentially inhibited by lactose and *N*-acetyl-D-galactosamine (GalNAc) (Olsnes *et al.*, 1982).
- Two lectin fractions with almost the same cytotoxic activity on MOLT-4 cells but with different carbohydrate affinities were isolated by affinity chromatography from the mistletoe drug: mistletoe lectin I with an affinity to D-galactose and GalNAc and mistletoe lectin II with an affinity to GalNAc. The lectin fractions and the mistletoe drug inhibited protein synthesis of MOLT-4 cells stronger than DNA synthesis (Olsnes *et al.*, 1982).
- Application of an aqueous extract from *Viscum album coloratum*, a Korean mistletoe significantly inhibited lung metastasis of tumor metastasis produced by highly metastatic murine tumor cells, B16-BL6 melanoma, colon 26-M3.1 carcinoma and L5178Y-ML25 lymphoma cells in mice. The antimetastatic effect resulted from the suppression of tumor growth and the inhibition of tumor-induced angiogenesis by inducing TNF-alpha (Yoon *et al.*, 1998).
- A peptide isolated from the *V. album* extract (Iscador) stimulated macrophages *in vitro* and *in vivo* and activated macrophages were found to have cytotoxic activity towards L-929 fibroblasts (Swiss Society for Oncology, 2001).
- Iscador Pini, an extract derived from *V. album* L. grown on pines and containing a non-lectin associated antigen, strongly induced proliferation of peripheral blood mononuclear cells (Cammarata and Cajelli, 1967).
- Polysaccharides are possibly involved in the pharmacological effects of *V. album* extracts, which are used in cancer therapy. The main polysaccharide of the green parts of *Viscum* is a highly esterified galacturonan whereas in *Viscum* 'berries' a complex arabinogalactan is predominant and interacting with the galactose-specific lectin (ML I) (Stein, 1999).
- Water-soluble polysaccharides of *V. album* were shown to exert a radioprotective effect which was a function of both the radiation dose and the drug dose and time of its injection. The maximum radioprotective efficacy of polysaccharides was observed after their injection 15 min before irradiation (Stein, 1999).

Antitumor activity

- The Korean mistletoe extract possesses antitumor activity *in vivo* and *in vitro*. Antiproliferative activities have been attributed to *Viscum album* C, *Viscum album* Qu *and Viscum album* M (trade name *Iscador*) on melanoma cell lines. *Viscum album* C contains *viscotoxin, alkaloids* and *lectins. Viscum album* Qu was extracted by Medac (Germany). *Viscum album* M is a preparation by the Institute Hiscia (Switzerland). The antiproliferative effect of the extracts on 11 melanoma cell lines obtained through the EORTC-MCG were tested in monolayer proliferation tests. In most of the melanoma cell lines tested, there was a significant antiproliferative effect of *V. album* C at a concentration of 100 $\mu g\,ml^{-1}$, whereas *V. album* M showed an antiproliferative effect at 1,000 $\mu g\,ml^{-1}$. The lectins isolated from *V. album* C, when compared with each other showed almost in all 11 melanoma cell lines tested a similar antiproliferative effect. It was seen at concentrations of 1–10 $ng\,ml^{-1}$. The antiproliferative effect of *viscotoxin* rises at concentrations of 0.5–1 $\mu g\,ml^{-1}$, whereas the antiproliferative effect of alkaloids begins at 10 $\mu g\,ml^{-1}$ (Yoon *et al.*, 1998).
- Iscador inhibited 20-methylcholanthrene-induced carcinogenesis in mice. Intraperitoneal administration of Iscador (1 $mg\,dose^{-1}$) twice weekly for 15 weeks could completely inhibit 20-methylcholanthrene-induced sarcoma in mice and protect these animals from tumour-induced death. Iscador was found to be effective even at lowered doses. After administration of 0.166, 0.0166 and 0.00166 $mg\,dose^{-1}$, 67, 50 and 17% of animals, respectively, did not develop sarcoma (Kuttan *et al.*, 1997).
- Patients with advanced breast cancer who were treated parenterally with Iscador showed an improvement in repair, possibly due to a stimulation of repair enzymes by lymphokines or cytokines secreted by activated leukocytes or an alteration in the susceptibility to exogenic agents resulting in less damage (Kovacs *et al.*, 1991).
- Macrophages from mice treated with *V. album* extract were shown to be active in inhibiting the proliferation of tumor cells in culture. These activated macrophages have now been shown to protect mice from dying of progressive tumors when injected intraperitoneally into the animals. Prophylactic as well as multiple treatments with macrophages activated with *V. album* extract seemed more effective than a single treatment. Thus, in addition to a direct cytotoxic effect of *V. album* extract, the activation of macrophages may contribute to the overall antitumor activity of the drug (Kuttan, 1993).
- Iscador was found to be cytotoxic to animal tumor cells such as Dalton's lymphoma ascites cells (DLA cells) and Ehrlich ascites cells *in vitro* and inhibited the growth of lung fibroblasts (LB cells), Chinese hamster ovary cells (CHO cells) and human nasopharyngeal carcinoma cells (KB cells) at very low concentrations. Moreover, administration of Iscador was found to reduce ascites tumors and solid tumors produced by DLA cells and Ehrlich ascites cells. The effect of the drug could be seen when the drug was given either simultaneously, after tumor development or when given prophylactically, indicating a mechanism of action very different from other chemotherapeutic drugs. Iscador was not found to be cytotoxic to lymphocytes (Luther *et al.*, 1977).
- The ML-I lectin from *V. album* has been shown to increase the number and cytotoxic activity of natural killer cells and to induce antitumor activity in animal models. The

same lectin inhibits cell growth and induces apoptosis (programmed cell death) in several cell types (Janssen *et al.*, 1993).

- In mice, an increased number of plaque-forming cells to sheep red blood cells (SRBC) followed the injection of Isorel (Novipharm, Austria) together with SRBC. Further, survival time of a foreign skin graft was shortened if Isorel was applied at the correct time. Finally, suppressed immune reactivity in tumorous mice recovered following Isorel injection. Isorel was further shown to be cytotoxic to tumor cells *in vitro*. Its application to tumor-bearing mice could prolong their life but without any therapeutic effect. However, a combination of local irradiation and Isorel was very effective: following 43 Gy of local irradiation to a transplanted methylcholanthrene-induced fibrosarcoma (volume about 240 mm^3) growing in syngeneic CBA/HZgr mice, the tumor disappeared in about 25% of the animals; the addition of Isorel increased the incidence of cured animals to over 65%. The combined action of Isorel, influencing tumor viability on the one hand and the host's immune reactivity on the other, seems to be favorable for its antitumor action *in vivo* (Pouckova *et al.*, 1986).

Anti-leukemic activity

- Mistletoe lectin I from *V. album* applied *in vitro* for 1 h in appropriate doses, caused irreversible inhibition of leukemic L1210 cell proliferation. The toxin appeared to be cytotoxic to normal bone marrow progenitor cells, as well as observed to the P-388 and L1210 leukemia cells.
- Iscador was found to reduce the leukocytopenia produced by radiation and cyclophosphamide treatment in animals. Weight loss due to radiation was considerable whereas weight loss due to cyclophosphamide was not altered. Hemoglobin levels also were not affected, indicating that treatment with the extract reduces lymphocytopenia and hence could be used along with chemotherapy and radiation therapy (Kutten *et al.*, 1993).

Other medical effects

- The 5-bromo-2′-deoxyuridine-induced sister chromatid exchange (SCE) frequency of amniotic fluid cells (AFC) remained stable after the addition of a therapeutical concentration of *V. album* (*Iscador P*) but decreased significantly after administration of high drug doses. As the proliferation index remained stable, even at extremely high drug concentrations, this effect could not be ascribed to a reduction of proliferation. No indications of cytogenetic damage or effects of mutagenicity were seen after the addition of the preparation. In addition, increasing concentrations of *V. album L.* extracts were shown to significantly reduce SCE frequency of phytohemagglutinin (PHA)-stimulated peripheral blood mononuclear cells (PBMC) of healthy individuals (Bussing *et al.*, 1995).
- The three mistletoe lectins. ML I, ML II and ML III, at concentrations from 0.02 to 20 pg ml^{-1} (100–10,000-fold lower than those showing toxic effects) were able to enhance the secretion of the cytokines tumor necrosis factor (TNF) alpha, interleukin (IL)-1 alpha, IL-1 beta and IL-6 by human monocytes several-fold over

control values were observed. The immunoactivating concentrations by the three lectins were found different for each donor. At toxic concentrations, the amounts of IL-1 alpha, IL-1 beta and to a less extent of TNF alpha in monocytes supernatants were particularly high (Ziska, 1998).

- The mistletoe lectin ML-A inactivates rat liver ribosomes by cleaving a *N*-glycosidic bond at A-4324 of 28S rRNA in the ribosomes, as it is characteristic of the common ribosome-inactivating proteins (RIPs) (Citores *et al.*, 1993).
- During a phase I/II study to determine the effect of *V. album* (Iscador) in HIV infection, 40 HIV-positive patients (with CD4-lymphocyte count > 200) were injected with 0.01 mg up to 10 mg subcutaneously twice a week over a period of 18 weeks. The extract was well tolerated and suggested to have anti-HIV activities (Gorter, 1994).

References

Barney, C.W., Hawksworth, F.G. and Geils, B.W. (1998) Hosts of *Viscum album*. *Eur. J. For. Path.* **28**, 187–208.

Büssing, A. (2000) *Mistletoe: The genus Viscum*. Harwood Academic Publishers.

Büssing, A., Schaller, G. and Pfuller, U. (1998) Generation of reactive oxygen intermediates (ROI) by the thionins from *Viscum album* L. *Anticancer Res.* **18**, 4291–6.

Büssing, A. and Schietzel, M. (1999) Apoptosis-inducing properties of *Viscum album* L. Extracts from different host trees, correlate with their content of toxic Mistletoe lectins. *Anticancer Res.* **19**, 23–8.

CA (Anonymous) (1983) Unproven methods of cancer management: *Iscador*. CA: a *Cancer J. Clinicians*, **33**, 186–8.

Cammarata, P.L. and Cajelli, E. (1967) Free amino acid content of *Viscum album* L. berries parasitizing the Pinus silvestris L. and Pinus nigra Arnold var. austriaca. *Boll. Chim. Farm.* Aug. **106**(8), 521–6.

Chen, P.M., Hsiao, K.I., Su, J.L., Liu, J. and Yang, L.L. (1992) Study of the activities of Chinese herb *Viscum alniformosanae* Part II: The components of conditioned medium produced by *Viscum alniformosanae*-stimulated mononuclear cells. *Am. J. Chin. Med.* **20**(3–4), 307–12.

Fink, J.M. (1988) *Third Opinion: An International Directory to Alternative Therapy Centers for the Treatment and Prevention of Cancer and Other Degenerative Diseases*. Second edn. Garden City Park, New York: Avery Publishing Group Inc., p.137.

Franz, H. (1986) Mistletoe lectins and their A and B chains. *Oncology* **43**(1), 23–34.

Grieve, M. (1994) *A Modern herbal*. Edited and introduced by Mrs. C.F. Leyel, Tiger books international, London.

Gomez, M.A, Saenz, M.T., Garcia, M.D., Ahumada, M.C. and De La Puerta, R. (1997) Cytostatic activity against Hep-2 cells of methanol extracts from *Viscum cruciatum* Sieber parasitic on Crataegus monogyna Jacq. and two isolated principles. *Phytother. Res.* **11**, 240–2.

Gorter, R. (1994) The European Mistletoe (*Viscum album*): new studies show significant results for AIDS and immune system problems. Institute for Oncological and Immunological Research.

Hauser, S. and Kast, A. (2001) Iscusin – preparations for pre- and postoperative treatment of malignant tumours. (BCCA Cancer Information Centre search file 701).

Hauser, S.P. (1993) Unproven methods in cancer treatment. *Curr. Opinion Oncol.* **5**, 646–54.

Heiny, B.M. and Benth, J. (1994) Mistletoe extract standardized for the galactoside-specific lectin (ML-1) induces B-endorphin release and immunopotentiation in breast cancer patients. *Anticancer Res.* **14**, 1339–42.

Janssen, O., Scheffler, A., Kabelitz, D. (1993) In vitro effects of mistletoe extracts and mistletoe lectins. Cytotoxicity towards tumor cells due to the induction of programmed cell death (apoptosis). *Arzneimittelforschung* **43**(11), 1221–7.

Kovacs, E., Hajto, T. and Hostanska, K. (1991) Improvement of DNA repair in lymphocytes of breast cancer patients treated with *Viscum album* extract (*Iscador*). *Eur. J. Cancer* **27**(12), 1672–6.

Kuttan, G., Menon, L.G., Antony, S. and Kuttan, R. (1997) Anticarcinogenic and antimetastatic activity of Iscador. *Anticancer Drugs* Apr. **8**(Suppl 1), S15–16.

Lukyanova, M., Chernyshov, P., Omelchenko, I., Slukvin, I., Pochinok, V., Antipkin, G., Voichenko, V., Heusser, P. and Schneiderman, G. (1992) Research on immune-suppressed children following the Chernobyl accident. Mistletoe effective for Chernobyl children. Ukrainian Institute for Pediatrics, Lukas Klinik, Switzerland.

Luther, P., Franz, H., Haustein, B. and Bergmann, K.C. (1977) Isolation and characterization of mistletoe extracts (*Viscum album L.*). II. Effect of agglutinating and cytotoxic fractions on mouse ascites tumor cells. *Acta Biol Med Ger* 36(1), 119–25.

Metzner, G., Franz, H., Kindt, A., Schumann, I. and Fahlbusch, B. (1987) Effects of lectin I from mistletoe (ML I) and its isolated A and B chains on human mononuclear cells: mitogenic activity and lymphokine release. *Pharmazie* May 42(5), 337–40.

Mueller, A.E. and Anderer, A.F. (1990) A *Viscum album* oligosaccharide activating human natural cytotoxicity is an interferon γ inducer. *Cancer Immunol Immunother.* 32, 221–7.

Olsnes, S., Stirpe, F., Sandvig, K. and Pihl, A. (1982) Isolation and characterization of *viscumin*, a toxic lectin from *Viscum album L.* (*Mistletoe*). *J. Biol. Chem.* 257(22), 13263–70.

Ontario Breast Cancer Information Exchange Project (1994) Guide to unconventional cancer therapies. First edn. Ontario Breast Cancer Information Exchange Project, Toronto. 76–79.

Rentea, R., Lyon, E. and Hunter, R. (1981) Biologic properties of iscador: a *Viscum album* preparation I. Hyperplasia of the thymic cortex and accelerated regeneration of hematopoietic cells following X-irradiation. *Lab. Invest.* Jan. 44(1), 43–8.

Samuelsson, G. (1992) Drugs of Natural Origin – A Textbook of Pharmacognosy. Third revised, enlarged and translated edition. Swedish Pharmaceutical Press.

Schink, M., Moser, D. and Mechelke, F. (1992) Two-dimensional isolectin patterns of the lectins from *Viscum album L.* (mistletoe). *Naturwissenschaften* Feb. 79(2), 80–1.

Stein, M.G., Edlund, U., Pfuller, U., Bussing, A. and Schietzel, M. (1999) Influence of polysaccharides from *Viscum album* L. on human lymphocytes, monocytes and granulocytes *in vitro*. *Anticancer Res.* 19, 3907–14.

Sweeney, E.C., Tonevitsky, A.G., Palmer, R.A., Niwa, H., Pfueller, U., Eck, J., Lentzen, H., Agapov, I.I. and Kirpichnikov, M.P. (1998) *Mistletoe* lectin I forms a double trefoil structure. *FEBS Lett.* 431(3), 367–70.

Swiss Society for Oncology. *Iscador*. (2001) (BCCA Cancer Information Centre search file 701).

Swiss Society for Oncology, Swiss Cancer League, Study Group on Unproven Methods in Oncology. *Helixor-mistletoe* preparations for treatment of cancer. Document UICC UMS010. (BCCA Cancer Information Centre search file 701).

U.S. Congress, Office of Technology Assessment. Unconventional cancer treatments. Washington, D.C.: U.S. Government Printing Office 1990 Sept. pp. 81–86.

Werner, M., Zanker, K.S. and Nikolai, G. (1998) Stimulation of T-cell locomotion in an *in vitro* assay by various *Viscum album L.* preparations (*Iscador*). *Int. J. Immunotherapy* XIV(3), 135–42.

Wagner, H., Jordan, E. and Feil, B. (1986) Studies on the standardization of *mistletoe* preparations. *Oncology* 43(1), 16–22.

Wilson, B.R. (1985). *Cancer Quackery Primer*. Dallas, Oregon.

Yoon, T.J., Yoo, Y.C., Kang, T.B., Baek, Y.J., Huh, C.S., Song, S.K., Lee, K.H., Azuma, I. and Kim, J.B. (1998) Prophylactic effect of Korean *mistletoe* (*Viscum album coloratum*) extract on tumor metastasis is mediated by enhancement of NK cell activity. *Int. J. Immunopharmacol* 20(4–5), 163–72.

Ziska, P., Franz, H. and Kindt, A. (1978) The lectin from *viscum album* L. purification by biospecific affinity chromatography. *Experientia*, 34(1), 123–4.

Useful addresses

Verein fuer Krebsforschung, 011 41 61 701 2323.

Prof. Dr Robert Gorter,
Institute for Oncological and Immunological Research,
011 49 303 976 3420 (Fax: 3422).

NOVIPHARM
A-9210 Portschach Klagenfurter Str 164, Austria.
Tel.: 04272 27510, Fax: 04272 3119.

Taxus baccata (Yew) (Taxaceae and Coniferae) — Antineoplastic agent

Location: Europe, North Africa and Western Asia. The important clinical efficacy of *taxol* has led to the drug supply crisis. As a result, NCI has developed plans to avert similar supply crisis in the future by initiating exploratory research projects for large-scale production.

Appearance (Figure 3.3)
Stem: a tree 1.2–1.5 m high, forming with age a very trunk covered with red-brown, peeling bark and topped with a rounded or wide-spreading head of branches.
Leaves: spirally attached to twigs, but by twisting of the stalks brought more or less into two opposed ranks, dark, glossy, almost black-green above, grey, pale-green or yellowish beneath, 15–45 cm long, 2–3 cm wide.
Flowers: unisexual, with the sexes invariably on different trees, produced in spring from the leaf axils of the proceeding summer's twigs. Male, a globose cluster of stamens; female, an ovule surrounded by small bracts, the so-called fruit bright red, sometimes yellow, juicy and encloses the seed.

Biology: Can be propagated by seed or cuttings. Seeds may require warm and cold stratification. Mature woodcuttings taken in winter can be rooted under mist.

Tradition: No tree is more associated with the history and legends of Great Britain. Before Christianity, it was a sacred tree favored by the Druids, who built their temples near these trees – a custom followed by the early christians. The association of the tree with places of worship still prevails. The wood was formerly much valued in archery for the making of long bows. The wood is said to resist the action of water and is very hard.

Part used: stem segments, needles 1–2 cm long, and roots.

Active ingredients:

- *Taxane* diterpenes, among them paclitaxel (earlier known as *taxol*), *cephalomannine*.
- Key precursors: *baccatin III, 10-desacetylbaccatin III, 9-dihydrobaccatin III, 13-Acetyl-9-dihydrobaccatin III, baccatin VI.*
- Related compounds, such as *taxotere*.

Figure 3.3 Taxus.

Particular value: *Taxol* research is being carried out on ovarian cancer, breast cancer, colon and gastric cancers, arthritis, Alzheimer's, as an aid in coronary and heart procedures and as an antiviral agent. The uses of yew in any form for any medical or health reason should only do after consulting a health care professional.

The status of taxus application in cancer therapy: *Taxol* (containing paclitaxel) is an anticancer drug, it was originally isolated from the Pacific Yew tree in the early 1960s, was recently approved by the Food and Drug Administration for use against ovarian cancer and has also shown activity against breast, lung and other cancers. This drug was also registered in Poland in 1996.

In 1958 the US NCI initiates a program to screen 35,000 plants species for anticancer activity. In 1963, Drs Monroe Wall and M.C. Wani of Research Triangle Institute, North Carolina subsequently find that an extract or the bark of Pacific yew tree has antitumor activity. Since that time its use as an anticancer drug has become well established (Cragg, 1998).

Human trials started in 1983. Despite a few deaths caused by unforeseen allergic reactions due to the form in which the drug was administered great promise was shown for women with previously incurable ovarian cancer. This led the NCI to issue a contract with Bristol Myers-Squibb (BMS), a pharmaceutical company based in the United States, for the clinical development of taxol (Rowinsky *et al.*, 1990).

Intense research on finding alternatives to taxol extracted from the bark of the Pacific yew is ongoing. Taxol has been chemically synthesized and semisynthetic versions have been developed using needles and twigs from other yew species grown in agricultural settings. This is reducing the pressure on natural stands of Pacific yew but bark is still being used for taxol production (Cragg *et al.*, 1993).

Precautions:

- Poisonous. Many cases of poisoning amongst cattle have resulted from eating parts of it. The fruit and seeds seem to be the most poisonous parts of the tree.
- In the treatment of cancer: reduction in white and red blood cells counts and infection. Other common side effects include hair loss, nausea and vomiting, joint and muscle pain, nerve pain, numbness in the extremities and diarrhea. Severe hypersensitivity can also occur, demonstrated by symptoms of shortness of breath, low blood pressure and rash. The likelihood of these reactions is lowered by the use of several kinds of medications that are given before the taxol infusion (NCI).

Indicative dosage and application: the doses of taxol given to most patients are

- $110\,\mathrm{mg\,m^{-2}}$ in 22%
- $135\,\mathrm{mg\,m^{-2}}$ in 48%
- $170\,\mathrm{mg\,m^{-2}}$ in 22%.

These doses are significantly lower, because of limited hematopoietic tolerance, than those previously demonstrated to be safe in minimally pre-treated or untreated patients (200–250 $\mathrm{mg\ m^{-2}}$).

Documented target cancers:

- Activity against the P-388, P-1534 and L-1210 murine leukemia models.
- Strong activity against the B16 melanoma system.
- Cytotoxic activity against KB cell culture system, Walker 256 carcinosarcoma, sarcoma 180 and Lewis lung tumors.
- Significant activity against several human tumor xenograft systems, including the MX-1 mammary tumor.

- Introduced to all ovarian cancer patients (meeting defined disease criteria).
- Responses in patients with metastatic breast cancer and in patients with other forms of advanced malignancy including lung cancer, cancer of the head and neck region and lymphomas.

Further details

Antitumor activity

- The antitumorous properties of paclitaxel are based on the ability to bind and to stabilize microtubules and block cell replication in the late G_2–M phase of the cell cycle. In 1979 it was demonstrated that taxol affects the *tubulin–microtubule equilibrium*: it decreases both the critical concentration of tubulin (to almost 0–1 $mg\,ml^{-1}$) and the induction time for polymerization, either in the presence or absence of GTP, MAPs and magnesium. Taken in conjunction with observations showing that *taxol* promotes the end-to-end joining of microtubules, these results point to a rather complex mechanism of action for *taxol* that is not yet completely understood (Cragg, 1998).
- Early studies with HeLa cells and BALB/c mouse fibroblasts treated with low concentrations of taxol (0.25 $\mu mol\,l^{-1}$), which produce minimal inhibition of DNA, RNA and protein synthesis, demonstrated that taxol blocks cell cycle traverse in the mitotic phases. Recently, taxol has been demonstrated to prevent transition from the G_0 phase to the S phase in fibroblasts during stimulation of DNA synthesis by growth factors and to delay traverse of sensitive leukemia cells in nonmitotic phases of the cell cycle. These findings indicate that the integrity of microtubules may be critical in the transmission of proliferative signals from cell-surface receptors to the nucleus. Proposed explanations that at least in part account for *taxol*'s inhibitory effects in nonmitotic phases include disruption of tubulin in the cell membrane and/or direct inhibition of the disassembly of the interphase cytoskeleton, which may upset many vital cell functions such as locomotion, intracellular transport and transmission of proliferative transmembrane signals.

Related species

- The plum yews (*Cephalotaxus harringtonia* Family: Cephalotaxaceae (plum yew family)) are similar to, and closely related to, the yews, family Taxaceae. Common Names: Japanese plum yew, Harrington plum yew, cow-tail pine, plum yew.

  The plum yews are evergreen, coniferous shrubs or small trees with flat, needle-like leaves arranged in two ranks on the green twigs and fleshy, plum-like seeds borne only on female plants. Japanese plum yew is a shrub or small tree, but most cultivars are quite a bit smaller. Japanese plum yew is native to Japan, Korea and eastern China, where it grows in the forest understory. Japanese plum yew has the potential to be a very useful landscape plant in the southern US. It is more tolerant of heat than the true yews (*Taxus*). It is produces *cephalomannine* a promising agent for cancer therapy.
- *Taxus brevifolia* can be regarded as the first source of taxol. It is common on the Olympic Peninsula in Washington and on Vancouver Island in British Columbia. The taxol supply needs for preclinical and early clinical studies were easily met by bark collections in Oregon between 1976 and 1985, from the bark of the tree. In 1988 it was demonstrated that the precursor, *10-desacetylbaccatin III*, isolated from the needles of the tree, can be converted to *taxol* and related active agents by a

relatively simple semisynthetic procedure, and alternative, more efficient processes for this conversion have recently been reported (Helfferich *et al.*, 1993).

- The taxol content of fresh needles of 35 different *Taxus* cultivars from different locations within the US has been analyzed. At least six contain amounts comparable to or higher than those found in the dried bark of *T. brevifolia*. These observations have resulted in the initiation of a study of the nursery cultivar, *Taxus* × *media Hicksii*, as a potential renewable large-scale source of taxol (Furmanova *et al.*, 1997).
- NCI and Program Resources, in collaboration with various organizations are undertaking analytical surveys of needles of a number of *Taxus* species. They include *T. baccata* from the Black Sea-Caucasus region of Georgia and Ukraine, and *T. cuspidata* from Siberian regions of Russia; *T. canadensis* from the Gaspe Peninsula of Quebec; *T. globosa* from Mexico, *T. sumatriensis* from the Philippines and various *Taxus* species from the US. In a number of samples, the *taxol* content of the needles is comparable to that of the dried bark of *T. brevifolia* (NCI, Cragg *et al.*, 1993).
- *Pestalotiopsis microspora* (an endophytic fungus) was isolated from the inner bark of a small limb of Himalayan yew, *T. wallachiana*, which has been shown to produce taxol in mycelial culture. Fungal *taxol* was evaluated in the standard 26 cancer cell line test and for its ability, when compared to authentic taxol, to inhibit cell division. The fungal compound found to be identical to authentic taxol (methods used: NMR, UV absorption and electrospray mass spectroscopy). It showed a pattern of activity comparable to that produced by standard authentic *taxanes* in the 26 cancer cell line test. In addition, its ability to induce mitotic arrest at a concentration of $37\,ng\,ml^{-1}$, consistent with a tubulin-stabilizing mode of action. The discovery that fungi make taxol increasingly adds to the possibility that horizontal gene transfer may have occurred between *Taxus* spp. and its corresponding endophytic organisms. This demonstration supports the idea that certain endophytic microbes of *Taxus* spp. may make and tolerate taxol in order to better compete and survive in association with these trees. Since *Taxus* spp. grow in places that are generally damp and shaded certain plant-pathogenic fungi (water molds) also prefer this niche (Strobel *et al.*, 1996).
- *Taxus marei* Hu ex Liu is a native Taiwan species sparsely distributed in mountainous terrain. Many are giant trees with a diameter at breast height greater than 100 cm and an estimated age of more than 1,000 years. Taxol concentration in the needles of these trees and selected superior trees with respect to high taxol and *10-desacetyl baccatin III* concentrations. It was found that rooted cutting (steckling) ramets of these trees also exhibited high taxol concentrations in mature needles, confirming that taxol yield is a heritable trait. Young needles from vegetatively propagated elite yew trees can serve as a renewable and economic tissue source for increasing taxol production. Micropropagation of mature *Taxus marei* was achieved using bud explants derived from approximately 1,000-year-old field grown trees. It might be a very useful tool to use for the mass propagation of superior yew trees and the production of high-quality (orthotropic) plantlets for nursery operation (Chang, 2001).

Antitumor activity

- Taxol has been shown to inhibit steroidogenesis in human Y-1 adrenocortical tumors and in MLTC-1 Leydig tumors by decreasing the intracellular transport of cholesterol to cholesterol side-chain cleavage enzymes. This effect appears to be related to perturbations in microtubule dynamics (Nicolaou *et al.*, 1994).

- Taxol has also been shown to inhibit specific functions in many nonmalignant tissues, which may be mediated through microtubule disruption. For example, in human neutrophils, taxol inhibits relevant morphological and biochemical processes, including chemotaxis, migration, cell spreading, polarization, generation of hydrogen peroxide and killing of phagocytosed microorganisms. Taxol also antagonizes the effects of microtubule-disrupting drugs on lymphocyte function and adenosine 3′,5′-cyclic monophosphate metabolism and inhibits the proliferation of stimulated human lymphocytes, but blast transformation is not affected during lymphocyte activation. Taxol has also been found to mimic the effects of endotoxic bacterial lipopolysaccharide on macrophages, resulting in a rapid decrement of receptors for tumor factor-α and TNF-α release. This finding suggests that an intracellular target affected by taxol may be involved in the actions of lipopolysaccharide on macrophages and other cells. Interestingly, taxol inhibits chorioretinal fibroblast proliferation and contractility in an *in vitro* model of proliferative vitreoretinopathy, a fact that may be relevant to the treatment of traction retinal detachment and proliferative vitreoretinopathy. Taxol inhibits, also, the secretory functions of many specialized cells. Examples include insulin secretion in isolated rat islets of Langerhans, protein secretion in rat hepatocytes and the nicotinic receptor-stimulated release of *catecholamines* from chromaffin cells of the adrenal medulla (Nicolaou *et al.*, 1994).

Related compounds

- *Taxotere* is a highly promising analog of *taxol* that has been synthesized. It promotes the assembly and stability of microtubules with potency approximately twice that of *taxol*. Recently, taxol and taxotere have been shown to compete for the same binding site. While most of the effects of taxotere mirror those of taxol, it appears that the microtubules formed by taxotere induction are structurally different from those formed by taxol induction. *Taxotere* is currently produced by attaching a synthetic sidechain to *10-desacetyl baccatin III*, which is readily available from the European yew *T. baccata*, in yields approaching 1 kg from 3.000 kg of needles (Hirasuna *et al.*, 1996).
- Cell culture has already been used to produce ^{14}C labeled *taxol* from ^{14}C sodium acetate. The USDA (United States Department of Agriculture) has received a patent for the production of taxol from cultured callus cells of *T. brevifolia*. They have licensed this process to Phyton Catalytic, who estimate that they will begin commercial production soon. The advantage of this system is that the major secretion product of the cells is taxol, which reduces the purification to an ether extraction of the medium. ESCA genetics has also announced technology for producing high levels of taxol in plant cell cultures, and they project large-scaled production in the near future. Additionally, callus cultures of *T. cuspidata* and *T. canadensis* have been sustained in a *taxol*-producing system for over two months. A fungus indigenous to *T. brevifolia*, that produces small amounts of taxol has recently been isolated and cultured (Helfferich *et al.*, 1993).
- As a target for chemical synthesis, taxol presents a plethora of potential problems. Perhaps most obvious is the challenge presented by the central B ring, an

eight-membered carbocycle. Such rings are notoriously difficult to form because of both entropic and enthalpic factors. The normally high transannular strain of an eight-membered ring is further increased in this case by the presence of the geminal dimethyl groups, which project into the interior of the B ring. Then the *trans*-fused C ring with its angular methyl group and another ring (A ring), which is a 1,3-C3 bridge, must be introduced. The A ring includes a somewhat problematic bridgehead alkene formally forbidden in a six-membered ring by Bredt's rule. If assembling the carbon skeleton alone is not a daunting enough task, one should consider the high degree of oxygenation that must be introduced in a manner which allows the differential protection of five alkoxy groups in a minimum of three orthogonal classes. Additionally, some of the functionality is quite sensitive to environmental conditions. The oxetane ring, for example, will open under acidic or nucleophilic conditions, and the 7-hydroxyl group, if left unprotected, will epimerize under basic conditions. Despite the many attempts to synthesize taxol, the molecule still remains inaccessible by total synthesis (Nicolaou *et al.*, 1994).

- *Taxol* is supplied as a sterile solution of $6\,mg\,ml^{-1}$ in 5 ml ampoules (30 mg per ampoule). Because of taxol's aqueous insolubility, it is formulated in 50% cremophor EL and 50% dehydrated alcohol. The contents of the ampoule must be diluted further in either 0.9% sodium chloride or 5% dextrose. During early phase I and II studies, taxol was diluted to final concentrations of 0.003–$0.60\,mg\,ml^{-1}$. These concentrations were demonstrated to be stable for 24 and 3 h, respectively, in early stability studies. This short stability period required the administration of large volumes of fluids and/or drug preparation at frequent intervals for patients receiving higher doses. In recent studies, concentrations of 0.3–$1.2\,mg\,ml^{-1}$ in either 5% dextrose or normal saline solution have demonstrated both chemical and physical stability for at least 12 h (Rowinsky *et al.*, 1990).
- Taxol and its relatives are emerging as yet another class of naturally occurring substances, like the enediyne antitumor antibiotics and the macrocyclic immunophilin ligands, that combine novel molecular architecture, important biological activity and fascinating mode of action.

References

Cragg, G.M. (1998) *Paclitaxel (Taxol)*: a success story with valuable lessons for Natural Product Drug discovery and development. John Wiley & Sons, Inc., New York.

Cragg, G.M., Schepartz, S.A., Suffness, M. and Grever, M.R. (1993) The *taxol* supply crisis. New NCI policies for handling the large-scale production of novel natural product anticancer and Anti-HIV agents. *J. Nat. Prod.* 56(10), 1657–68.

Chang, S.H., Ho, C.K., Chen, Z.Z. and Tsay, J.Y. (2001) Micropropagation of *Taxus mairei* from mature trees. *Plant Cell Rep.* 20, 496–502.

Furmanowa, M., Glowniak, K., Syklowska-Baranek, K., Zgorka, G. and Jozefczyk, A. (1997) Effect of picloram and methyl jasmonate on growth and *taxane* accumulation in callus culture of *Taxus* X *media var. Hatfieldii. Plant Cell, Tissue Organ Culture* 49, 75–79.

Grieve M. (1994) *A Modern Herbal.* Edited and introduced by Mrs C.F. Leyel, Tiger books international, London.

Helfferich, C. (1993) *Taxol* Revisited Article, *Alaska Science Forum*, 1126.

Hirasuna, T.J., Pestchanker, L.J., Srinivasan, V. and Shuler, M.L. (1996) *Taxol* production in suspension cultures of *Taxus baccata. Plant Cell, Tissue Organ Culture* 44, 95–102.

Ketchum, R.E.B. and Gibson, D.M. (1996) Pactitaxel production in suspension cell cultures of *Taxus. Plant Cell, Tissue Organ Culture* 46, 9–16.

Ketchum, R.E.B., Gibson, D.M. and Greenspan Gallo, L. (1995) Media optimization for maximum biomass production in cell cultures of pacific yew. *Plant Cell, Tissue Organ Culture*, 42, 185–193.

Luo, J.P., Mu Q. and Gu, Y.-H. (1999) Protoplast culture and paclitaxel production by *Taxus yunnanensis. Plant Cell, Tissue Organ Culture* 59, 25–29.

Nicolaou, K.C., Dai, W.M. and Guy, R.K. (1994) Chemistry and Biology of *Taxol* Angew. *Chem. Int. Ed. Engl.* 33, 15–44.

Rowinsky, E.K., Cazenave, L.A. and Donehower, R.C. (1990) Taxol: a novel investigational antimicrotubule agent. Review. *J. Natnl Cancer Inst.*, 82(15), 1247–1259.

Samuelsson, G. (1992) *Drugs of Natural Origin – A textbook of Pharmacognosy*. Third revised, enlarged and translated edition. Swedish Pharmaceutical Press.

Strobel, G., Yang, X., Sears, J., Kramer, R., Sidhu, R.S. and Hess, W.M. (1996) *Taxol* from Pestalotiopsis microspora, an endophytic fungus of *Taxus wallachiana. Microbiology* 142, 435–440.

Sho-saiko-to, Juzen-taiho-to

Sho-saiko-to (SST) and Juzen-taiho-to (JTT) are not plants but Japanese modified Chinese herbal medicines, or *Kampo*. Juzen-taiho-to was formulated by Taiping Hui-Min Ju (Public Welfare Pharmacy Bureau) in Chinese Song Dynasty in AD 1200. It is prepared by extracting a mixture of ten medical herbs (*Rehmannia glutinosa, Paeonia lactiflora, Liqusticum wallichii, Angelica sinesis, Glycyrrhiza uralensis, Poria cocos, Atractylodes macrocephala, Panax ginseng. Astragalus membranaceus* and *Cinnamomum cassia*) that tone the blood and vital energy, and strengthen health and immunity. (Aburada *et al.*, 1983). This potent and popular prescription has traditionally been used against anemia, anorexia, extreme exhaustion, fatigue, kidney and spleen insufficiency and general weakness, particularly after illness. TT is the most effective biological response modifier among 116 Chinese herbal formulates (Hisha *et al.*, 1997). Animal models and clinical studies have revealed that it demonstrates extremely low toxicity ($LD_{50} > 15\,g\,kg^{-1}$ of murine), self-regulatory and synergistic actions of its components in immunomodulatory and immunopotentiating effects (by stimulating hemopoietic factors and interleukins production in association with NK cells, etc.), potentiates therapeutic activity in chemotherapy (mitomycin, cisplatin, cyclophosphamide and fluorouracil) and radiotherapy, inhibits the recurrence of malignancies, prolongs survival, as well as ameliorate and/or prevents adverse toxicities (GI disturbances such as anorexia, nausea, vomiting, hematotoxicity, immunosuppression, leukopenia, thrombocytopenia, anemia and nephropathy, etc.) of many anticancer drugs (Horie *et al.*, 1994; Ikehara *et al.*, 1992; Ohnishi *et al.*, 1998).

Liver metastasis: the effect of the medicine was assayed after the inoculation of a liver-metastatic variant (L5) of murine colon 26 carcinoma cells into the portal vein. (Ohnishi *et al.*, 1998). Oral administration of JTT for 7 days before tumor inoculation resulted in dose-dependent inhibition of liver tumor colonies and significant enhancement of survival rate as compared with the untreated control, without side effects. JTT significantly inhibited the experimental liver metastasis of colon 26-L5 cells in mice pretreated with anti-asialo GM1 serum and untreated normal mice, whereas it did not inhibit metastasis in 2-chloroadenosine-pretreated mice or T-cell-deficient nude mice. Oral administration of Juzen-taiho-to activated peritoneal exudate macrophages (PEM) to become cytostatic against the tumor cells. These results show that oral administration of Juzen-taiho-to inhibited liver metastasis of colon 26-L5 cells, possibly through a mechanism mediated by the activation of macrophages and/or T-cells in the host immune system. Thus, Juzen-taiho-to may be efficacious for the prevention of cancer metastasis.

Both SST and JTT suppressed the activities of thymidylate synthetase and thymidine kinase involved in *de novo* and salvage pathways for pyrimidine nucleotide synthesis, respectively, in mammary tumors of SHN mice with the reduction of serum prolactin level. These results indicate that SST and JTT may have the antitumor effects on mammary tumors (Sakamoto *et al.*, 1994).

Juzen-taiho-to also improves the general condition of cancer patients receiving chemotherapy and radiation therapy. Oral administration of TJ-48 accelerates recovery from hemopoietic injury induced by radiation and the anticancer drug mitomycin C. The effects are found to be due to its stimulation of spleen colony-forming units. It has been suggested that the administration of TJ-48 should be of benefit to patients receiving chemotherapy, radiation therapy or bone marrow transplantation.

In combination with an anticancer drug UFT (5-fluorouracil derivative), it prevented the body weight loss and the induction of the colonic cancer in rats treated with a chemical carcinogen 1,2-dimethylhydrazine (DMH), and suppressed markedly the activity of thymidylate synthetase (TS) involved in the *de novo* pathway of pyrimidine synthesis in colonic cancer induced by DMH (Sakamoto *et al.*, 1991).

The combination of TJ-48 and mitomycin C (MMC) produced significantly longer survival in p-388 tumor-bearing mice than MMC alone, and TJ-48 decreased the diverse effects of MMC such as leukopenia, thrombopenia and weight loss.

Immunostimulation: In mice, TJ-48 augmented antibody production and activated macrophage by oral administration of TJ-48, but reduced the MMC-induced immunosuppression in mice. TJ-48 showed a mitogenic activity in splenocytes but not in thymocytes, and an anti-complementary activity was also observed. Anti-complementary activity and mitogenic activity were both observed in high-molecular polysaccharide fraction but not in low-molecular weight fraction (Satomi *et al.*, 1989). Of several polysaccharide fractions in TJ-48, only pectic polysaccharide fraction (F-5-2) showed potent mitogenic activity. F-5-2 was also shown to have the highest anti-complementary activity. However, the polygalacturonan region is essential for the expression of the mitogenic activity, but that the contribution of polygalacturonan region to the anti-complementary activity is less. F-5–2 activates complement via alternative complement pathway and induces the proliferation of B cells but does not differentiate those cells from antibody producing cells.

Contribution to the prevention of the lethal and marked side effects of recombinant human TNF (rhTNF) and lipopolysaccharide (LPS) without impairing their antitumor activity. These drugs are thought to decrease the oxygen radicals and stabilize the cell membranes, with a deep relation to the arachidonic cascade. The release of prostaglandins and leukotriene B4 was suppressed by pretreatment with Shosaiko-to (Yano *et al.*, 1994). Thromboxane B2 was transiently increased, followed by suppression. After pretreatment with Hochu-ekki-to or Juzen-taiho-to, suppression of leukotriene B4 could not be observed. The release of prostaglandin D2 was suppressed in mice pretreated with SST, JTT or Ogon (*Scutellariae Radix*) but it increased following pretreatment with Hochu-ekki-to. Chemicals that could prevent the lethality of rhTNF and LPS also revealed suppression of prostaglandins, leukotriene B4 and thromboxane B2. In general, drugs that prevented the lethality of rhTNF and LPS without impairing the antitumor activity could inhibit the release of leukotriene B4 and/or prostaglandin D2 (Sugiyama *et al.*, 1995). rhTNF could activate the arachidonic cascade in combination with LPS. The lethality of rhTNF and LPS could be prevented by pretreatment with Japanese modified traditional Chinese medicines and the crude drug, Ogon.

In BDF1-mice which were implanted with P-388 leukemic cells, JTX prolonged significantly the average survival days of MMC-treated group. In tumor-free BDF1-mice, JTX improved the leukopenia and the body weight loss which were caused by MMC. Additionally, JTX delayed the appearance of deaths by lethal doses of MMC. These results indicate that JTX enhances the antitumor activity of MMC and lessens the adverse effects of it. JTX may be useful for patients undertaking MMC treatment.

TJ-48 has the capacity to accelerate recovery from hematopoietic injury induced by radiation and the anticancer drug MMC. The effects are found to be due to its stimulation of spleen colony-forming unit (CFU-S) counts on day 14.

Compound isolation: n-Hexane extract from TJ-48 shows a significant immunostimulatory activity. The extract is further fractionated by silica gel chromatography and HPLC in order to identify its active components. 1H-NMR and GC-EI-MS indicate that the active fraction is composed of free fatty acids (oleic acid and linolenic acid). When 27 kinds of free fatty acids (commercially available) are tested using the HSC proliferating assay, oleic acid, elaidic acid and linolenic acid are found to have potent activity. The administration of oleic acid to MMC-treated mice enhances CFU-S counts on days 8 and 14 to twice the control group. These findings strongly suggest that fatty acids contained in TJ-48 actively promote the proliferation of HSCs. Although many mechanisms seem to be involved in the stimulation of HSC proliferation, we speculate that at least one of the signals is mediated by stromal cells, rather than any direct interaction with the HSCs.

The inhibitory effect of JTT on progressive growth of a mouse fibrosarcoma is partly associated with prevention of gelatin sponge-elicited progressive growth, probably mediated by endogenous factors including antioxidant substances, in addition to the augmentation of host-mediated antitumor activity (Ohnishi *et al.*, 1996).

Juzen-taiho-to could be an effective drug for protecting against the side effects (nephrotoxicity, immunosuppression, hepatic toxicity and gastrointestinal toxicity) induced by carboplatin in the clinic as well as by cisplatin.

Sodium L-malate, $C_4H_4Na_2O_5$, was found to exhibit protective effects against both nephrotoxicity (ED_{50}: 0.4 mg kg^{-1}, p.o.) and bone marrow toxicity (ED_{50}: 1.8 mg/kg^{-1}, p.o.), without reducing the antitumor activity of *cis*-diamminedichloroplatinum (II) (CDDP) (Sugiyama *et al.*, 1994). These findings indicate that *Angelicae Radix* and its constituent sodium L-malate could provide significant protection against CDDP-induced nephrotoxicity and bone marrow toxicity without reducing the antitumor activity.

Water-soluble related compounds of the herbal medicine SST dose-dependently inhibited the proliferation of a human hepatocellular carcinoma cell line (KIM-1) and a cholangiocarcinoma cell line (KMC-1). Fifty percent effective doses on day 3 of exposure to SST were 353.5 +/− 32.4 μg ml^{-1} for KIM-1 and 236.3 +/− 26.5 μg ml^{-1} for KMC-1. However, almost no suppressive effects were detected in normal human peripheral blood lymphocytes or normal rat hepatocytes (Hano *et al.*, 1994). Sho-saiko-to suppressed the proliferation of the carcinoma cell lines significantly more strongly than did each of its major related compounds, that is, saikosaponin a, c and d, ginsenoside Rb1 and Rg1, glycyrrhizin, baicalin, baicalein and wogonin, or another herbal medicine, JTT ($P < 0.05$ or 0.005). Because such related compounds are barely soluble in water, there could be synergistic or additive effects of the related compounds in SST. Morphological, DNA, and cell cycle analyses revealed two possible modes of action of SST to suppress the proliferation of carcinoma cells: (a) it induces apoptosis in the early period of exposure; and (b) it induces arrest at the G_0/G_1 phase in the late period of exposure.

The effect of Shi-Quan-Da-Bu-Tang (TJ-48) on hepatocarcinogenesis induced by *N*-nitrosomorpholine (NNM) was investigated in male Sprague–Dawley rats. (Tatsuta *et al.*, 1994). Rats were given drinking water containing NNM for 8 weeks, and also from the start of the experiment, regular chow pellets containing 2.0% or 4.0% TJ-48 until the end of the experiment. Preneoplastic and neoplastic lesions staining for the placental type of glutathione-*S*-transferase (GST-P) or γ-glutamyl transpeptidase (GGT) were examined histochemically. In week 15, quantitative histological analysis showed that prolonged administration of either 2.0% or 4.0% TJ-48 in the diet significantly reduced the size, volume and/or number of GST-P-positive and GGT-positive hepatic lesions. This treatment also caused a significant increase in the proportion of interleukin-2 receptor-positive lymphocytes among the lymphocytes infiltrating the tumors as well as a significant decrease in the labeling index of preneoplastic lesions. These findings indicate that TJ-48 inhibits the growth of hepatic enzyme-altered lesions, and suggests that its effect may be in part due to activation of the immune system.

References

Aburada, M., Takeda, S., Ito, E., Nakamura, M. and Hosoya, E. (1983) Protective effects of juzentaihoto, dried decoctum of 10 Chinese herbs mixture, upon the adverse effects of mitomycin C in mice. *J. Pharmacobiodyn* 6(12), 1000–4.

Hisha, H., Yamada, H., Sakurai, M.H., Kiyohara, H., Li, Y., Yu, C., Takemoto, N., Kawamura, H., Yamaura, K., Shinohara, S., Komatsu, Y., Aburada, M. and Ikehara, S. (1997) Isolation and identification of hematopoietic stem cell-stimulating substances from Kampo (Japanese herbal) medicine, Juzen-taiho-to. *Blood* 90(3), 1022–30.

Horie, Y., Kato, K., Kameoka, S., Hamano, K. (1994) Bu ji (hozai) for treatment of postoperative gastric cancer patients. *Am. J. Chin. Med.* 22, (3–4), 309–19.

Horii, A., Kyo, M., Asakawa, M., Yasumoto, R. and Maekawa, M. (1991) Multidisciplinary treatment for bladder carcinoma–biological response modifiers and kampo medicines. *Urol Int.* 471, 108–12.

Ikehara, S., Kawamura, H., Komatsu, Y., Yamada, H., Hisha, H., Yasumizu, R., Ohnishi-Inoue, Y., Kiyohara, H., Hirano, M. and Aburada, M. (1992) Effects of medicinal plants on hemopoietic cells. *Adv. Exp. Med. Biol.* 319, 319–30.

Onishi, Y., Yamaura, T., Tauchi, K., Sakamoto, T., Tsukada, K., Nunome, S., Komatsu, Y. and Saiki, I. (1998) Expression of the anti-metastatic effect induced by Juzen-taiho-to is based on the content of Shimotsu-to constituents. *Biol. Pharm. Bull.* 21(7), 761–5.

Ohnishi, Y., Fujii, H., Hayakawa, Y., Sakukawa, R., Yamaura, T., Sakamoto, T., Tsukada, K., Fujimaki, M., Nunome, S., Komatsu, Y. and Saiki, I. (1998) Oral administration of a Kampo (Japanese herbal) medicine Juzen-taiho-to inhibits liver metastasis of colon 26-L5 carcinoma cells. *Jpn. J. Cancer Res.* 89(2), 206–13.

Ohnishi, Y., Fujii, H., Kimura, F., Mishima, T., Murata, J., Tazawa, K., Fujimaki, M., Okada, F., Hosokawa, M. and Saiki, I. (1996) Inhibitory effect of a traditional Chinese medicine, Juzen-taiho-to, on progressive growth of weakly malignant clone cells derived from murine fibrosarcoma. *Jpn. J. Cancer Res.* 87(10), 1039–44.

Sakamoto, S., Furuichi, R., Matsuda, M., Kudo, H., Suzuki, S., Sugiura, Y., Kuwa, K., Tajima, M., Matsubara, M. and Namiki, H. (1994) Effects of Chinese herbal medicines on DNA-synthesizing enzyme activities in mammary tumors of mice. *Am. J. Chin. Med.* 22(1), 43–50.

Sakamoto, S., Kudo, H., Kuwa, K., Suzuki, S., Kato, T., Kawasaki, T., Nakayama, T., Kasahara, N. and Okamoto, R. (1991) Anticancer effects of a Chinese herbal medicine, juzen-taiho-to, in combination with or without 5-fluorouracil derivative on DNA-synthesizing enzymes in 1,2-dimethylhydrazine induced colonic cancer in rats. *Am. J. Chin. Med.* 19(3–4), 233–41.

Satomi, N., Sakurai, A., Iimura, F., Haranaka, R. and Haranaka, K. (1989) Japanese modified traditional Chinese medicines as preventive drugs of the side effects induced by tumor necrosis factor and lipopolysaccharide. *Mol. Biother.* 1(3), 155–62.

Sugiyama, K., Ueda, H. and Ichio, Y. (1995) Protective effect of juzen-taiho-to against carboplatin-induced toxic side effects in mice. *Biol. Pharm. Bull.* 18(4), 544–8.

Sugiyama, K., Ueda, H., Ichio, Y. and Yokota, M. (1995) Improvement of cisplatin toxicity and lethality by juzen-taiho-to in mice. *Biol. Pharm. Bull.* 18(1), 53–8.

Sugiyama, K., Ueda, H., Suhara, Y., Kajima, Y., Ichio, Y. and Yokota, M. (1994) Protective effect of sodium L-malate, an active constituent isolated from Angelicae radix, on cis-diamminedichloroplatinum(II)-induced toxic side effect. *Chem. Pharm. Bull.* (Tokyo) 42(12), 2565–8.

Tatsuta, M., Iishi, H., Baba, M., Nakaizumi, A. and Uehara, H. (1994) Inhibition by shi-quan-da-bu-tang (TJ-48) of experimental hepatocarcinogenesis induced by *N*-nitrosomorpholine in Sprague-Dawley rats. *Eur. J. Cancer* 30(1), 74–8.

Yamada, H. (1989) Chemical characterization and biological activity of the immunologically active substances in Juzen-taiho-to. *Gan To Kagaku Ryoho* 16(4 Pt 2-2), 1500–5.

Yano, H., Mizoguchi, A., Fukuda, K., Haramaki, M., Ogasawara, S., Momosaki, S. and Kojiro, M. (1994) The herbal medicine sho-saiko-to inhibits proliferation of cancer cell lines by inducing apoptosis and arrest at the G0/G1 phase. *Cancer Res.* 54(2), 448–54.

Zee-Cheng, R.K. (1992) Shi-quan-da-bu-tang (ten significant tonic decoction), SQT. A potent Chinese biological response modifier in cancer immunotherapy, potentiation and detoxification of anticancer drugs. *Methods Find Exp. Clin. Pharmacol.* 14(9), 725–36.

3.2.2. Promising candidates for the future: plant species with a laboratory-proven potential

***Acronychia oblongifolia* (Acronychia) (Rutaceae)**	**Cytotoxic**

Location: In all types of rainforest.

Appearance (Figure 3.4)
Stem: 12 m high.
Leaves: 4–12 cm long and emit a pleasant smell when crushed. Oil dots are visible and numerous, and the leaf blade is very glossy.
Flowers: they are produced on the bare stems and behind the foliage.
Parts used: bark, stem.

Active ingredients

- Flavonols: *5,3′-dihydroxy-3,6,7,8,4′-pentamethoxyflavone, 5-hydroxy-3,6,7,8,3′,4′-hexamethoxyflavone, digicitrin, 3-O-demethyldigicitrin, 3,5,3′-trihydroxy-6,7,8,4′-tetramethoxyflavone* and *3,5-dihydroxy-6,7,8,3′,4′-pentamethoxyflavone*.
- Alkaloids: *1,2,3-trimethoxy-10-methyl-acridone, 1,3,4-trimethoxy-10-methyl-acridone, des-N-methyl acronycine, normelicopine* and *noracronycine*.

Documented target cancers

- human nasopharyngeal carcinoma
- tubulin inhibitor.

Figure 3.4 Acronychia sp.

Further details

Related species

- *Acronychia porteri* contains various *flavonols* (see above) which showed activity against (KB) human nasopharyngeal carcinoma cells (IC_{50} 0.04 μg ml^{-1}) and inhibited tubulin assembly into microtubules (IC_{50} 12 μM) (Lichius *et al.*, 1994).
- *Acronychia pedunculata*: The bark contains *acrovestone* and *bauerenol*, two crystalline substances (Wu *et al.*, 1989; Zhu *et al.*, 1989).
- *Acronychia baueri* (Rutaceae): the bark contains the alkaloids, *1,2,3-trimethoxy-10-methyl-acridone, 1,3,4-trimethoxy-10-methyl-acridone, des-N-methyl acronycine, normelicopine* and *noracronycine* (Svoboda *et al.*, 1966).
- *Acronychia laurifolia BL*: contains *acronylin*, a phenolic compound (Biswas *et al.*, 1970).
- *Acronychia haplophylla*: This plant contains the alkaloids *acrophylline* and *acrophyllidine* (Lahey *et al.*, 1968).

References

Biswas, G.K. and Chatterjee, A. (1970) Isolation and structure of acronylin: a new phenolic compound from *Acronychia laurifolia* BL. *Chem. Ind.* 16(20), 654–5.

Chowrashi, B.K., Mukherjea, B. and Sikder, S. (1976) Some central effects of *Acronychia laurifolia* Linn (letter). *Indian J. Physiol. Pharmacol.* 20(4), 250–1.

Funayama, S. and Cordell, G.A. (1984) Chemistry of acronycine IV. Minor constituents of acronine and the phytochemistry of the genus *Acronychia*. *J. Nat. Prod.* 47(2), 285–91.

Lahey, F.N. and McCamish, M. (1968) Acrophylline and acrophyllidine. Two new alkaloids from *Acronychia haplophylla*. *Tetrahedron Lett.*, 12, 1525–7.

Lichius, J.J., Thoison, O., Montagnac, A., Pais, M., Gueritte-Voegelein, F., Sevenet, T., Cosson, J.P. and Hadi, A.H. (1994) Antimitotic and cytotoxic flavonols from *Zieridium pseudobtusifolium* and *Acronychia porteri*. *J. Nat. Prod.*, 57(7), 1012–6.

Svoboda, G.H., Poore, G.A., Simpson, P.J. and Boder, G.B. (1966) Alkaloids of *Acronychia Baueri* Schott I. Isolation of the alkaloids and a study of the antitumor and other biological properties of acronycine. *J. Pharm. Sci.*, 55(8), 758–68.

Wu, T.S., Wang, M.L., Jong, T.T., McPhail, A.T., McPhail, D.R. and Lee, K.H. (1989) X-ray crystal structure of acrovestone, a cytotoxic principle from *Acronychia pedunculata*. *J. Nat. Prod.*, 52(6), 1284–9.

Zhou, F.X. and Min, Z.D. (1989) Studies on the chemical constituents of *Acronychia pedunculata* (L.) Mig. *Chung Kuo Chung Yao Tsa Chih.* 14(2), 30–1, 62.

Agrimonia pilosa **(Agrimony) (Rosaceae)** — **Immunomodulator Cytotoxic**

Location: Of Chinese origin, it is found in most places – on hedge-banks, meadows, open woods and roadsides – though not in the far north.

Appearance
Stem: erect and cylindrical, hairy, 50–150 cm high, mostly unbranched.
Root: long, woody and black.
Leaves: 7.7–20 cm long, pinnate with to other leaflets.
Flowers: small, yellow, on terminal spikes, emitting an apricot-like odor. Fruits bear hairy spines. Fruit deeply grooved.
In bloom: June–September.

Tradition: One of the most famous "magic" herbs, it has been used against wounds of various causes and for the prevention and cure of liver disorders. The Chinese *A. pilosa* is known as *xian he cao*.

Part used: Root

Active ingredients: *agrimoniin* (tannin), unidentified components of methanolic extract.

Particular value: Its use presents a relatively low risk of side effects.

Precautions: Avoid use in case of constipation.

Indicative dosage and application: *agrimoniin*: intraperitoneal injection with $10\,mg\,kg^{-1}$.

Documented target cancers

- *Agrimoniin* is capable of inducing interleukin-1.
- The methanol extract from roots of the plant helps to prolong the life span of mammary carcinoma-bearing mice while inhibiting tumor growth.
- Is cytotoxic to tumor cells, normal cells are far less affected.

Further details

Related compounds

- An antimutagenic activity against benzo[*a*]pyrene (B[*a*]P) was marked in the presence of *A. pilosa* extracts (boiled for 2 h in a water bath) whereas that against 1,6-dinitropyrene (1,6-diNP) and 3,9-dinitrofluoranthene (3,9-diNF) varied from 20%

to 86%. The observed differences in inhibition might be due to the inactivation of metabolic enzymes (Horikawa *et al.*, 1994).

- A significant amount of interleukin-1 (IL-1) beta in the culture supernatant of the human peripheral blood mononuclear cells was stimulated with *agrimoniin* (Miyamoto, 1988). *Agrimoniin* induced IL-1 beta secretion dose- and time-dependently (Murayama, 1992). The adherent peritoneal exudate cells from mice intraperitoneally injected with agrimoniin ($10\,mg\,kg^{-1}$) also secreted IL-1 four days later. These results suggested that *agrimoniin* is a novel cytokine inducer.

Antitumor activity

- To evaluate the antitumor activity of *A. pilosa*, the effects of the methanol extract from roots of the plant (AP-M) on several transplantable rodent tumors were investigated. AP-M inhibited the growth of S-180 solid type tumors (Miyamoto, 1987). On the other hand, the prolongation of life span induced by AP-M on S-180 ascites type tumor-bearing mice was markedly minimized or abolished by the pretreatment with cyclophosphamide. AP-M showed considerably strong cytotoxicity on MM-2 cells *in vitro*, but the effect was diminished to one-tenth by the addition of serum to the culture. Against the host animals, the peripheral white blood cells in mice were significantly increased from 2 to 5 days after the i.p. injection of AP-M. On day 4 after the injection of AP-M, the peritoneal exudate cells, which possessed the cytotoxic activity on MM-2 cells *in vitro*, were also increased to about 5-fold relative to those in the non-treated control. The spleen of the mice was enlarged, and the spleen cells possessed the capacity to uptake 3H-thymidine. However, AP-M did not show direct migration activity like other mitogens against spleen cells from non-treated mice (Miyamoto, 1987). These results indicate that the roots of *A. pilosa* contain some antitumor constituents, and possible mechanisms of the antitumor activity may include host-mediated actions and direct cytotoxicity.

References

Horikawa, K., Mohri, T., Tanaka, Y. and Tokiwa, H. (1994) Moderate inhibition of mutagenicity and carcinogenicity of benzo[a]pyrene, 1,6-dinitropyrene and 3,9-dinitrofluoranthene by Chinese medicinal herbs. *Mutagenesis* 9(6), 523–6.

Kimura, Y., Takido, M. and Yamanouchi, S. (1968) Studies on the standardization of crude drugs. XI. Constituents of *Agrimonia pilosa var. japonica Yakugaku Zasshi* 88(10), 1355–7.

Koshiura, R., Miyamoto, K., Ikeya, Y. and Taguchi, H. (1985) Antitumor activity of methanol extract from roots of Agrimonia pilosa Ledeb. *Jpn. J. Pharmacol.* 38(1), 9–16.

Min, B.S., Kim, Y.H., Tomiyama, M., Nakamura, N., Miyashiro, H., Otake, T. and Hattori, M. (2001) Inhibitory effects of Korean plants on HIV-1 activities. *Phytother. Res.* 15(6), 481–6.

Murayama, T., Kishi, N., Koshiura, R., Takagi, K., Furukawa, T. and Miyamoto, K. (1992) Agrimoniin, an antitumor tannin of *Agrimonia pilosa* Ledeb., induces interleukin-1. *Anticancer Res.* 12(5), 1471–4.

Miyamoto, K., Kishi, N. and Koshiura, R. (1987) Antitumor effect of agrimoniin, a tannin of *Agrimonia pilosa* Ledeb., on transplantable rodent tumors. *Jpn. J. Pharmacol.* 43(2), 187–95.

Miyamoto, K., Kishi, N., Murayama, T., Furukawa, T. and Koshiura, R. (1988) Induction of cytotoxicity of peritoneal exudate cells by agrimoniin, a novel immunomodulatory tannin of *Agrimonia pilosa* Ledeb. *Cancer Immunol. Immunother.* 27(1), 59–620.

Pei, Y.H., Li, X., Zhu, T.R. and Wu, L.J. (1990) Studies on the structure of a new flavanonol glucoside of the root-sprouts of *Agrimonia pilosa* Ledeb *Yao Xue Xue Bao* 5(4), 267–70.

Angelica archangelica L. (Angelica) (Umbelifereae) — Cytotoxic

Location: Of Syria origin, native in cold and moist places in Scotland, and in countries further north (Lapland, Iceland). It can be easily found, as it is largely cultivated in some places.

Appearance (Figure 3.5)
Stem: stout, fluted, 1.3–2 m high and hollow.
Root: long, spindle-shaped, thick and fleshy with large heavy specimens.
Leaves: bright green, composed of numerous small leaflets, divided into three principal groups each of which is subdivided into three lesser groups. Edges are finely toothed or serrated.
Flowers: small and numerous, yellowish or greenish, grouped into large, globular umbels.
In bloom: July.

Tradition: It was well known for its protection against contagion, for purifying the blood and for curing every conceivable malady, such as poisons, agues and all infectious maladies.

Part used: root, leaves, seeds.

Active ingredients

- *Pyranocoumarins: decursin, archangelici, and 8(S),9(R)-9-angeloyloxy-8,9-dihydrooroselol.*
- *Chalcones: 4-hydroxyderricin, xanthoangelol and ashitaba-chalcone.*
- *Polysaccharide: uronic acid.*

Precautions: Should not be given to patients who have tendency towards diabetes, because it increases sugar in the urine.

Documented target cancers: Skin cancer (mouse), Ehrlich tumors (mouse), and the stimulation of the uptake of tritiated thymidine into murine and human spleen cells.

Figure 3.5 Angelica archangelica.

Further details

Related compounds

- *Pyranocoumarins decursin* is cytotoxic against various human cancer cell lines, possibly due to protein kinase C activation. Relatively low cytotoxicity against normal fibroblasts.
- *Polysaccharide*: cytotoxic, immunostimulating.

Related species

- *Angelica gigas*: roots contain the cytotoxic *pyranocoumarin decursin* (also found in *A. decursiva*) Fr. et Sav. (Ahn *et al.*, 1996).
- *Angelica sinensis*: the rhizome contains a low molecular weight (3 kd) *polysaccharide* composed partly of uronic acid. It shows strong antitumor activity on Ehrlich Ascites tumor-bearing mice. It also exhibits immunostimulating activities, both *in vitro* and *in vivo* (Choy *et al.*, 1994).
- *Angelica keiskei*: roots contain two angular furanocoumarins, *archangelicin* and *8(S),9(R)-9-angeloyloxy-8,9-dihydrooroselol* as well as three chalcones, *4-hydroxyderricin, xanthoangelol and ashitaba-chalcone* which can suppress 12-*O*-tetradecanoylphorbol-13-acetate (TPA)-stimulated $^{32}P_i$-incorporation into phospholipids of cultured cells. In addition, *4-hydroxyderricin* and *xanthoangelol* have antitumor-promoting activity in mouse skin carcinogenesis induced by 7,12-dimethylbenz[*a*]anthracene (DMBA) plus TPA, possibly due to the modulation of calmodulin involved systems (Okuyama *et al.*, 1991).
- *Angelica acutiloba* is one of the main components of the oriental Kampo-prescription, Shi-un-kou (in which other two constituents are *Lithospermum erythrorhizon* and *Macrotomia euchroma)*. The drug exhibits inhibitory activity on Epstein–Barr virus activation and skin tumor formation in mice. Roots contain an immunostimulating polysaccharide (AIP) consisting of uronic acid, hexose and peptide (Kumazawa *et al.*, 1982).
- *Angelica radix* is another oriental herb whose administration in mice is associated with an increased production of the TNF, possibly through stimulation of the reticuloendothelial system (RES) (Haranaka *et al.*, 1985).

References

Ahn, K.S., Sim, W.S. and Kim, I.H., (1996) Decursin: a cytotoxic agent and protein kinase C activator from the root of *Angelica gigas. Planta Med.* 62(1), 7–9.

Choy, Y.M., Leung, K.N., Cho, C.S., Wong, C.K. and Pang, P.K. (1994) Immunopharmacological studies of low molecular weight polysaccharide from *Angelica sinensis. Am. J. Chin. Med.* 22(2), 137–45.

Haranaka, K., Satomi, N., Sakurai, A., Haranaka, R., Okada, N. and Kobayashi, M. (1985) Antitumor activities and tumor necrosis factor producibility of traditional Chinese medicines and crude drugs. *Cancer Immunol. Immunother.* 20(1), 1–5.

Konoshima, T., Kozuka, M., Tokuda, H. and Tanabe, M. (1989) Anti-tumor promoting activities and inhibitory effects on Epstein–Barr virus activation of Shi-un-kou and its constituents *Yakugaku Zasshi.* 109(11), 843–6.
Kumazawa, Y., Mizunoe, K. and Otsuka, Y. (1982) Immunostimulating polysaccharide separated from hot water extract of *Angelica acutiloba* Kitagawa (Yamato tohki). *Immunology* 47(1), 75–83.
Okuyama, T., Takata, M., Takayasu, J., Hasegawa, T., Tokuda, H., Nishino, A., Nishino, H. and Iwashima, A. (1991) Anti-tumor-promotion by principles obtained from *Angelica keiskei. Planta Med.* 57(3), 242–6.

***Annona cherimola* (Annona) (Annonaceae) — Cytotoxic**

Location: Central America (Ecuador, Colombia and Bolivia)

Appearance (Figure 3.6)
Stem: 5–10 m high, erect, low brunched.
Leaves: briefly deciduous, alternate, 2-ranked, with minutely hairy petioles 0.8–1.5 cm long, ovate to elliptic or ovate-lanceolate.
Flowers: fragrant, solitary or in groups of 2 or 3, on short hairy stalks along the branches, 3 outer greenish petals and 3 smaller, inner pinkish petals.
In bloom: Spring, summer, autumn, winter.

Part used: fruit.

Active ingredients: *Annonaceous acetogenins* (*lactones*), *alkaloids*.

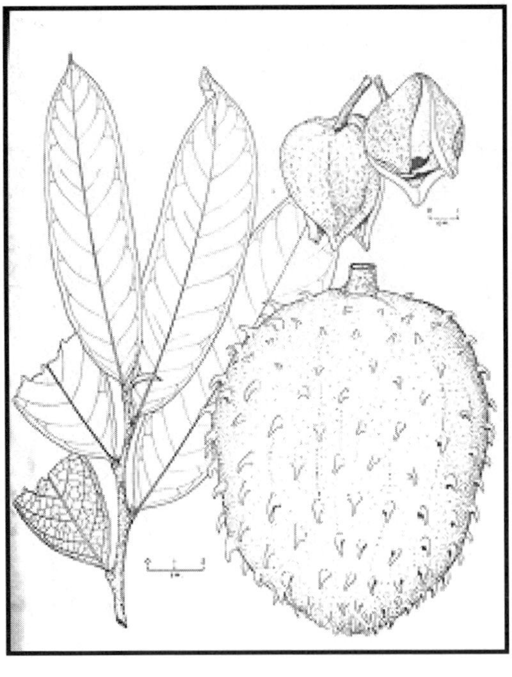

Figure 3.6 Annona sp.

Documented target cancers

- Prostate adenocarinoma
- pancreatic carcinoma cell line (human)
- sarcoma.

Further details

Related species

- *Annona muricata*: leaves contain two Annonaceous acetogenins, muricoreacin and murihexocin C., showing significant cytotoxicities among human tumor cell lines with selectivities to the prostate adenocarinoma (PC-3) and pancreatic carcinoma (PACA-2) cell lines (Kim *et al.*, 1998).
- *Annona senegalensis* is used against sarcomas (Durodola *et al.*, 1975a,b).
- *Annona purpurea* contains alkaloids (Sonnet *et al.*, 1971).
- *Annona reticulata*: seeds contain the cytotoxic *gamma-lactone acetogenin, cis-/trans-isomurisolenin*, along with *annoreticuin, annoreticuin-9-one, bullatacin, squamocin, cis-/trans-bullatacinone* and *cis-/trans-murisolinone* (Chang, 1998).

Related compounds

- The bark of *A. squamosa* yielded three new mono-tetrahydrofuran (THF) ring acetogenins, each bearing two flanking hydroxyls and a carbonyl group at the C-9 position. These compounds were isolated using the brine shrimp lethality assay as a guide for the bioactivity-directed fractionation. (2,4-*cis* and *trans*)-*Mosinone A* is a mixture of ketolactone compounds bearing a *threo/trans/threo* ring relationship and a double bond two methylene units away from the flanking hydroxyl. The other two new acetogenins differ in their stereochemistries around the THF ring; *mosin B* has a *threo/trans/erythro* configuration across the ring, and *mosin C* possesses a *threo/cis/threo* relative stereochemistry. Also found was *annoreticuin-9-one*, a known acetogenin that bears a *threo/trans/threo* ring configuration and a C-9 carbonyl and is new to this species. The structures were elucidated based on spectroscopic and chemical methods. Compounds 1–4 all showed selective cytotoxic activity against the human pancreatic tumor cell line, PACA-2, with potency 10–100 times that of Adriamycin (Hopp *et al.*, 1997).
- Activity-guided fractionation of the stem bark of *A. senegalensis* gave four bioactive ent-kaurenoids. Compound 2 showed selective and significant cytotoxicity for MCF-7 (breast cancer) cells (ED_{50} 1.0 $\mu g\,ml^{-1}$), and 3 and 4 exhibited cytotoxic selectivity for PC-3 (prostate cancer) cells but with weaker potencies (ED_{50} 17–18 $\mu g\,ml^{-1}$). The structure of the new compound, 3, was deduced from spectral evidence (Fatope *et al.*, 1996).
- The bark extracts of *A. squamosa* yielded a new bioactive acetogenin, *squamotacin* (1), and the known compound, *molvizarin*, which is new to this species. Compound 1 is

identical to the potent acetogenin, *bullatacin*, except that the adjacent bis- THF rings and their flanking hydroxyls are shifted two carbons toward the γ-lactone ring. Compound 1 showed cytotoxic activity selectively for the human prostate tumor cell line (PC-3), with a potency of over 100 million times that of Adriamycin (Hopp *et al.*, 1997).

- Bioactivity-directed fractionation of the seeds of *A. muricata* L. (Annonaceae) resulted in the isolation of five new compounds: *cis-annonacin, cis-annonacin-10-one, cis-goniothalamicin, arianacin* and *javoricin*. Three of these) are among the first *cis* mono-THF ring acetogenins to be reported. NMR analyses of published model synthetic compounds, prepared cyclized formal acetals, and prepared Mosher ester derivatives permitted the determinations of absolute stereochemistries. Bioassays of the pure compounds, in the brine shrimp test, for the inhibition of crown gall tumors, and in a panel of human solid tumor cell lines for cytotoxicity, evaluated relative potencies. Compound 1 was selectively cytotoxic to colon adenocarcinoma cells (HT-29) in which it was 10,000 times the potency of adriamycin (Rieser *et al.*, 1996).
- In a continuing activity-directed search for new antitumor compounds, using brine shrimp lethality test (BST), mixtures of three additional pairs of bis-THF ketolactone acetogenins were isolated from the ethanol extract of the bark of *A. bullata* Rich. (Annonaceae). Compared with (2,4-*cis* and *trans*)-*bullatacinone*, these new compounds each have one more aliphatic OH group at a different position on the hydrocarbon chain and thus, were named *(2,4-cis and trans)-10-hydroxybullatacinone (1 and 2), (2,4-cis and trans)-12-hydroxybullatacinone (3 and 4), and (2,4-cis and trans)-29-hydroxybullatacinone*. These mixtures all showed potent activities in the BST and exhibited cytotoxicities comparable to those of adriamycin against human solid tumor cells in culture with selectivities exhibited especially toward the breast cancer cell line (MCF-7) (Gu *et al.*, 1993).
- From *A. bullata*, three more pairs of new ketolactone *Annonaceous acetogenins* were isolated by bioactivity-directed isolation. They are hydroxylated adjacent bis-THF acetogenins and are named *(2,4-cis and trans)-32-hydroxybullatacinone (1 and 2), (2,4-cis and trans)-31-hydroxybullatacinone (3 and 4), and (2,4-cis and trans)-30-hydroxybullatacinone*. The structures were elucidated by analysis of the ^{1}H- and ^{13}C-NMR spectra of 1–6 and their acetates and the MS of their tri-trimethylsilyl (TMSi) derivatives as compared with *bullatacinone*. This is the first time that *Annonaceous acetogenins* with OH groups at successive positions near the end of the aliphatic chain have been reported. All of the new compounds showed potent activities in the BST and against human solid tumor cells in culture, with selectivities exhibited especially toward the colon cancer cell line (HT-29) (Gu *et al.*, 1994).
- Structural work and chemical studies are reported for several cytotoxic agents from the plants *Annona densicoma, Annona reticulata, Claopodium crispifolium, Polytrichum obioense*, and *Psorospermum febrifugum*. Studies are also reported based on development of a mammalian cell culture benzo[*a*]pyrene metabolism assay for the detection of potential anticarcinogenic agents from natural products (Cassady *et al.*, 1990).

References

Cassady, J.M., Baird, W.M. and Chang, C.J. (1990) Natural products as a source of potential cancer chemotherapeutic and chemopreventive agents. *J. Nat. Prod.* 53(1), 23–41.

Chang, F.R., Chen, J.L., Chiu, H.F., Wu, M.J. and Wu, Y.C. (1998) Acetogenins from seeds of *Annona reticulata*. *Phytochemistry* 47(6), 1057–61.

Durodola, J.I. (1975a) Viability and transplantability of developed tumour cells treated *in vitro* with antitumour agent C/M2 isolated from a herbal cancer remedy – of *Annona senegalensis*. *Planta Med.* 28(4), 359–62.

Durodola, J.I. (1975b) Antitumour effects against sarcoma 180 ascites of fractions of *Annona senegalensis*. *Planta Med.* 28(1), 32–6.

Fatope, M.O., Audu, O.T., Takeda, Y., Zeng, L., Shi, G., Shimada, H. and McLaughlin, J.L. (1996) Bioactive ent-kaurene diterpenoids from *Annona senegalensis*. *J. Nat. Prod.* 59(3), 301–3.

Gu, Z.M., Fang, X.P., Hui, Y.H. and McLaughlin, J.L. (1994) 10-, 12-, and 29-hydroxybullatacinones: new cytotoxic Annonaceous acetogenins from *Annona bullata* Rich (Annonaceae). *Nat. Toxins.* 2(2), 49–55.

Gu, Z.M., Fang, X.P., Miesbauer, L.R, Smith, D.L. and McLaughlin, J.L. (1993) 30-, 31-, and 32-hydroxybullatacinones: bioactive terminally hydroxylated annonaceous acetogenins from *Annona bullata*. *J. Nat. Prod.* 56(6), 870–6.

Hopp, D.C., Zeng, L., Gu, Z.M., Kozlowski, J.F. and McLaughlin, J.L. (1997) Novel mono-tetrahydrofuran ring acetogenins, from the bark of *Annona squamosa*, showing cytotoxic selectivities for the human pancreatic carcinoma cell line, PACA-2. *J. Nat. Prod.* 60(6), 581–6.

Hopp, D.C., Zeng, L., Gu, Z. and McLaughlin, J.L. (1996) Squamotacin: an annonaceous acetogenin with cytotoxic selectivity for the human prostate tumor cell line (PC-3). *J. Nat. Prod.* 59(2), 97–9.

Kim, G.S., Zeng, L., Alali, F., Rogers, L.L., Wu, F.E., Sastrodihardjo, S. and McLaughlin, J.L. (1998) Muricoreacin and murihexocin C, mono-tetrahydrofuran acetogenins, from the leaves of *Annona muricata*. *Phytochemistry* 49(2), 565–71.

Rieser, M.J., Gu, Z.M., Fang, X.P., Zeng, L., Wood, K.V. and McLaughlin, J.L. (1996) Five novel mono-tetrahydrofuran ring acetogenins from the seeds of *Annona muricata*. *J. Nat. Prod.* 59(2), 100–8.

Sonnet, P.E. and Jacobson, M. (1971) Tumor inhibitors. II. Cytotoxic alkaloids from *Annona purpurea*. *J. Pharm. Sci.* 60(8), 1254–6.

***Brucea antidysenterica* (Brucea) (Simaroubaceae)** (Figure 3.7) — **Cytotoxic**

Location: China, Japan.

Part used: stem.

Active ingredients

- Cytotoxic: *Bruceoside C, bruceanic acid A* and its methyl ester 2 (new), *bruceanic acid B, C and D*.
- Quassinoid glucosides: *bruceosides D, E* and *F, bruceantinoside C* and *yadanziosides G* and *N, bruceanic acids*.
- Alkaloids: *1,11-dimethoxycanthin-6-one, 11-hydroxycanthin-6-one* and *canthin-6-one*.

Indicative dosage and application: Tested in human carcinoma cells at:

- 250 $\mu g\,ml^{-1}$ showed 42% growth inhibition.
- 500 $\mu g\,ml^{-1}$ showed 56% growth inhibition.

Figure 3.7 Brucea.

The 50% of the results are visible after the first 7 h.

Documented target cancers: Leukemia and non-small-cell lung, colon, CNS, melanoma and ovarian cancer.

- *Bruceanic acid D* is cytotoxic against P-388 lymphocytic leukemia cells.
- *Bruceanic acid A* against KB and TE 671 tumor cells, brain metastasis, in lung cancer with radiotherapy.
- *Bruceoside C* is used against KB, A-549, RPMI and TE-671 tumor cells.
- The three above-mentioned alkaloids are cytotoxic and are used as anti-leukemic alkaloids.

Further details

Antitumor activity

- The fruit of *Brucea javanica* contains *quassinoid glucosides*, which show selective cytotoxicity in the leukemia and non-small cell lung, colon, CNS, melanoma and ovarian cancer, cell lines with $\log GI_{50}$ values ranging from −4.14 to −5.72. A fruit-derived emulsion inhibited human squamous cell carcinoma cells. At a dose of 250 $\mu g\,ml^{-1}$ at 96 h after drug exposure, it showed 42% growth inhibition, and at 500 $\mu g\,ml^{-1}$ inhibited 56% of the cell growth. The effect of more than 50% of the growth inhibition was evident at more than 7 h after drug exposure. In the analysis of the mechanism of the drug using a flow cytometry, the arrest in G_1 phase of cell cycle was found during incubation of cancer cells with drug (Fukamiya *et al.*, 1992).
- The 10% *Brucea javanica* emulsion has synergetic with radiotherapy in treating brain metastasis in lung cancer. Median survival (15 months) of the patients treated was prolonged for 50% (Wang, 1992).

- In addition, the venous emulsion of BJOE had strong action against the elevation of intracranial pressure produced by SNP ($P < 0.01$) while oral emulsion had mild action against it, which was similar to the clinical observation exhibiting improvement of clinical manifestations after application of BJOE on intracranial hypertension caused by brain metastasis from lung cancer (Wang, 1992; Lu *et al.*, 1994).

Related compounds

- The stem of *Brucea antidysenterica* contains *bruceanic acid A* and its methyl ester 2, as well as the *bruceanic acids B, C,* and *D*. It also contains three cytotoxic, *quassinoid glycosides, bruceantinoside C* and the *yadanziosides G* and *N* (Toyota *et al.*, 1990).
- These species also contains three cytotoxic anti-leukemic alkaloids, *1,11-dimethoxy-canthin-6-one, 11-hydroxycanthin-6-one* and *canthin-6-one*.

References

Fukamiya, N., Okano, M., Miyamoto, M., Tagahara, K. and Lee, K.H. (1992) Antitumor agents, 127. Bruceoside C, a new cytotoxic quassinoid glucoside, and related compounds from *Brucea javanica. J. Nat. Prod.* 55(4), 468–75.

Fukamiya, N., Okano, M., Aratani, T., Negoro, K., McPhail, A.T., Ju-ichi, M. and Lee, K.H. (1986) Antitumor agents, 79. Cytotoxic anti-leukemic alkaloids from *Brucea antidysenterica. J. Nat. Prod.* 49(3), 428–34.

Fukamiya, N., Okano, M., Tagahara, K., Aratani, T., Muramoto, Y. and Lee, K.H. (1987) Antitumor agents, 90. Bruceantinoside C, a new cytotoxic quassinoid glycoside from *Brucea antidysenterica. J. Nat. Prod.* 50(6), 1075–9.

Kupchan, S.M., Britton, R.W., Ziegler, M.F. and Sigel, C.W. (1973) Bruceantin, a new potent anti-leukemic simaroubolide from *Brucea antidysenterica. J. Org. Chem.* 38(1), 178–9.

Lu, J.B., Shu, S.Y. and Cai, J.Q. (1994) Experimental study on effect of *Brucea javanica* oil emulsion on rabbit intracranial pressure. *Chung Kuo Chung Hsi I Chieh Ho Tsa Chih.* 14(10), 610–1.

Ohnishi, S., Fukamiya, N., Okano, M., Tagahara, K. and Lee, K.H. (1995) Bruceosides D, E, and F, three new cytotoxic quassinoid glucosides from *Brucea javanica. J. Nat. Prod.* 58(7), 1032–8.

Okano, M., Lee, K.H. and Hall, I.H. (1981) Antitumor agents. 39. Bruceantinoside-A and -B, novel anti-leukemic quassinoid glucosides from *Brucea antidysenterica. J. Nat. Prod.* 44(4), 470–4.

Phillipson, J.D. and Darwish, F.A. (1981) Bruceolides from Filjian *Brucea javanica. Planta Med.* 41(3), 209–20.

Phillipson, J.D. and Darwish, A. (1979) TLX-5 lymphoma cells in rapid screening for cytotoxicity in *Brucea* extracts. *Planta Med.* 35(4), 308–15.

Sakaki, T., Yoshimura, S., Tsuyuki, T., Takahashi, T. and Honda, T. (1986) Yadanzioside P, a new anti-leukemic quassinoid glycoside from *Brucea javanica* (L.) Merr with the 3-O-(beta-D-glucopyranosyl)bruceantin structure. *Chem. Pharm. Bull.* (Tokyo) 34(10), 4447–50.

Toyota, T., Fukamiya, N., Okano, M., Tagahara, K., Chang, J.J. and Lee, K.H. (1990) Antitumor agents, 118. The isolation and characterization of bruceanic acid A, its methyl ester, and the new bruceanic acids B, C, and D, from *Brucea antidysenterica. J. Nat. Prod.* 53(6), 1526–32.

Wang, Z.Q. (1992) Combined therapy of brain metastasis in lung cancer. *Chung Kuo Chung Hsi I Chieh Ho Tsa Chih.* 12(10), 609–10.

Xuan, Y.B., Yasuda, S., Shimada, K., Nagai, S. and Ishihama, H. (1994) Growth inhibition of the emulsion from to *Brucea javanica* cultured human carcinoma cells. *Gan To Kagaku Ryoho.* 21(14), 2421–5.

Bursera simaruba (Bursera) (Burseraceae) — Cytotoxic, Antitumor

Location: Central and northern South America.

Appearance (Figure 3.8)
Stem: 6–17 m high, with reddish bark that reveal a smooth and sinuous gray underbark, thick trunk, large irregular branches.
Leaves: 10–28 cm long with 3–7 oval or elliptic leaflets, each 2,5–5 cm long.
Flowers: small, inconspicuous, with 3–5 greenish petals, blooming in elongate racemes.
In bloom: Winter.

Part used: stem, leaves.

Active ingredients (*lignans*): *deoxypodophyllotoxin, beta-peltatin methyl ether, picro-beta-peltatin methyl ether* and *dehydro-beta-peltatin methyl ether*.

Documented target cancers

- lymphocytic leukemia, human epidermoid carcinoma of the nasopharynx.
- *Lignans*: *deoxypodophyllotoxin* (KB, PS test systems), 5′- *desmethoxydeoxypodophyllotoxin* (*morelensin*) (KB test system).
- *Sapelins A and B*: PS system.

Further details

Related compounds

- The stem of *Bursera permollis* contains four cytotoxic lignans: deoxypodophyllotoxin, *beta*-peltatin methyl ether, picro-beta-peltatin methyl ether and dehydro-*beta*-peltatin methyl ether (Wickramaratne *et al.*, 1995). Deoxypodophyllotoxin and another lignan, 5′-desmethoxydeoxypodophyllotoxin, were also isolated from the

Figure 3.8 Bursera.

dried exudate of *B. morelensis* (Jolad *et al.*, 1977b), *B. microphylla* also contains deoxypodophyllotoxin (Bianchi *et al.*, 1968).

- The leaves of *B. klugii* contain non-polar substances, such as sapelins A and B, which showed activity against two test systems, the P-388 lymphocytic leukemia (3PS) and the human epidermoid carcinoma of the nasopharynx (9KB) (Jolad *et al.*, 1977a).
- The isolation and identification from Burseraceae are reported.
- The existence of lignans with antitumor activity in *B. schlechtendalii* has been reported (McDoniel *et al.*, 1972).

References

Bianchi, E., Caldwell, M.E. and Cole, J.R. (1968) Antitumor agents from *Bursera microphylla* (Burseraceae) I. Isolation and characterization of deoxypodophyllotoxin. *J. Pharm. Sci.* 57(4), 696–7.

Jolad, S.D., Wiedhopf, R.M. and Cole, J.R. (1977a) Cytotoxic agents from *Bursera klugii* (Burseraceae) I: isolation of sapelins A and B. *J. Pharm. Sci.* 66(6), 889–90.

Jolad, S.D., Wiedhopf, R.M. and Cole, J.R. (1977b) Cytotoxic agents from *Bursera morelensis* (Burseraceae): deoxypodophyllotoxin and a new lignan, 5′-desmethoxydeoxypodophyllotoxin. *J. Pharm. Sci.* 66(6), 892–3.

McDoniel, P.B. and Cole, J.R. (1972) Antitumor activity of *Bursera schlechtendalii* (Burseraceae): isolation and structure determination of two new lignans. *J. Pharm. Sci.* 61(12), 1992–4.

Wickramaratne, D.B., Mar, W., Chai, H., Castillo, J.J., Farnsworth, N.R., Soejarto, D.D., Cordell, G.A., Pezzuto, J.M. and Kinghorn, A.D. (1995) Cytotoxic constituents of *Bursera permollis*. *Planta Med.* 61(1), 80–1.

Cassia acutifolia (Cassia, Senna) (Leguminosae) — Cytotoxic

Location: Egypt, Nubia, Arabia and Sennar.

Appearance
Stem: erect, smooth, pale green with long spreading branches, 0.70 m high.
Leaves: bearing leaflets in four or five pairs, 1 inch long, lanceolate or obovate, brittle, grayish-green, of a faint, peculiar odor and mucilaginous, sweetish taste.
Flowers: small, yellow.

Parts used: dried leaflets, pods.

Active ingredients (*Bitetrahydroanthracene* derivative): *torosaol-III, Pyranosides, Polysaccharides, Piperidine.*

Documented target cancers: KB cells, solid Sarcoma-180 (mice).

Further details

It has been found that contains Related compounds that are cytotoxic and DNA damaging.

Related species

- *Cassia torosa* Cav.: The flowers contain *torosaol-III*, *physcion*, *5,7′-physcionanthrone-physcion, 5,7′-biphyscion, torosanin-9,10-quinone, 5,7-dihydroxy-chromone, naringenin* and *chrysoeriol*. Dimeric *tetrahydroanthracenes* exhibited cytotoxic activity against KB cells in the tissue culture (Kitanaka *et al*., 1994).

- *Cassia angustifolia* L.: The leaves contain water-soluble polysaccharides, including *L-rhamnose, L-arabinose, D-galactose, D-galacturonic acid* and derivatives thereof, exhibiting activity against the solid Sarcoma-180 in CD1 mice (Muller *et al*., 1987).
- *Cassia leptophylla* contains the DNA-damaging compound piperidine.

References

Kitanaka, S. and Takido, M. (1994) Bitetrahydroanthracenes from flowers of *Cassia torosa* Cav. *Chem. Pharm. Bull.* (Tokyo) 42(12), 2588–90.

Kwon, B.M., Lee, S.H., Choi, S.U., Park, S.H., Lee, C.O., Cho, Y.K., Sung, N.D. and Bok, S.H. (1998) Synthesis and *in vitro* cytotoxicity of cinnamaldehydes to human solid tumor cells. *Arch. Pharm. Res.* 21(2), 147–52.

Lee, C.W., Hong, D.H., Han, S.B., Park, S.H., Kim, H.K., Kwon, B.M. and Kim, H.M. (1999) Inhibition of human tumor growth by 2′-hydroxy- and 2′-benzoyloxycinnamaldehydes. *Planta Med.* 65(3), 263–6.

Messana, I., Ferrari, F., Cavalcanti, M.S. and Morace, G. (1991) An anthraquinone and three naphthopyrone derivatives from *Cassia pudibunda*. *Phytochemistry* 30(2), 708–10.

Muller, B.M., Kraus, J. and Franz, G. (1989) Chemical structure and biological activity of water-soluble polysaccharides from *Cassia angustifolia* leaves. *Planta Med.* 55(6), 536–9.

***Chelidonium majus* L. (Chelodonium, Celandine) (Papaveraceae)** — Immunomodulatory

Location: found by old walls, on waste ground and in hedges, nearly always in the neighborhood of human habitations.

Appearance
Stem: slender, round and slightly hairy, 0.5–1 m high, much branched.
Root: thick, fleshy.
Leaves: yellowish-green, much paler, almost grayish below, graceful in form and slightly hairy, 15–30 cm long, 5–7.5 cm wide, deeply divided as far as the central rib, so as to form usually two pairs of opposite leaflets with rounded teeth edges.
Flowers: arranged at the ends of the stems in loose umbels.
In bloom: summer.

Tradition: It was used as a drug plant since the Middle Ages and Dioscorides and Pliny mention it. It was used to take away specks from the eye and to stop incipient suffusions. It is useful, also, as alterative, diuretic, purgative, in jaundice, eczema and scrofulous diseases.

Part used: the whole herb.

Active ingredients (*Alkaloids*): *chelidonine* and its semisynthetic compound; *Tris(2-({5bS-(5ba,6b,12ba)})-5b,6,7,12b,13,14-Hexahydro-13-methyl)({1,3}-benzodioxolo{5,6-c}-1,3-dioxolo{4,5-i} phenanthridinium-6-ol-Ethaneaminyl) Phosphinesulfide 6HCl (Ukrain)*.

Particular value: Although *Ukrain*, of high concentrations is cytostatic for malignant cells and may suppress the growth of cancer, is not cytostatic of normal concentrations.

Indicative dosage and application

- Every second day in a dose of 10 mg per injection. Each patient receives 300 mg of the drug (30 injections).
- In lung cancer it is used in an intravenous injection every three days. One course consisted of 10 applications of 10 mg each.

Documented target cancers: It has been reported that the herb extract of *Chelidonium majus* showed preventive effects on glandular stomach tumor development in rats treated with *N-methyl-N′-nitro-N nitrosoguanidine* (MNNG) and hypertonic sodium chloride. The incidence of forestomach neoplastic lesions (papillomas and squamous cell carcinomas) also showed a tendency to decrease with the herbal extract treatment (Bruller, 1992).

Further details

Related compounds

- *Ukrain*, is a semi-synthetic thiophosphoric acid compound of alkaloid *chelidonine* isolated from *Chelidonium majus L.* Its full chemical name is *Tris(2-({5bS-(5ba,6b,12ba)})-5b,6,7,12b,13,14-Hexahydro-13-methyl}({1,3}-benzodioxolo{5,6-c}-1,3-dioxolo{4,5-i}phenanthridinium-6-ol-Ethaneaminyl) Phosphinesulfide 6HCl.* Ukrain causes a regression of tumors and metastases in many oncological patients. More than 400 documented patients with various carcinomas in different stages of development have been treated with Ukrain. J.W. Nowicky produced Ukrain for the first time in 1978. (Austrian Patent No. 354644, Vienna, January 25, 1980.) Ukrain can be immunologically effective in lung cancer patients and can improve human cellular response (Nowicky *et al.*, 1991).

Antitumor activity

- Ukrain was applied as an i.v. injection every three days on nine men (aged 42–68 years, mean 57 years) with histologically proven lung cancer, previously untreated. One course consisted of 10 applications of 10 mg each. The treatment was generally well tolerated. The results showed an increase in the proportion of total T-cells, and a significant decrease in the percentage of T-suppressor cells. There were no signs of activation of NK, T-helper and B-cells. The restoration of cellular immunity was accompanied by an improvement in the clinical course of the disease. This effect was particularly pronounced in patients who responded to further chemotherapy. Objective tumor regression was seen in 44.4% of treated patients. Four out of nine patients (44.4%) died of progressive disease during the course of this study (Staniszewski *et al.*, 1992).

- Thirty-six stage III cancer patients were treated with *Ukrain*. The drug was injected intravenously every second day in a dose of 10 mg per injection. Each patient received 300 mg of the drug (30 injections). The cytostatic effect of Ukrain was monitored clinically and by ultrasonography (USG) and computer tomography (CT), as well as by determination of CEA and CA-125 in the sera of patients with rectal and ovarian cancers, respectively. The influence of Ukrain on immune parameters was evaluated by monoclonal antibodies (MAb) to CD2, CD4, CD8 and CD22. The influence of Ukrain on immune parameters in cancer patients was matched with its effect on these parameters in 20 healthy volunteer controls. The results obtained indicate that Ukrain, in a concentration not cytostatic in normal cells, is cytostatic for malignant ones, may suppress the growth of cancer. The compound also has immunoregulatory properties, regulating the T lymphocyte subsets (Steinacker *et al.*, 1996).
- The effect of Ukrain on the growth of Balb/c syngenic mammary adenocarcinoma was assessed. Intravenous, but not subcutaneous or intraperitoneal, administration of this drug was found to be effective in delaying tumor growth in an actual therapeutic protocol initiated five days after tumor implantation. No untoward side effects were observed using these *in vivo* treatment modalities. Ukrain's *in vivo* effects against the development of mammary tumors may be due, at least in part, to its ability to restore macrophage cytolytic function.
- Ukrain is an effective biological response modifier augmenting, by up to 48-fold, the lytic activity of splenic lymphocytes obtained from alloimmunized mice. The lytic activities of IL-2-treated spleen cells and peritoneal exudate lymphocytes were also significantly increased by the addition of Ukrain to the cell mediated lysis (CML) assay medium. The highest Ukrain-induced enhancement of splenic lymphocytolytic activity *in vitro* was found to occur at day 18 after alloimmunization was dose-dependent and specific for the immunizing P815 tumor cells. Since *Ukrain* was present only during the CML assays, its mode of action is thought to be via direct activation of the effector cells' lytic mechanism(s). The effect of Ukrain on the growth of Balb/c syngenic mammary adenocarcinoma was also evaluated. Intravenous, but not subcutaneous or intraperitoneal, administration of this drug was found to be effective in delaying tumor growth in an actual therapeutic protocol initiated five days after tumor implantation. No deleterious side effects were observed using these *in vivo* treatment modalities. The role of macrophages in the observed retardation of tumor development was investigated, using PEM in cytotoxicity assays. Previous studies showed that PEM of mammary tumor-bearing mice lose their capacity to kill a variety of tumor target cells including the *in vitro* cultured homologous tumour cells (DA-3). Pretreatment of PEM from normal mice with 2.5 μM Ukrain for 24 h, followed by stimulation with either IFN-γ or with LPS plus IFN-γ enhanced their cytotoxic activity. Treatment of PEM from tumour-bearing mice with 2.5 μM Ukrain and LPS results in a reversal of their defective cytotoxic response against DA-3 target cells. Furthermore, Ukrain alone, in the absence of a secondary signal, induced the activation of tumoricidal function of PEM from tumor-bearing, but not from normal, mice. These data indicate that Ukrain's *in vivo* effects against the development of mammary tumors may be due, at least in part, to its ability to restore macrophage cytolytic function (Sotomayor *et al.*, 1992).

Other medical activity

- For the treatment of AIDS patients with Kaposi's sarcoma, *Ukrain* was injected i.v. in the dose of 5 mg every other day for a total of 10 injections. During treatment the Kaposi's sarcoma lesions diminished in size, showed decoloration and no lesion appeared in the 30-day interval after the beginning of treatment. Both patients tolerated Ukrain well and showed an improved immunohematological status: an increase in total leukocytes, T-lymphocytes and T-suppressor numbers. In one case T-helper lymphocytes were also increased (Voltchek *et al.*, 1996).

References

Bruller, W. (1992) Studies concerning the effect of Ukrain *in vivo* and *in vitro*. *Drugs Exp. Clin. Res.* 18, 13–6.

Ciebiada, I., Korczak, E., Nowicky, J.W. and Denys, A. (1996a) Estimation of direct influence of Ukrain preparation on influenza viruses and the bacteria *E. coli* and *S. aureus*. *Drugs Exp. Clin. Res.* 22(3–5), 219–23.

Ciebiada, I., Korczak, E., Nowicky, J.W. and Denys, A. (1996b) Does the Ukrain preparation protect mice against lethal doses of bacteria? *Drugs Exp. Clin. Res.* 22(3–5), 207–11.

Ebermann, R., Alth, G., Kreitner, M. and Kubin, A. (1996) Natural products derived from plants as potential drugs for the photodynamic destruction of tumor cells. *J. Photochem., Photobiol. B.* 36(2), 95–7.

Kim, D.J., Ahn, B., Han, B.S. and Tsuda, H. (1997) Potential preventive effects of *Chelidonium majis* L. (Papaveraceae) herb extract on glandular stomach tumor development in rats treated with N-methyl-N′-nitro-N nitrosoguanidine (MNNG) and hypertonic sodium chloride. *Cancer Lett.* 112(2), 203–8.

Liepins, A. and Nowicky, J.W. (1996) Modulation of immune effector cell cytolytic activity and tumour growth inhibition *in vivo* by Ukrain (NSC 631570). *Drugs Exp. Clin. Res.* 22(3–5), 103–13.

Liepins, A. and Nowicky, J.W. (1992) Activation of spleen cell lytic activity by the alkaloid thiophosphoric acid derivative: Ukrain. *Int. J. Immunopharmacol.* 14(8), 1437–42.

Lohninger, A. and Hamler, F. (1992) Chelidonium *majus* L. (Ukrain) in the treatment of cancer patients. *Drugs Exp. Clin. Res.* 18, 73–7.

Malaveille, C., Friesen, M., Camus, A.M., Garren, L., Hautefeuille, A., Bereziat, J.C., Ghadirian, P., Day, N.E. and Bartsch, H. (1982) Mutagens produced by the pyrolysis of opium and its alkaloids as possible risk factors in cancer of the bladder and oesophagus. *Carcinogenesis* 3(5), 577–85.

Nowicky, J.W., Manolakis, G., Meijer, D., Vatanasapt, V. and Brzosko, W.J. (1992) Ukrain both as an anticancer and immunoregulatory agent. *Drugs Exp. Clin. Res.* 18, 51–4.

Nowicky, J.W., Staniszewski, A., Zbroja-Sontag, W., Slesak, B., Nowicky, W. and Hiesmayr, W. (1991) Evaluation of thiophosphoric acid alkaloid derivatives from *Chelidonium majus* L. ("Ukrain") as an immunostimulant in patients with various carcinomas. *Drugs Exp. Clin. Res.* 17(2), 139–43.

Ranadive, K.J., Gothoskar, S.V. and Tezabwala, B.U. (1973) Testing carcinogenicity of contaminants in edible oils. II. Argemone oil in mustard oil. *Indian J. Med. Res.* 61(3), 428–34.

Ranadive, K.J., Gothoskar, S.V. and Tezabwala, B.U. (1972) Carcinogenicity of contaminants in indigenous edible oils. *Int. J. Cancer.* 10(3), 652–66.

Shi, G.Z. (1992) Blockage of *Glyrrhiza uralensis* and *Chelidonium majus* in MNNG induced cancer and mutagenesis. *Chung Hua Yu Fang I Hsueh Tsa Chih.* 26(3), 165–7.

Sotomayor, E.M., Rao, K., Lopez, D.M. and Liepins, A. (1992) Enhancement of macrophage tumouricidal activity by the alkaloid derivative Ukrain. *In vitro* and *in vivo* studies. *Drugs Exp. Clin. Res.* 18, 5–11.

Slesak, B., Nowicky, J.W. and Harlozinska, A. (1992) *In vitro* effects of *Chelidonium majus* L. alkaloid thiophosphoric acid conjugates (Ukrain) on the phenotype of normal human lymphocytes. *Drugs Exp. Clin. Res.* **18**, 17–21.
Staniszewski, A., Slesak, B., Kolodziej, J., Harlozinska-Szmyrka, A. and Nowicky, J.W. (1992) Lymphocyte subsets in patients with lung cancer treated with thiophosphoric acid alkaloid derivatives from *Chelidonium majus* L. (Ukrain). *Drugs Exp. Clin. Res.* **18**, 63–7.
Steinacker, J., Kroiss, T., Korsh, O.B. and Melnyk, A. (1996) Ukrain therapy in a frontal anaplastic grade III astrocytoma (case report). *Drugs Exp. Clin. Res.* 22(3–5), 275–7.
Voltchek, I.V., Liepins, A., Nowicky, J.W. and Brzosko, W.J. (1996) Potential therapeutic efficacy of Ukrain (NSC 631570) in AIDS patients with Kaposi's sarcoma. *Drugs Exp. Clin. Res.* 22(3–5), 283–6.
Xian, M.S., Hayashi, K., Lu, J.P. and Awai, M. (1989) Efficacy of traditional Chinese herbs on squamous cell carcinoma of the esophagus: histopathologic analysis of 240 cases. *Acta Med. Okayama.* 43(6), 345–51.

Cinnamomum camphora **(Cinnamomum, Camphor tree) (Lauraceae)** — **Cytotoxic Immunomodulator**

Location: East Asia. It can be found in most sub-tropical countries, as it can be cultivated successfully there.

Appearance (Figure 3.9)
Stem: 20–40 m, many branched, evergreen.
Leaves: evergreen with oval oblong blades.
Flowers: white, small and clustered.
In bloom: Spring.

Tradition: Chinese use the camphor oil exudes in the process of extracting camphor for many centuries. It was mentioned by Marko Polo in the thirteenth century and Camoens in 1571, who called it the "balsam of disease". Very useful in complaints of stomach and bowels, in spasmodic cholera and flatulent colic.

Part used: gum.

Active ingredients (*Cinnamaldehydes*): *2′-Hydroxycinnamaldehyde* (*HCA*) and *2′-benzoxy-cinnamaldehyde* (*BCA*).

Precautions: In large doses it is very poisonous. Should be used cautiously in certain heart disease.

Documented target cancers: Human cancer cells lines, SW-620 human tumor xenograft.

Further details

Other medical effects

- The species are cytotoxic (the key functional group of the cinnamaldehyde-related compounds in the antitumor activity is the propenal group) (Ling and Liu, 1996).
- Immunomodulation is effected due to the inhibition of farnesyl protein transferase. RAS activation, which is accompanied with its farnesylation, has been known to be

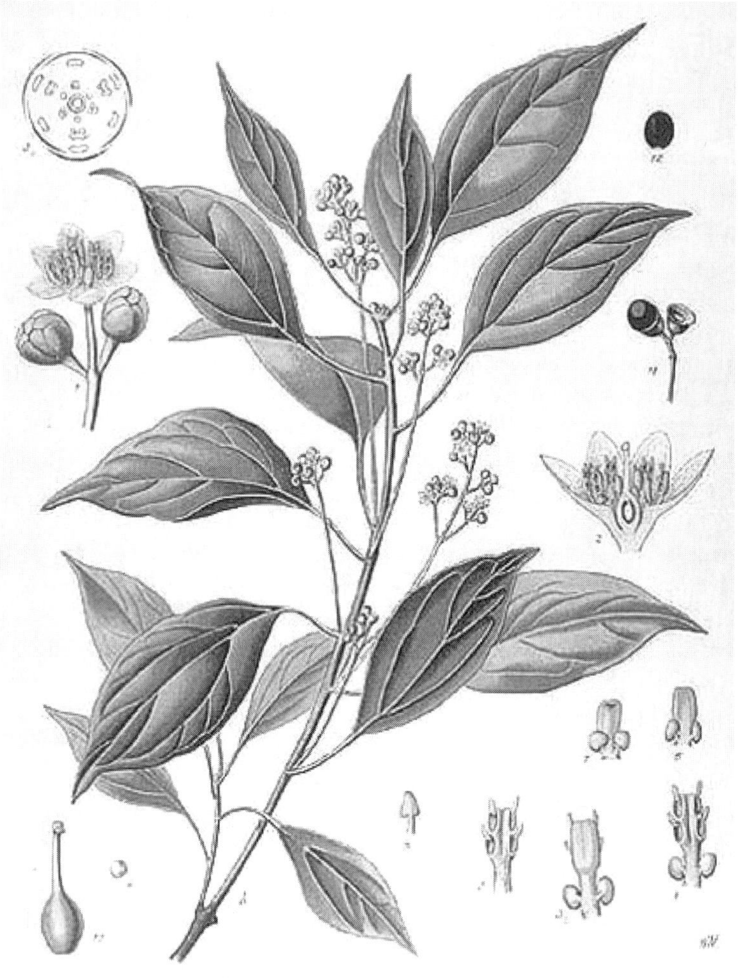

Figure 3.9 *Cinnamomum camphora*.

important in immune cell activation as well as in carcinogenesis. Extracts inhibit the lymphoproliferation and induce a T-cell differentiation through the blockade of early steps in signaling pathway leading to cell growth.

Related species

- *Cinnamomum cassia* Blume (Lauraceae): the bark contains *2′-hydroxycinnamaldehyde* which reacts with benzoyl chloride in order to give *2′-benzoyloxycinnamaldehyde*

(Lee *et al.*, 1999). Both compounds strongly inhibited *in vitro* growth of 29 kinds of human cancer cells and *in vivo* growth of SW-620 human tumor xenograft without the loss of body weight in nude mice.

Related compounds

- Two kinds of *cinnamaldehyde* derivative, HCA and *BCA*, were studied for their immunomodulatory effects. These compounds were screened as anticancer drug candidates from stem bark of *Cinnamomum cassia* for their inhibitory effect on activity (Lee *et al.*, 1999). Treatment of these cinnamaldehydes to mouse splenocyte cultures induced suppression of lymphoproliferation following both Con A and LPS stimulation in a dose-dependent manner. A dose of 1 μM of HCA and BCA inhibited the Con A-stimulated proliferation by 69% and 60%, and the LPS-induced proliferation by 29% and 21%, respectively. However, the proliferation induced by PMA plus ionomycin was affected by neither HCA nor BCA treatment. Decreased levels of antibody production by HCA or BCA treatment were observed in both SRBC-immunized mice and LPS-stimulated splenocyte cultures. The exposure of thymocytes to HCA or BCA for 48 h accelerated T-cell differentiation from CD4 and CD8 double positive cells to CD4 or CD8 single positive cells. The inhibitory effect of cinnamaldehyde on lymphoproliferation was specific to the early phase of cell activation, showing the strongest inhibition of Con A- or LPS-stimulated proliferation when added concomitantly with the mitogens. In addition, the treatment of HCA and BCA to splenocyte cultures attenuated the Con A-triggered progression of cell cycle at G_1 phase with no inhibition of S–G_2/M phase transition. Although cinnamaldehyde treatment had no effect on the IL-2 production by splenocyte cultures stimulated with Con A, it inhibited markedly and dose-dependently the expression of IL-2Rα and IFN-γ. Taken together, the results in this study suggest both HCA and BCA.

References

Balachandran, B. and Sivaramkrishnan, V.M. (1995) Induction of tumours by Indian dietary constituents. *Indian J. Cancer* 32(3), 104–9.

Chen, C.H., Yang, S.W. and Shen, Y.C. (1995) New steroid acids from Antrodia cinnamomea, a fungal parasite of *Cinnamomum micranthum*. *J. Nat. Prod.* 58(11), 1655–61.

Choi, J., Lee, K.T., Ka, H., Jung, W.T., Jung, H.J. and Park, H.J. (2001) Constituents of the essential oil of the *Cinnamomum cassia* stem bark and the biological properties. *Arch Pharm Res.* 24(5), 418–23.

Haranaka, R., Hasegawa, R., Nakagawa, S., Sakurai, A., Satomi, N. and Haranaka, K. (1988) Antitumor activity of combination therapy with traditional Chinese medicine and OK432 or MMC. *J. Biol. Response. Mod.* 7(1), 77–90.

Haranaka, K. Satomi, N., Sakurai, A., Haranaka, R., Okada, N. and Kobayashi, M. (1985) Antitumor activities and tumor necrosis factor producibility of traditional Chinese medicines and crude drugs. *Cancer Immunol. Immunother.* 20(1), 1–5.

Ikawati, Z., Wahyuono, S. and Maeyama, K. (2001) Screening of several Indonesian medicinal plants for their inhibitory effect on histamine release from RBL-2H3 cells. *J. Ethnopharmacol.* 75(2–3), 249–56.

Kwon, B.M., Lee, S.H., Choi, S.U., Park, S.H., Lee, C.O., Cho, Y.K., Sung, N.D. and Bok, S.H. (1998) Synthesis and *in vitro* cytotoxicity of cinnamaldehydes to human solid tumor cells. *Arch. Pharm. Res.* 21(2), 147–52.

Lee, C.W., Hong, D.H., Han, S.B., Park, S.H., Kim, H.K., Kwon, B.M. and Kim, H.M. (1999) Inhibition of human tumor growth by 2′-hydroxy- and 2′-benzoyloxycinnamaldehydes. *Planta Med.* 65(3), 263–6.
Ling, J. and Liu, W.Y. (1996) Cytotoxicity of two new ribosome-inactivating proteins, cinnamomin and camphorin, to carcinoma cells. *Cell Biochem. Funct.* 14(3), 157–61.
Mihail, R.C. (1992) Oral leukoplakia caused by cinnamon food allergy. *J. Otolaryngol.* 21(5), 366–7.
Sakamoto, S., Yoshino, H., Shirahata, Y., Shimodairo, K. and Okamoto, R. (1992) Pharmacotherapeutic effects of kuei-chih-fu-ling-wan (keishi-bukuryo-gan) on human uterine myomas. *Am. J. Chin. Med.* 20(3–4), 313–7.
Sedghizadeh, P.P. and Allen, C.M. (2002) White plaque of the lateral tongue. *J. Contemp. Dent. Pract.* 15, 3(3), 46–50.
Westra, W.H., McMurray, J.S., Califano, J., Flint, P.W. and Corio, R.L. (1998) Squamous cell carcinoma of the tongue associated with cinnamon gum use: a case report. *Head Neck* 20(5), 430–3.
Zee-Cheng, R.K. (1992) Shi-quan-da-bu-tang (ten significant tonic decoction), SQT. A potent Chinese biological response modifier in cancer immunotherapy, potentiation and detoxification of anticancer drugs. *Methods Find Exp. Clin. Pharmacol.* 14(9), 725–36. Review.

Chrysanthemum

See in Glycyrriza under Further details.

Colchicum autumnale (Meadow saffron) (Liliaceae) Cytotoxic

Synonyms: Autumn Crocus, Naked Ladies.

Location: In Southern and Central Europe, in meadows and deciduous woods.
Appearance
Root: scaly corm, up to 7 cm.
Leaves: basal, linear-lanceolate, up to 40 cm long.
Flowers: long-tubed purple or white, directly emerging from the underground corm. They share a resemblance to the flowers of *Crocus sativus*, but they possess 6 anthers.
Fruit: oval capsule.
In bloom: August–October.

Tradition: Considered to be the Hermodactyls of the Arabians, it has been used against rheumatism and gout.

Part used: Root, seeds.

Active ingredients: *colchicine* (alkaloid) and related compounds, such as *thiocolchicine* and *thioketones*.

Particular value: It is used as anti-rheumatic, cathartic, emetic.

Precautions: Extremely poisonous. Colchicine acts upon all secretive organs, such as the bowels and kidneys.

Documented target cancers

- Colchicine and several of its analogues show good antitumor effect in mice infected with P388 lymphocytic leukemia (Kupchan *et al.*, 1973).
- High antitubulin effects of derivatives of 3-demethylthiocolchicine, methylthio ethers of natural colchicinoids and thioketones derived from thiocolchicine (Muzaffar *et al.*, 1990).
- Treatment of esophageal cancer with colchamine (Vitkin, 1969).

Further details

Other medical effects

- *Colchicum autumnale:* It is also, considered to have cytostatic effects bibliography.
- *Colchicine* can cause induction of chromosome (loss and gain): The fruit fly *Drosophila melanogaster* is one of the standard systems used for mutagen screening. The colchicine-containing drugs Colchicum-Dispert and Colchysat Burger were fed at extremely low concentrations (1 : 300 000 and 1:50 000 respectively) to *Drosophila* females. Among their offspring a remarkably high frequency of aneuploid individuals (XO and XXY flies) were found. These aneuploids correspond karyotypically to the human Ullrich-Turner (XO) and Klinefelter's (XXY) syndromes and result from chromosome loss (XO) and chromosome gain (XXY). The maximum aneuploidy frequency observed after colchicine feeding was 24 times the control value. Depending on their size the aneuploidy frequencies are as great as those obtained by X-ray-irradiation with some hundred or some thousand R (Traut and Sommer, 1976).

Antitumor activity

- Esterification of the phenolic group in 3-demethylthiocolchicine and exchange of the *N*-acetyl group with other *N*-acyl groups or a *N*-carbalkoxy group afforded many compounds which showed superior activity over the parent drug as inhibitors of tubulin polymerization and of the growth of L1210 murine leukemia cells in culture (Muzaffar *et al.*, 1990). A comparison of naturally occurring colchicum alkaloids with thio isosters, obtained by replacing the OMe group at C(10) with a SCH3 group, showed the thio ethers to be invariably more potent in these assays. The comparison included 3-demethylthiodemecolcine prepared from 3-demethylthiocolchicine by partial synthesis. Thiation of thiocolchicine with Lawesson's reagent afforded novel thiotropolones which exhibited high antitubulin activity. Their structures are fully secured by spectral data. Colchicine and several of its analogues show good antitumor effect in mice infected with P388 lymphocytic leukemia, and all of them show high affinity for tubulin and inhibit tubulin polymerization at low concentration (Muzaffar *et al.*, 1990). Consequently, antitubulin assays with this class of compounds can serve as valuable prescreens for the initial evaluation of potential antitumor drugs.

Related species

- *Colchicum speciosum*: It concerns as a tumor inhibitor, with anti-leukemic activity (Kupchan *et al.*, 1973).

References

Brncic, N., Viskovic, I., Peric, R., Dirlic, A., Vitezic, D. and Cuculic, D. (2001) Accidental plant poisoning with *Colchicum autumnale*: report of two cases. *Croat. Med. J.* 42(6), 673–5.

Danel, V.C., Wiart, J.F., Hardy, G.A., Vincent, F.H. and Houdret, N.M. (2001) Self-poisoning with *Colchicum autumnale L.* flowers. *J. Toxicol. Clin. Toxicol.* 39(4), 409–11.

Haupt, H. (1996) Toxic and less toxic plants 29 *Kinderkrankenschwester* 15(9), 337–8.

Klintschar, M., Beham-Schmidt, C., Radner, H., Henning, G. and Roll, P. (1999) Colchicine poisoning by accidental ingestion of meadow saffron (*Colchicum autumnale*): pathological and medicolegal aspects. *Forensic. Sci. Int.* 20, 106, (3), 191–200.

Kupchan, S.M., Britton, R.W., Chiang, C.K., NoyanAlpan, N. and Ziegler, M.F. (1973) Tumor inhibitors. 88. The antileukemia principles of *Colchicum speciosum. Lloydia*, 36(3), 338–40.

Lindholm, P., Gullbo, J., Claeson P., Goransson, U., Johansson, S., Backlund, A., Larsson, R. and Bohlin, L. (2002) Selective cytotoxicity evaluation in anticancer drug screening of fractionated plant extracts. *J. Biomol. Screen.* 7(4), 333–40.

Muzaffar, A., Brossi, A., Lin, C.M. and Hamel, E. (1990) Antitubulin effects of derivatives of 3-demethylthiocolchicine, methylthio ethers of natural colchicinoids, and thioketones derived from thiocolchicine. Comparison with colchicinoids. *J. Med. Chem.* 33(2), 567–71.

Rueffer, M. and Zenk, M.H. (1998) Microsome-mediated transformation of O-methylandrocymbine to demecolcine and colchicine. *FEBS Lett.* 30, 438 (1–2), 111–3.

Schrader, A., Schulz, O., Volker, H. and Puls, H. (2001) Recent plant poisoning in ruminants of northern and eastern Germany. *Communication from the practice for the practice Berl Munch Tierarztl Wochenschr.* 114(5–6), 218–21.

Traut, H. and Sommer, U. (1976) The induction of chromosome loss and gain by colchicines. *MMW Munch Med. Wochenscr.* 3, 118, (36), 1113–6.

Van and Os, F.H. (1970) Plants with cytostatic effect Farmaco. *Science* 25(6), 455–83. Review.

Vitkin, B.S. (1969) Treatment of esophageal cancer with colchamine. *Vopr. Onkol.* 15(11), 90–2.

Weinberger, A. and Pinkhas, J. (1980) The history of colchicine. *Korot* 7(11–12), 760.

Yamada, M., Kobayashi, Y., Furuoka, H. and Matsui, T. (2000) Comparison of enterotoxicity between autumn crocus (*Colchicum autumnale L.*) and colchicine in the guinea pig and mouse: enterotoxicity in the guinea pig differs from that in the mouse. *J. Vet. Med. Sci.* 62(8), 809–13.

Yamada, M., Matsui, T., Kobayashi, Y., Furuoka, H., Haritani, M., Kobayashi, M. and Nakagawa, M. (1999) Supplementary report on experimental autumn crocus (*Colchicum autumnale* L.) poisoning in cattle: morphological evidence of apoptosis. *J. Vet. Med. Sci.* 61(7), 823–5.

Yamada, M., Nakagawa, M., Haritani, M., Kobayashi, M., Furuoka, H. and Matsui, T. (1998) Histopathological study of experimental acute poisoning of cattle by autumn crocus (*Colchicum autumnale* L.). *J. Vet. Med. Sci.* 60(8), 949–52.

***Crocus sativus* (Saffron) (Iridaceae) — Cytotoxic Chemopreventive**

Synonyms: Crocus, Saffron Crocus, Krokos (Greek), Zaffer (Arabian).

Location: Wild forms are found in Italy, Greece, the Balkans, Eastern Asia (mainly Iran). From Europe to Asia, it can be found, in meadows or (mostly) in cultivation.

Appearance
Root: corm.
Leaves: short and linear, with a white-pale central nerve, up to 30 cm long.
Flowers: long-tubed pale violet, directly emerging from the underground corm, with 3 yellow anthers and red-orange styles, up to 10 cm long.
In bloom: September–November.

Biology: The plant is perennial, with five forms existing in the wild state. Fruit setting requires cross-fertilization. Corms must not be left to grow in the same ground for too long (longer than three years).

Tradition: Already known in ancient times, saffron is referred to as Karkom in the Song of Solomon (iv. 14). The luxury yellow dye traditionally derived from the plant has been mentioned in various Greek myths, along with its scent and flavor.

Part used: Flower stigmas.

Active ingredients: crocin, crocetin, picrocrocin and safranal (carotenoids).

Particular value: It is used as carminative, emmenagogue, diaphoretic for children and for chronic hemorrhage of the uterus in adults.

Indicative dosage and application

- Oral administration of 200 mg/kg^{-1} body weight of the extract increased the life span of S-180, EAC, DLA tumor-bearing mice to 111.0%, 83.5% and 112.5%, respectively. The same extract was found to be cytotoxic to P38B, S-180, EAC and DLA tumor cells *in vitro* (potential use of saffron as an anticancer agent).
- Intraperitoneal administration of *Nigella sativa* (100 mg/kg^{-1} body wt) and oral administration of *Crocus sativus* (100 mg/kg^{-1} body wt) 30 days after subcutaneous administration of MCA (745 nmol + 2 days) restricted tumor incidence to 33.3% and 10%, respectively, compared with 100% in MCA-treated controls.

Documented target cancers

- Crocin, safranal and picrocrocin inhibit the growth of human cancer cells *in vitro* (Escribano *et al.*, 1996).
- Saffron extract (dimethyl-crocetin) possesses anticarcinogenic, anti-mutagenic and immunomodulating effects: dose-dependent cytotoxic effect to carcinoma, sarcoma and leukemia cells *in vitro*, delayed ascites tumor growth and increased the life span of the treated mice compared to untreated controls by 45–120%. In addition, it delayed the onset of papilloma growth, decreased incidence of squamous cell carcinoma and soft tissue sarcoma in treated mice (Salomi *et al.*, 1991).
- Crocetin has a dose-dependent inhibitory effect on DNA and RNA synthesis in isolated nuclei and suppressed the activity of purified RNA polymerase II. Also, crocetin causes a dose-dependent inhibition of nucleic acid and protein synthesis (Abdullaev, 1994). (Cell lines: HeLa (cervical epitheloid carcinoma), A549 (lung adenocarcinoma) and VA13 (SV-40 transformed fetal lung fibroblast) cells.)
- Antitumor activity against intraperitoneally transplanted sarcoma-180 (S-180), Ehrlich ascites carcinoma (EAC) and Dalton's lymphoma ascites (DLA) tumors in mice (Nair *et al.*, 1991).

Further details

Related compounds

- Doses inducing 50% cell growth inhibition (LD_{50}) on HeLa cells were 2.3 mg ml^{-1} for an ethanolic extract of saffron dry stigmas, 3 mM for crocin, 0.8 mM for safranal and 3 mM for picrocrocin. Crocetin did not show any cytotoxic effect (Escribano *et al.*, 1996).
- Cells treated with crocin exhibited wide cytoplasmic vacuole-like areas, reduced cytoplasm, cell shrinkage and pyknotic nuclei, suggesting apoptosis induction

(Abdullaev, 1994). Considering its water-solubility and high inhibitory growth effect, crocin is the more promising saffron compound to be assayed as a cancer therapeutic agent.

- Saffron (dimethyl-crocetin) disrupts DNA–protein interactions for example, *topoisomerases II*, important for cellular DNA synthesis (significant inhibition in the synthesis of nucleic acids but not protein synthesis) (Abdullaev, 1994).

Antitumor activity

- The effects of carotenoids of *Crocus sativus* L. (saffron) on cell proliferation and differentiation of HL-60 cells have been studied and compared with those of all-*trans* retinoic acid. Results demonstrated that the doses inducing 50% inhibition of cell growth were 0.12 μM for all-*trans* retinoic acid (ATRA) and for carotenoids of saffron 0.8 μM for dimethylcrocetin (DMCRT), 2 μM for crocetin (CRT) and 2 μM for crocins (CRCs). At 5 μM, all these compounds induced differentiation of HL-60 cells, at 85% for ATRA, 70% for DMCRT, 50% for CRT and 48% for CRCs. In these experiments, leukemic cells were cultured for 5 days in the absence or in the presence of up to 5 μM ATRA or seminatural and natural carotenoids. Since retinoids have a potential application as chemopreventive agents in humans, their toxicity as an important limiting factor for their use in treatment should be extensively explored. The seminatural (DMCRT and CRT) and natural carotenoids (CRCs) of *Crocus sativus* L. are not provitamin A precursors and could therefore be less toxic than retinoids, even at high doses (Tarantilis *et al.*, 1994).
- Topical application of *Nigella sativa* and *Crocus sativus* extracts (common food spices) inhibited two-stage initiation/promotion [dimethylbenz[a]anthracene (DMBA)/croton oil] skin carcinogenesis in mice. A dose of 100 mg kg^{-1} body wt of these extracts delayed the onset of papilloma formation and reduced the mean number of papillomas per mouse, respectively. The possibility that these extracts could inhibit the action of 20-methylcholanthrene (MCA)-induced soft tissue sarcomas was evaluated by studying the effect of these extracts on MCA-induced soft tissue sarcomas in albino mice (Salomi *et al.*, 1991).

References

Abdullaev, F.I. (1994) Inhibitory effect of crocetin on intracellular nucleic acid and protein synthesis in malignant cells. *Toxicol. Lett.* 70, 2, 243–51.

Escribano, J., Alonso, G.L., Coca-Prados, M. and Fernandez, J.A. (1996) Crocin, safranal and picrocrocin from saffron (*Crocus sativus* L.) inhibit the growth of human cancer cells *in vitro*. *Cancer Lett.* 27, 100, (1–2), 23–30.

Nair, S.C., Kurumboor, S.K. and Hasegawa, J.H. (1995) Saffron chemoprevention in biology and medicine: a review. *Cancer Biother.* 10(4), 257–64.

Nair, S.C., Pannikar, B. and Panikkar, K.R. (1991) Antitumour activity of saffron (*Crocus sativus*). *Cancer Lett.* 57(2), 109–14.

Salomi, M.J., Nair, S.C. and Panikkar, K.R. (1991) Inhibitory effects of *Nigella sativa* and saffron (*Crocus sativus*) on chemical carcinogenesis in mice. *Nutr. Cancer.* 16(1), 67–72.

Tarantilis, P.A., Morjani, H., Polissiou, M. and Manfait, M. (1994) Inhibition of growth and induction of differentiation of promyelocytic leukemia (HL-60) by carotenoids from *Crocus sativus* L. *Anticancer Res.* 14(5A), 1913–8.

Dendropanax arboreus (Dendropanax) (Araliaceae) — Cytotoxic

Appearance
Stem: spines absent.
Root: stilt roots absent.
Leaves: spiral, not scale-like, simple, trinerved at base, coriaceous, symmetric at the base, palmately lobed, smooth margined.
Flowers: bisexual, stalked, round.

Active ingredients

- *Falcarinol, dehydrofalcarinol, diyenne, falcarindiol, dehydrofalcarindiol*; and
- two novel *polyacetylenes: dendroarboreols A* and *B*.

Further details

Related compounds

The major compound responsible for the *in vitro* cytotoxicity was *falcarinol*. Several other known compounds were isolated and found to be cytotoxic, including *dehydrofalcarinol*, a *diyenne, falcarindiol* and *dehydrofalcarindiol*. In addition, two novel *polyacetylenes, dendroarboreols A* and *B*, were isolated and characterized by standard and inverse-detected NMR methods (Bernart *et al*., 1996).

References

Arikawa, J., Nogita, T., Murata, Y. and Kawashima, M. (1998) Contact dermatitis due to *Dendropanax trifidus Makino. Contact Dermatitis* 38(5), 291–2.

Bernart, M.W., Cardellina, J.H., Balaschak, M.S., Alexander, M.R., Shoemaker, R.H. and Boyd, M.R. (1996) Cytotoxic falcarinol oxylipins from *Dendropanax arboreus. J. Nat. Prod.* 59(8), 748–53.

Huang, J.Y., Liu, C.M., Qi P.L. and Liu, K.M. (1989) Studies on the antiarrhythmic effects of leaves of *Dendropanax chevalieri* (Vig.) *Merr. Et Chun Zhongguo Zhong Yao Za Zhi* 14(6), 367–70, 384.

Lans, C., Harper, T., Gearges, K. and Bridgewater, E. (2001) Medicinal and ethnoveterinary remedies of hunters in Trinidad. *BMC Complement Altern Med.* 1(1)10.

Moriarity, D.M., Huang, J., Yancey, C.A., Zhang, P., Setzer, W.N., Lawton, R.O., Bates, R.B. and Caldera, S. (1998) Lupeol is the cytotoxic principle in the leaf extract of *Dendropanax cf. querceti. Planta Med.* 64(4), 370–2.

Oka, K., Saito, F., Yasuhara, T. and Sugimoto, A. (1997) The major allergen of *Dendropanax trifidus* Makino. *Contact Dermatitis* 36(5), 252–5.

Oka, K. and Saito, F. (1999) Allergic contact dermatitis from *Dendropanax trifidus. Contact Dermatitis* 41(6), 350–1.

Oka, K., Saito, F., Yasuhara, T. and Sugimoto, A. (1999) The allergens of *Dendropanax trifidus* Makino and *Fatsia japonica* Decne. et Planch. and evaluation of cross-reactions with other plants of the *Araliaceae* family. *Contact Dermatitis* 40(4), 209–13.

Setzer, W.N., Green, T.J., Whitaker, K.W., Moriarity, D.M., Yancey, C.A., Lawton, R.O. and Bates, R.B. (1995) A cytotoxic diacetylene from *Dendropanax arboreus Planta Med.* 61(5), 470–1.

Eriophyllum

See in *Eupatorium* under Active ingredients.

Ervatamia divaricata (Ervatamia) (Apocynaceae) Cytotoxic

Location: Southeast Asia.

Appearance
Stem: round, many branches, 0.5–3 m.
Leaves: single, green, the surfaces of which are smooth and with raised veins. The length is 6–15 cm and its width is 2–4 cm.
Flowers: snow white, 1–5 cm diameter, fragrant. The flower stalk protrudes from the leaves and bears 1 or 2 flowers.

Part used: root, steam and leaf.

Active ingredients: *Vinca alkaloids* (*conophylline*).

Documented target cancers

- *Conophylline* inhibits the growth of K-ras-NRK cells, but this inhibition is reversible.
- The alkaloid also inhibits the growth of K-ras-NRK and K-ras-NIH3T3 tumors transplanted into nude mice.
- On the other hand, it shows no effect on survival of the mice loaded with L1210 leukemia.

Further details

Other species

- *Ervatamia heyneana*: the whole plant contains unidentified factors with anticancer properties (Chitnis *et al.*, 1971).
- *Ervatamia microphylla* contains *conophylline, a vinca alkaloid*, isolated from the plant (Umezawa *et al.*, 1996).

References

Chitnis, M.P., Khandalekar, D.D., Adwankar, M.K. and Sahasrabudhe, M.B. (1971) Anticancer activity of the extracts of root, stem & leaf of *Ervatamia heyneana*. *Indian J. Exp. Biol.* 9(2), 268–70.

Johnson, R.K., Chitnis, M.P., Embrey, W.M. and Gregory, E.B. (1978) *In vivo* characteristics of resistance and cross-resistance of an adriamycin-resistant subline of P388 leukemia. *Cancer Treat. Rep.* 62(10), 1535–47.

Umezawa, K., Taniguchi, T., Toi, M., Ohse, T., Tsutsumi, N., Yamamoto, T., Koyano, T. and Ishizuka, M. (1996) Growth inhibition of K-ras-expressing tumours by a new vinca alkaloid, conophylline, in nude mice. *Drugs Exp. Clin. Res.* 22(2), 35–40.

Eupatorium cannabinum **(Agrimony (Hemp)) (Compositae)** **Antitumor**

Synonyms: Holy Rope, St John's Herb.

Location: Common on the banks of rivers, sides of ditches, at the base of cliffs on the seashore, and in other damp places in most parts of Britain and Europe.

Appearance
Stem: round, growing from 60 to 150 cm, with short branches, reddish in color, covered with downy hair, woody below.
Root: woody.
Leaves: the root-leaves are on long stalks. The stem-leaves have only very short footstalks. All the leaves bear distinct, short hairs.
Flowers: flower heads being arranged in crowded masses of a dull lilac color at the top of the stem or branches. Each little composite head consists of about five or six florets.
In bloom: late in Summer and Autumn.

Tradition: It has the reputation of being a good wound herb, whether bruised or made into an ointment with lard. They used it as a strong purgative and emetic, and for curing dropsy.

Part used: herb.

Active ingredients

- Sesquiterpene lactones: *eupatoriopicrin* (EUP) (*E. cannabinum*), *eupaserrin* and *deacetyleupaserrin* (*E. semiserratum*), *eupacunin* (*E. cuneifolium*), lactones from *E. rotundifolium*.
- γ-lactones: *germacranolides* (*E. semiserratum* and *Eriophyllum confertiflorum*).
- Eupatolide (*E. formosanum* HAY)
- Flavones: *eupatorin* and *5-hydroxy-3′,4′,6,7-tetramethoxyflavone* (*E. altissimum*).

Particular value: Herbalists recognize its cathartic, diuretic and anti-scorbutic properties and consider it a good remedy for purifying the blood.

Precautions: Cytotoxicity.

Indicative dosage and application

- Growth inhibition of the Lewis lung carcinoma and the F10 26 fibrosarcoma, was found after i.v. injection of 20 or $40\,mg\,kg^{21}$ EUP (in mice C57B1), at a tumor volume of $500\,\mu l$.

Documented target cancers

- Anti-leukemic: *eupaserrin* and *deacetyleupaserrin, germacranolides*, flavones.
- Antitumor: *eupatoriopicrin, eupatolide, flavones*.
- Cytotoxic: *flavones*.

Further details

Related compounds

- The sesquiterpene lactone EUP from *Eupatorium cannabinum* L. has been shown to be cytotoxic in a glutathione (GSH)-dependent way, through the induction of DNA damage in tumor cells. The amount of EUP, requested to demonstrate DNA damage after a 24 h post-incubation period lay within the concentration range that was effective in the clonogenic assay (1–10 $\mu g\,ml^{-1}$). Glutathione depletion of the cells to about 99%, by use of buthionine sulphoximine (BSO), enhanced the extent of DNA damage (Woerdenbag *et al.*, 1989).

- *Germacranolides*: the α,β-unsaturated ester side chain adjacent to the γ-lactone and either a primary or secondary allylic alcohol or both demonstrates an *in vivo* anti-leukemic activity (Kupchan *et al.*, 1978).
- Flavones showed confirmed activity in the P-388 lymphocytic leukemia assay in mice, and the chloroform solubles showed both cytotoxic activity in the 9KB carcinoma of the nasopharynx cell culture assay and antitumor activity in the P-388 lymphocytic leukemia assay (Dobberstein *et al.*, 1977).

Related species

- *E. rotundifolium* is a native of new England and Virginia.

References

Dobberstein, R.H., Tin-wa, M., Fong, H.H., Crane, F.A. and Farnsworth, N.R. (1977) Flavonoid constituents from *Eupatorium altissimum* L. (*Compositae*). *J. Pharm. Sci.* 66(4), 600–2.

Elsasser-Beile, U., Willenbacher, W., Bartsch, H.H., Gallati, H., Schulte Monting, J., von Kleist, S. (1996) Cytokine production in leukocyte cultures during therapy with Echinacea extract. *J. Clin. Lab. Anal.* 10(6), 441–5.

Kupchan, S.M., Ashmore, J.W. and Sneden, A.T. (1978) Structure–activity relationships among *in vivo* active germacranolides. *J Pharm. Sci.* 67(6), 865–7.

Kupchan, S.M., Fujita, T., Maruyama, M. and Britton, R.W. (1973) The isolation and structural elucidation of eupaserrin and deacetyleupaserrin, new anti-leukemic sesquiterpene lactones from *Eupatorium semiserratum*. *J. Org. Chem.* 38(7), 1260–4.

Kupchan, S.M., Maruyama, M., Hemingway, R.J., Hemingway, J.C., Shibuya, S., Fujita, T., Cradwick, P.D., Hardy, A.D. and Sim, G.A. (1971) Eupacunin, a novel anti-leukemic sesquiterpene lactone from *Eupatorium cuneifolium*. *J. Am. Chem. Soc.* 93(19), 4914–6.

Kupchan, S.M., Kelsey, J.E., Maruyama, M., Cassady, J.M., Hemingway, J.C. and Knox, J.R. (1969) Tumor inhibitors. XLI. Structural elucidation of tumor-inhibitory sesquiterpene lactones from *Eupatorium rotundifolium*. *J. Org. Chem.* 34(12), 3876–83.

Lee, K.H., Huang, H.C., Huang, E.S. and Furukawa, H. (1972) Antitumor agents. II. Eupatolide, a new cytotoxic principle from *Eupatorium formosanum* HAY. *J. Pharm. Sci.* 61(4), 629–31.

Woerdenbag, H.J., van der Linde, J.C., Kampinga, H.H., Malingre, T.M. and Konings, A.W. (1989) Induction of DNA damage in Ehrlich ascites tumour cells by exposure to eupatoriopicrin. *Biochem. Pharmacol.* 38(14), 2279–83.

Woerdenbag, H.J., Lemstra, W., Malingre, T.M. and Konings, A.W. (1989) Enhanced cytostatic activity of the sesquiterpene lactone eupatoriopicrin by glutathione depletion. *B. J. Cancer* 59(1), 68–75.

***Fagara macrophylla* (Fagara) (Rutaceae)** — **Cytotoxic Anti-leukemic**

Location: Africa.

Part used: roots.

Active ingredients

- *Alkaloids*: *nitidine chloride, 6-oxynitidine, 6-methoxy-5,6-dihydronitidine* (*Fagara macrophylla*).
- *Fagaronine* (Fine) (*Fagara xanthoxyloides*).

Indicative dosage and application

- *Alkaloids*: *nitidine chloride* and *6-methoxy-5,6-dihydronitine* are used at doses of 30–50 $mg\,kg^{-1}$.
- *Fagaronine* is used at a concentration of $3 \times 10-6\,mol\,l^{-1}$ at day 4.

Documented target cancers

- The alkaloids *nitidine chloride* and *6-methoxy-5,6-dihydronitine* are about equipotent in P-388 mouse leukemia, giving high *T/C* values of 240–260% (Wall *et al.*, 1987).
- *Fagaronine* (Fine) inhibits cell proliferation of human erythroleukemia K562 cells by 50% at a concentration of $3 \times 10^{-6}\,mol\,l^{-1}$ at day 4 (more informations in Further details) (Comoe *et al.*, 1988).

Further details

Related compounds

- The known alkaloids nitidine chloride (1), 6-oxynitidine (2) and 6-methoxy-5,6-dihydronitidine (3) have been isolated from *Fagara macrophylla*. Compound 3 was the major product and was shown to be an artifact. The alkaloids 1 and 3 have been interconverted by treatment of 1 under basic conditions or 3 under acidic conditions. On sublimation 1 and 3 formed *8,9-dimethoxy-2,3-methylenedioxybenzo{c}phenanthridine* which could then be converted to *5,6-dihydronitidine*. The alkaloids 1 and 3 are about equipotent in P-388 mouse leukemia, giving high *T/C* values of 240–260% at doses of 30–50 $mg\,kg^{-1}$. The other compounds were inactive. The structural requirement for antitumor activity in the phenanthridine series is the ability to form a C-6 iminium ion (Wall *et al.*, 1987).
- Fagaronine (Fine) is an anti-leukemic drug extracted from the root of *Fagara xanthoxyloides* Lam. (Rutaceae). Fine inhibits cell proliferation of human erythroleukemia K562 cells by 50% at a concentration of $3 \times 10^{-6}\,mol\,l^{-1}$ on day 4. It stimulates incorporation of labelled macromolecular thymidine on day 1, but decreases incorporation on days 2, 3 and 4. Fine induces a cell accumulation in G_2 and late-S phases (Messmer *et al.*, 1972).

References

Comoe, L., Carpentier, Y., Desoize, B. and Jardillier, J.C. (1988) Effect of fagaronine on cell cycle progression of human erythroleukemia K562 cells. *Leuk. Res.* 12(8), 667–72.

Comoe, L., Kouamouo, J., Jeannesson, P., Desoize, B., Dufour, R., Yapo, E.A. and Jardillier, J.C. (1987) Cytotoxic effects of root extracts of *Fagara zanthoxyloides* Lam. (Rutaceae) on the human erythroleukemia K562 cell line. *Ann. Pharm. Fr.* 45(1),79–86.

Messmer, W.M., Tin-Wa, M., Fong, H.H., Bevelle, C., Farnsworth, N.R., Abraham, D.J. and Trojanek, J. (1972) Fagaronine, a new tumor inhibitor isolated from *Fagara zanthoxyloides* Lam. (Rutaceae). *J. Pharm. Sci.* 61(11), 1858–9.

Wall, M.E., Wani, M.C. and Taylor, H. (1987) Plant antitumor agents, 27. Isolation, structure, and structure activity relationships of alkaloids from *Fagara macrophylla*. *J. Nat. Prod.* 50(6), 1095–9.

Ficus carica L. (Ficus) (Urticaceae) — Anti-leukemic

Location: Indigenous to Persia, Asia Minor and Syria, wild in most of the Mediterranean countries.

Appearance (Figure 3.10)
Stem: 6–7 m high.
Root: free from stagnant water, sheltered from cold.
Leaves: broad, rough, deciduous, deeply lobed.
Flowers: concealed within the body of the fruit.
In bloom: July–August.

Part used: Seeds, fruit.

Active ingredients: *Lectins* (*Ficus cunia*).

Documented target cancers: It is used for different types of leukemia (chronic myeloid leukemia, acute myeloblastic leukemia, acute lymphoblastic leukemia and chronic lymphocytic leukemia) (Agrawal *et al.*, 1990; Guyot *et al.*, 1986).

Figure 3.10 Ficus carica.

Further details

Related species

- The seeds of *Ficus cunia* contain a lectin with a molecular weight of 3300–3500, which can agglutinate white blood cells (leukocytes and mononuclear cells) from patients with different types of leukemia (as mentioned above) (Ray *et al.*, 1993).

Related compounds

- A lectin, isolated from the seeds of *Ficus cunia* and purified by affinity chromatography on fetuin-Sepharose, was homogeneous in PAGE, GPC, HPLC, and immunodiffusion, and had molecular weight of 3200–3500. In SDS-PAGE and HPLC in the absence and presence of 2-mercaptoethanol, the lectin gave a single band or peak corresponding to M(r) 3300–3500, thus indicating it to be a monomer. The lectin agglutinated human erythrocytes regardless of blood group, bound to Ehrlich ascites cells and to human rat spermatozoa, and was thermally stable; Ca^{2+} enhanced its activity. The lectin is a metalloprotein that was inactivated by dialysis with EDTA followed by acetic acid, but reactivated by the addition of Ca^{2+}. The lectin contained 2.0% of carbohydrates, large proportions of acidic amino acids, but little methionine. In hapten-inhibition assays, chitin oligosaccharides linked β-GlcNAc} and *N*-acetyl-lactosamine were inhibitors of which *N,N′*-tetra-acetylchitotetraose was the most potent. Among the macromolecules tested that contain either multiple *N*-acetyl-lactosamine and/or linked β-GlcNAc, asialofetuin glycopeptide was the most potent inhibitor. Thus, an *N*-acetyl group and substitution at C-1 of D-GlcN are necessary for binding (Ray *et al.*, 1993).
- Semipurified saline extracts of seeds from *Crotolaria juncea, Cassia marginata, Ficus racemosa, Cicer arietinum (L-532), Gossipium indicum (G-27), Melia composita, Acacia lenticularis, Meletia ovalifolia, Acacia catechu* and *Peltophorum ferrenginium* were tested for leukoagglutinating activity against whole leukocytes and mononuclear cells from patients with chronic myeloid leukemia, acute myeloblastic leukemia, acute lymphoblastic leukemia, chronic lymphocytic leukemia, various lymphoproliferative/hematologic disorders and normal healthy subjects. In addition, bone marrow cells from three patients undergoing diagnostic bone marrow aspiration and activated lymphocytes from mixed lymphocyte cultures (MLC) were also tested. All the seed extracts agglutinated white blood cells from patients with different types of leukemia. But none of them reacted with peripheral blood cells of normal individuals, patients with various lymphoproliferative/hematologic disorders or cells from MLC. Leukoagglutination of leukemic cells with each of the seed extracts was inhibited by simple sugars. Only in one instance, cells from bone marrow of an individual who had undergone diagnostic bone marrow aspiration for a non-malignant condition were agglutinated. It is felt that purification of these seed extracts may yield leukemia-specific lectins (Agrawal *et al.*, 1990).

References

Agrawal, S. and Agarwal, S.S. (1990) Preliminary observations on leukaemia specific agglutinins from seeds. *Indian J. Med. Res.* 92, 38–42.

Guyot, M., Durgeat, M. and Morel, E. (1986) Ficulinic acid A and B, two novel cytotoxic straight-chain acids from the sponge *Ficulina ficus*. *J. Nat. Prod.* 49(2), 307–9.

Peraza-Sanchez, S.R., Chai, H.B., Shin, Y.G., Santisuk, T., Reutrakul, V., Farnsworth, N.R., Cordell, G.A., Pezzuto, J.M. and Kinghorn, A.D. (2002) Constituents of the leaves and twigs of *Ficus hispida*. *Planta Med.* 68(2), 186–8.

Ray, S., Ahmed, H., Basu, S. and Chatterjee, B.P. (1993) Purification, characterisation, and carbohydrate specificity of the lectin of *Ficus cunia*. *Carbohydr. Res.* 7, 242, 247–63.

Rubnov, S., Kashman, Y., Rabinowitz, R., Schlesinger, M. and Mechoulam, R. (2001) Suppressors of cancer cell proliferation from fig (*Ficus carica*) resin: isolation and structure elucidation. *J. Nat. Prod.* 64(7), 993–6.

Simon, P.N., Chaboud, A., Darbour, N., Di Pietro, A., Dumontet, C., Lurel, F., Raynaud J. and Barron, D. (2001) Modulation of cancer cell multidrug resistance by an extract of *Ficus citrifolia*. *Anticancer Res.* 21(2A), 1023–7.

Garcinia hombrioniana (Garcinia) (Guttifereae) — Cytotoxic, Antitumor

Location: Riverine and coastal alluvial regions. Malaysia, Brunei.

Appearance (Figure 3.11)
Stem: 10 m high, numerous branches.
Leaves: tertiary branches hold much of the leaves.
Flowers: in clusters of not more than five small flowers.

Tradition: Very powerful drastic hydragogue, cathartic, very useful in dropsical conditions.

Part used: gum resin.

Active ingredients: *Garonolic acids*.

Figure 3.11 Garcinia fruit.

Precautions: Full dose is rarely given alone, as it causes vomiting, nausea and griping. In high dose it can cause death.

Documented target cancers

- *Garcinia hunburyi* (Gamboge), when steam processed (0.15 MPa, 126°C for 30 min) is cytotoxic on K562 tumor cells (Lu *et al.*, 1996).

Further details

Related species

- The technology for processing steamed *Garcinia hunburyi* with high pressure was synthetically selected by using orthogonal experimental design, based on the indexes of anti-inflammatory, bacteriocidal, antitumour effects and gambagic acid content. The result shows that the best way is to steam for 0.5 h at 126°C (Ye *et al.*, 1996).

Antitumor activity

- The cytotoxicity of different processed products of Gamboge on K562 tumor cell was observed. The result showed that the antitumor action of *Garcinia hunburyi* processed by steaming (0.15 MPa, 126°C for 30 min) was the strongest (Lu *et al.*, 1996).

Other medical effects

- However, there is a possibility that the Nigerian cola plant (*Garcinia*) may be a cause of human cancer in countries where kola nuts are widely consumed as stimulants (e.g. via chewing), because of their content of primary and secondary amines, and their relative methylating potential due to nitrosamide formation (Atawodi *et al.*, 1995).

References

Atawodi, S.E., Mende, P., Pfundstein, B., Preussmann, R. and Spiegelhalder, B. (1995) Nitrosatable amines and nitrosamide formation in natural stimulants: *Cola acuminata, C. nitida* and *Garcinia cola. Food Chem. Toxicol.* 33(8), 625–30.

Lu, Y., Wang, G. and Ye, D. (1996) Comparison of cytotoxicity of different processed products of gamboge on K562 tumor cells. *Chung Kuo Chung Yao Tsa Chih.* 21(2), 90–1, 127.

Ye, D. and Kong, L. (1996) Selection of technology for processing steamed *Garcinia hunburyi* with high pressure. *Chung Kuo Chung Yao Tsa Chih.* 21(8), 472–3.

Glycyrrhiza glabra **L. (Glycyrrhiza, Liquorice) (Leguminosae) — Antitumor**

Location: It can be found in Southeast Europe, Southwest Asia. It is cultured in Spain, Italy, UK and USA.

Appearance
Stem: graceful, with light, spreading, pinnate foliage, presenting an almost feathery appearance from a distance.
Root, double: the one part consisting of a vertical or tap root, often with several branches penetrating to a depth of 1–1.5 m, the other of horizontal rhizomes or stolons, thrown out of the root below the surface of the ground.
Leaves: leaflets.
Flowers: from the axils of the leaves spring racemes or spikes of papilionaceous small pale blue, violet, yellowish-white or purplish, followed by small pods.
In bloom: summer.

Tradition: Very common in use in South Italy for stomach disorders, cough and also as a sweeter.

Part used: root and stolons.

Active ingredients: *Glycyrrhizic acid, glycyrrhetinic acid, flavonoids, triterpenoids.*

Documented target cancers

- Prevention, skin cancer, leukemia.
- Some triterpenoids from *Glycyrrhiza* spp. were effective against adriamycin (ADM)-resistant P-388 leukemia cells (P-388/ADM), which were resistant to multiple anticancer drugs (Hasegawa *et al.*, 1995).

Further details

Related species

- *Glycyrrhiza uralensis* is one of the main related compounds of Hua-sheng-ping (*Chrysanthemum morifolium, Glycyrrhiza uralensis, Panax notoginseng*), which has many medicinal uses (Yu, 1993).

Related compounds

- *Glycyrrhizic acid*, the active ingredients in licorice, and its metabolite *carbenoxolone* are members of short-chain dehydrogenase reductase (SDR) enzymes. The SDR family includes over 50 proteins from human, mammalian, insect and bacterial sources (Duax *et al.*, 1997).

Other medical activity

- *Glycyrrhiza uralensis*: Extracts have strong antimutagenic properties, indicated for syndromes such as Spleen–Stomach Asthenic Cold and has been proved to be an effective prescription for precancerous lesions. An important component is *glycyrrhetinic acid*, which can protect rapid DNA damage and decrease the unscheduled DNA synthesis induced by benzo(alpha)pyrene (Chen *et al.*, 1994).

- *Glycyrrhizae inflata*: Extracts contain 6 flavonoids with significant antioxidant effects, showing anti-promoting effects on two-stage carcinogenesis in mouse skin induced by DMBA plus croton oil. The TPA enhanced $32P_i$-incorporation into phospholipid fraction in HeLa cells was inhibited, and the micronuclei in mouse bone marrow cells induced by cytoxan were also depressed (Agarwal *et al.*, 1991).

References

Agarwal, R., Wang, Z.Y. and Mukhtar, H. (1991) Inhibition of mouse skin tumor-initiating activity of DMBA by chronic oral feeding of glycyrrhizin in drinking water. *Nutr. Cancer* 15(3–4), 187–93.

Biglieri, E.G. (1995) My engagement with steroids: a review. *Steroids* 60(1), 52–8.

Chen, X. and Han, R. (1995) Effect of glycyrrhetinic acid on DNA damage and unscheduled DNA synthesis induced by benzo(alpha)pyrene. *Chin. Med. Sci J.* 10(1), 16–9.

Chen, X.G. and Han, R. (1994) Effect of glycyrrhetinic acid on DNA damage and unscheduled DNA synthesis induced by benzo (a) pyrene. *Yao Hsueh Hsueh Pao.* 29(10), 725–9.

Duax, W.L. and Ghosh, D. (1997) Structure and function of steroid dehydrogenases involved in hypertension, fertility, and cancer. *Steroids* 62(1), 95–100.

Fu, N., Liu, Z. and Zhang, R. (1995) Anti-promoting and anti-mutagenic actions of G9315. *Chung Kuo I Hsueh Ko Hsueh Yuan Hsueh Pao* 17(5), 349–52.

Hasegawa, H., Sung, J.H., Matsumiya, S., Uchiyama, M., Inouye, Y., Kasai, R., Yamasaki, K. (1995) Reversal of daunomycin and vinblastine resistance in multidrug-resistant P388 leukemia *in vitro* through enhanced cytotoxicity by triterpenoids. *Planta Med.* 61(5), 409–13.

Horn, B. (1986) Rarities in family practice. Consequences for education and continuing education. *Schweiz Rundsch. Med. Prax.* 75(44), 1323–7.

Liu, X.R., Han, W.Q. and Sun, D.R. (1992) Treatment of intestinal metaplasia and atypical hyperplasia of gastric mucosa with xiao wei yan powder. *Chung Kuo Chung Hsi I Chieh Ho Tsa Chih.* 12(10), 602–3, 580.

Montanari, G., Zanaletti, F., Quadrelli, G.C., Di Battista, F., Ongis, G.A., Fraschini, A., Zinzalini, G. and Abbiati, C. (1988) Arterial hypertension with hypokalemia. *Minerva Med.* 79(3), 209–14.

Shackleton, C.H. (1993) Mass spectrometry in the diagnosis of steroid-related disorders and in hypertension research. *J. Steroid Biochem. Mol. Biol.* 45(1–3), 127–40.

Shi, G.Z. (1992) Blockage of *Glycyrrhiza uralensis* and *Chelidonium majus* in MNNG induced cancer and mutagenesis. *Chung Hua Yu Fang I Hsueh Tsa Chih.* 26(3), 165–7.

Takahashi, K., Yoshino, K., Shirai, T., Nishigaki, A., Araki, Y. and Kitao, M. (1988) Effect of a traditional herbal medicine (shakuyaku-kanzo-to) on testosterone secretion in patients with polycystic ovary syndrome detected by ultrasound. *Nippon Sanka Fujinka Gakkai Zasshi.* 40(6), 789–92.

Yamashiki, M., Nishimura, A., Huang, X.X., Nobori, T., Sakaguchi, S. and Suzuki, H. (1997) Effects of the Japanese herbal medicine "Sho-saiko-to" (TJ-9) on *in vitro* interleukin-10 production by peripheral blood mononuclear cells of patients with chronic hepatitis C. *Hepatology* 25(6), 1390–7.

Yu, X.Y. (1993) A prospective clinical study on reversion of 200 precancerous patients with hua-sheng-ping. *Chung Kuo Chung Hsi I Chieh Ho Tsa Chih* 13(3), 147–9.

Wang, Z.Y., Agarwal, R., Khan, W.A. and Mukhtar, H. (1992) Protection against benzo[a]pyrene- and *N*-nitrosodiethylamine-induced lung and forestomach tumorigenesis in A/J mice by water extracts of green tea and licorice. *Carcinogenesis* 13(8), 1491–4.

White, P.C., Mune, T. and Agarwal, A.K. (1997) 11 β-Hydroxysteroid dehydrogenase and the syndrome of apparent mineralocorticoid excess. *Endocr Rev.* 18(1), 135–56.

Webb, T.E., Stromberg, P.C., Abou-Issa, H., Curley, R.W. Jr and Moeschberger, M. (1992) Effect of dietary soybean and licorice on the male F344 rat: an integrated study of some parameters relevant to cancer chemoprevention. *Nutr. Cancer* 18(3), 215–30.

Zee-Cheng, R.K. (1992) Shi-quan-da-bu-tang (ten significant tonic decoction), SQT. A potent Chinese biological response modifier in cancer immunotherapy, potentiation and detoxification of anticancer drugs. *Methods Find Exp. Clin. Pharmacol.* 14(9), 725–36.

***Goniothalamus* sp. (Annonaceae)** **Cytotoxic**

Location: Malaysia, China.

Active ingredients

- *Acetogenins: gardnerilins A and B*;
- *Styrylpyrone(SPD), goniodiol-7-monoacetate*;
- *Acetogenin lactones: goniothalamicin, annonacin.*

Indicative dosage and application: Doses used in rat mammary tumors with good effects were: 2, 10 and 50 mg kg^{-1}.

Documented target cancers: Antiestrogen (mice), breast cancer, cytotoxic, 9ASK (astrocytoma) and weakly active against 3PS murine leukemia.

Further details

Related species

- *Goniothalamus gardneri*: the roots contain the C35 acetogenins *gardnerilins A and B* (Chen *et al.*, 1998).
- *Goniothalamus amuyon*, and other *Goniothalamus* species contain the *styrylpyrone, goniodiol-7-monoacetate* [6*R*-(7*R*,8*R*-dihydro-7-acetoxy-8-hydroxystyryl)-5,6-dihydro-2-pyrone] (Wu *et al.*, 1991).
- The stem bark of *Goniothalamus giganteus* Hook. Thomas contains the *γ-lactone goniothalamicin, a tetrahydroxy-mono-tetrahydrofuran fatty acid*, along with *annonacin.*

Antitumor activity

- The estrogen antagonism: agonism ratio for SPD is much higher than *Tamoxifen*, which is indicative of the breast cancer antitumor activity as seen in compounds such as MER-25. Pretreatment assessment on 1 mg kg^{-1} BW SPD and Tam showed that SPD is not a very good, estrogen antagonist compared to Tam, as it was unable to revert the estrogenicity effect of estradiol benzoate (EB) on immature rat uterine weight. Antitumor activity assessment for SPD exhibited significant tumor growth retardation in DMBA-induced rat mammary tumors at all doses employed (2, 10 and 50 mg kg^{-1}) compared to the controls. This compound was found to be more potent than Tam (2 and 10 mg kg^{-1}) and displayed greater potency at a dose of 10 mg kg^{-1}. It caused complete remission of 33.3% of tumors but failed to prevent onset of new tumors. However, SPD administration at 2 mg kg^{-1} caused 16.7% complete remission and partial remission. It also prevented the onset of new tumors throughout the experiment (Hawariah and Stanslas, 1998).

Related compounds

- *Goniodiol-7-monoacetate* showed potent (ED_{50} values less than 0.1 $\mu g\, ml^{-1}$) cytotoxicities against KB, P-388, RPMI, and TE671 tumor cells (Wu *et al.*, 1991).
- *Goniothalamicin* is cytotoxic and insecticidal and inhibits the formation of crown gall tumors on potato discs. *Annonacin*, the only other reported mono-tetrahydrofuran acetogenin, was also isolated, which is active against 9ASK (astrocytoma) and weakly active against 3PS murine leukemia (Alkofahi *et al.*, 1988).

References

Alkofahi, A., Rupprecht, J.K., Smith, D.L., Chang, C.J. and McLaughlin, J.L. (1988) Goniothalamicin and annonacin: bioactive acetogenins from *Goniothalamus giganteus* (Annonaceae). *Experientia* 44(1), 83–5.

Chen, Y., Jiang, Z., Chen, R.R. and Yu, D.Q. (1998) Two linear acetogenins from *Goniothalamus gardneri*. *Phytochemistry* 49(5), 1317–21.

Hawariah, A. and Stanslas, J. (1998) Antagonistic effects of styrylpyrone derivative (SPD) on 7,12-dimethylbenzanthracene-induced rat mammary tumors. *In Vivo* 12(4), 403–10.

Wu, Y.C., Duh, C.Y., Chang, F.R., Chang, G.Y., Wang, S.K., Chang, J.J., McPhail, D.R., McPhail, A.T. and Lee, K.H. (1991) The crystal structure and cytotoxicity of goniodiol-7-monoacetate from *Goniothalamus amuyon*. *J. Nat. Prod.* 54(4), 1077–81.

Gossypium herbaceum L. (GOSSYPIUM, Cotton root) (Malvaceae) — Cytotoxic

Location: Asia Minor, cultivated in USA. and Egypt, Mediterranean, India.

Appearance (Figure 3.12)
Stem: 0.5–2 m high, branching stems.
Root: the root bark consists of thin flexible bands covered with a brownie yellow. periderm, odor not strong, tastes slightly acid.
Leaves: palmate, hairy, green, lobes lanceolate and acute.
Flowers: yellow with a purple spot in the center.
In bloom: August–September.

Tradition: One of the well-known Chinese medicine used as an anticancer crude drug.

Part used: bark of root.

Active ingredients: *Catechin*.

Indicative dosage and application: It is used as crude extract, mixed with other herbs, usually oral intake.

Documented target cancers: Murine B16 melanoma and L1210 lymphoma cells.

Figure 3.12 Gossypium herbaceum.

Further details

Related species

- *Gossypium indicum* has a moderate antimutagenic activity against benzo[*a*]pyrene. Its aqueous–alcoholic extracts from unripe cotton balls are well known for their antitumor activity. The hydrophilic fractions contain certain amounts of *catechin* and its derivatives, which are responsible for the antitumor activities of the herb (Choi *et al.*, 1998).

References

Choi, J.J., Yoon, K.N., Lee, S.K., Lee, Y.H., Park, J.H., Kim, W.Y., Kim, J.K. and Kim, W.K. (1998) Antitumor activity of the aqueous-alcoholic extracts from unripe cotton ball of *Gossypium indicum*. *Arch Pharm. Res.* 21(3), 266–72.

Hoffmann, K., Kaspar, K., Gambichler, T. and Altmeyer, P. (2000) *In vitro* and *in vivo* determination of the UV protection factor for lightweight cotton and viscose summer fabrics: a preliminary study. *J. Am. Acad. Dermatol.* 43(6), 1009–16.

Lee, H. and Lin, J.Y. (1988) Antimutagenic activity of extracts from anticancer drugs in Chinese medicine. *Mutat. Res.* 204(2), 229–34.

MacFarlane, D. and Goldberg, L.H. (1999) Use of the cotton-tipped applicator in lower eyelid surgery. *Dermatol Surg.* 25(4), 326–7.

Hannoa chlorantha **(Hannoa) (Simaroubaceae)** **Anti-leukemic**

Location: Africa.

Tradition: *Hannoa chlorantha* and *Hannoa klaineana* (Simaroubaceae) are used in traditional medicine of Central African countries against fevers and malaria.

Part used: stem bark, root bark.

Active ingredients
Quassinoids (*15-desacetylundulatone*), 14-hydroxychaparrinone, chaparrinone 15-*O*-β-D-glucopyranosyl-21-hydroxy-glaucarubolone was found to be more toxic while 6-α-tigloyloxy-glaucarubol and 21-hydroxyglaucarubolone was found inactive.

Documented target cancers: P-388 cells mouse lymphocytic leukemia, colon 38 adenocarcinoma.

Further details

Other medical activity

- *Hannoa chlorantha* and *Hannoa klaineana*: Apart from their documented antimalaria activity, stem bark extracts from *H. klaineana* and *H. chlorantha* are also cytotoxic against P-388 cells mouse lymphocytic leukemia cells. This activity is due to the presence of 14-hydroxychaparrinone (and, in a lesser degree, chaparrinone) from *H. klaineana* (Francois *et al.*, 1998). In addition, the quassinoid *15-desacetylundulatone* isolated from the root bark of *Hannoa klaineana*, was found active against P-388 and colon 38 adenocarcinoma, while 15-*O*-β-D-glucopyranosyl-21-hydroxy-glaucarubolone were found to be more toxic while 6-α-tigloyloxy-glaucarubol and 21-hydroxyglaucarubolone were found inactive (Francois *et al.*, 1998).

References

Francois, G., Diakanamwa, C., Timperman, G., Bringmann, G., Steenackers, T., Atassi, G., Van Looveren, M., Holenz, J., Tassin, J.P., Assi, L.A., Vanhaelen-Fastre, R. and Vanhaelen, M. (1998) Antimalarial and cytotoxic potential of four quassinoids from *Hannoa chlorantha* and *Hannoa klaineana*, and their structure–activity relationships. *Int. J. Parasitol.* **28**(4), 635–40.

Lumonadio, L., Atassi, G., Vanhaelen, M. and Vanhaelen-Fastre, R. (1991) Antitumor activity of quassinoids from *Hannoa klaineana. J. Ethnopharmacol.* 31(1), 59–65.

Polonsky, J. and Bourguignon-Zylber, N. (1965) Study of the bitter constituents of the fruit of Hannoa klaineana (Simarubaceae): chaparrinone and klaineanone *Bull. Soc. Chim. Fr.* **10**, 2793–9.

***Helenium microcephalum* (Sneezeweed) (Compositae) Cytotoxic**

Other common names: smallhead sneezeweed, red and gold sneezeweed.

Location: Of North America origin, it is found in mountain meadows and moist places. It can be easily cultivated.

Appearance
Stem: stout, 20–90 cm high.
Leaves: alternate, lance-shaped, up to 2–2.5 cm long.
Flowers: yellow-orange flowerheads.
In bloom: June–September or generally during the warm season of the year.

Biology: *Helenium* is a perennial plant, growing well on moist but well drained soil and requiring full sun. It can be easily propagated by seed and by dividing clumps every 3–4 years.

Tradition: Species of the genus *Helenium* are long valued daisy-like ornamentals used in cutting and butterfly gardens for late summer color ('Helen's Flower').

Part used: Whole plant.

Active ingredients
Helenalin (a sesquiterpene lactone), microhelenin-E (1) and -F (2) (nor-pseudoguaianolides).

Precautions
The plant and related species (such as *H. hoopesii*) are very poisonous. Helenalin has a documented acute toxicity. Reported effects on liver, kidney and lung include depression, appetite loss, weak irregular pulse, weakness, stiffness, nasal discharge, bloat, "spewing sickness", vomiting, foaming at the mouth, coughing, green nasal discharge, diarrhea, and photosensitization. Death may occur rapidly, 4–24 h of ingestion, or over a longer period in chronic cases.

Indicative dosage and application

- The oral median lethal dose of helenalin for 5 mammalian species is between 85 and 105 $mg\,kg^{-1}$.
- In a study, they used a single i.p. dose of helenalin in male mice 43 $mg\,kg^{-1}$ and they continue for the next three days with i.p. injection of 25 mg helenalin kg^{-1}.

Documented target cancers

- Helenalin, a sesquiterpene lactone found in species of the plant genus *Helenium*, inhibits the proliferation of cancer cells.
- Microlenin acetate, a dimeric sesquiterpene lactone, has a significant anti-leukemic activity.

Further details

Antitumour activity

- *Helenalin* causes a marked potentiation of the increases in intracellular free Ca^{2+} concentration ($[Ca^{2+}]_i$) produced by mitogens such as vasopressin, bradykinin, and platelet-derived growth factor in Swiss mouse 3T3 fibroblasts. Removing external Ca^{2+} partly attenuated the increased $[Ca^{2+}]_i$ responses caused by helenalin. The increased $[Ca^{2+}]_i$ responses occurred at concentrations of helenalin that inhibited cell proliferation. At higher concentrations, helenalin inhibited the $[Ca^{2+}]_i$ responses. No change in resting $[Ca^{2+}]_i$ was caused by helenalin even at high concentrations. Other helenalin analogues also increased the $[Ca^{2+}]_i$ response (Powis *et al.*, 1994). Helenalin did not inhibit protein kinase C (PKC) and PKC appeared to play a minor role in the effects of helenalin on $[Ca^{2+}]_i$ responses in intact cells. Studies with saponin-permeabilized HT-29 human colon carcinosarcoma cells indicated that helenalin caused an increased accumulation of Ca^{2+} into nonmitochondrial stores and that the potentiating effect of helenalin on mitogen-stimulated $[Ca^{2+}]_i$ responses was due in part to an increase in the inositol-(1,4,5)-trisphosphate-mediated release of Ca^{2+} from these stores.

Related compounds

- Two new nor-pseudoguaianolides, microhelenin-E (1) and -F (2), were isolated from Texas *Helenium microcephalum* and their structures elucidated on the basis of physicochemical data and spectral evidence (Kasai *et al.*, 1982). Microhelenin-E demonstrated significant *in vitro* and *in vivo* cytotoxic and anti-leukemic activities against KB tissue cell culture ($ED_{50} = 1.38\,\mu g\,ml^{-1}$) P-388 lymphocytic leukemia growth in BDF1 male mice (T/C-166% at $8\,mg\,kg^{-1}$per day), respectively.
- The antitumor sesquiterpene lactones microhelenins-A, B and C, microlenin acetate and plenolin were isolated from *Helenium microcephalum*. The structures and stereochemistry of these lactones were determined by physical methods as well as by chemical transformations and correlations (Lee *et al.*, 1976). Microlenin acetate is probably the first novel dimeric sesquiterpene lactone demonstrated to have significant anti-leukemic activity.
- The known compound isohelenalin and a new anti-leukemic sesquiterpene lactone, isohelenol were isolated from *Helenium microcephalum* (Sims *et al.*, 1979).

Cytotoxic activity

- Studies with smallhead sneezeweed indicated that helenalin, is the only significant toxic constituent present. The oral median lethal dose of helenalin for 5 mammalian species was between 85 and $105\,mg\,kg^{-1}$.
- The acute toxicity of helenalin was examined in male BDF1 mice. The 14-day LD_{50} for a single ip dose of helenalin in male mice was $43\,mg\,kg^{-1}$. A single i.p. injection of $25\,mg\,kg^{-1}$ helenalin increased serum alanine aminotransferase (ALT), lactate dehydrogenase (LDH), urea nitrogen (BUN), and sorbitol dehydrogenase within 6 h of treatment (Chapman *et al.*, 1988). Multiple helenalin exposures, i.p. injection of $25\,mg\,kg^{-1}$ for 3 days, increased differential polymorphonuclear leukocyte counts and decreased lymphocyte counts. Serum ALT, BUN and cholesterol levels were also increased by multiple helenalin exposures at $25\,mg\,kg^{-1}$ per day. Helenalin significantly reduced liver, thymus and spleen relative weights and histologic evaluation revealed substantial effects of multiple helenalin exposures on lymphocytes of the thymus, spleen and mesenteric lymph nodes. No helenalin-induced histologic changes were observed in the liver or kidney. Multiple helenalin exposures (25 mg kg^{-1}per day) significantly inhibited hepatic microsomal enzyme activities (aminopyrine demethylase and aniline hydroxylase) and decreased microsomal cytochromes P-450 and b5 contents. Three concurrent days of diethyl maleate (DEM) pretreatment ($3.7\,mmol\,kg^{-1}$, 0.5 h before helenalin treatment) significantly increased the toxicity of helenalin exposure. These results indicate that the hepatic microsomal drug metabolizing system and lymphoid organs are particularly

vulnerable to the effects of helenalin. In addition, helenalin toxicity is increased by DEM pretreatments, which have been shown to decrease GSH concentrations.

- Helenalin (25 mg kg^{-1}) administered to immature male ICR mice caused a rapid decrease in hepatic GSH levels and was lethally toxic to greater than 60% of the animals within 6 days. L-2 Oxothiazolidine 4-carboxylate (OTC), a compound that elevates cellular GSH levels, administered to ice 6 or 12 h before helenalin protected against hepatic GSH depletion and the lethal toxicity of these toxins. OTC administered at the same time as the sesquiterpene lactones was not protective, suggesting that the critical events against which GSH is protective occur within the first 6 h. In primary rat hepatocyte cultures, helenalin (4–16 μM) caused a rapid lethal injury as determined by the release of lactate dehydrogenase. Cotreatment of cultures with *N*-acetylcysteine at high concentrations (4 mM) afforded significant protection against lethal injury by both toxins (Merrill *et al.*, 1988). In contrast, BCNU, which inhibits glutathione reductase, or diethylmaleate, which depletes hepatocellular GSH, potentiated the hepatotoxicity of helenalin in monolayer rat hepatocytes. These studies suggest that the *in vivo* and *in vitro* toxicity of helenalin is strongly dependent on hepatic GSH levels, which helenalin rapidly depletes at very low concentrations.

References

Kasai, R., Shingu, T., Wu, R.Y., Hall, I.H. and Lee, K.H. (1982) Antitumor agents 57. The isolation and structural elucidation of *microhelenin-E*, a new anti-leukemic nor-pseudoguaianolide, and microhelenin-F from *Helenium microcephalum*. *J. Nat. Prod.* 45(3), 317–20.

Lee, K.H., Imakura, Y. and Sims, D., (1976) Antitumor agents XVII; Structure and stereochemistry of *microhelenin-A*, a new antitumor sesquiterpene lactone from *Helenium microcephalum*. *J. Pharm. Sci.* 65(9), 1410–2.

Merrill, J.C., Kim, H.L., Safe, S., Murray, C.A. and Hayes, M.A. (1988) Role of glutathione in the toxicity of the sesquiterpene lactones hymenoxon and helenalin. *J. Toxicol. Environ. Health* 23(2), 159–69.

Sims, D., Lee, K.H. and Wu, R.Y. (1979) Antitumor agents 37. The isolation and structural elucidation of isohelenol, a new anti-leukemic sesquiterpene lactone, and isohelenalin from *Helenium microcephalum*. *J. Nat. Prod.* 42(3), 282–6.

***Hypericum perforatum* L. (Hypericum (St John's Wort)) (Hypericaceae)** — Cytotoxic

Location: Britain and throughout Europe and Asia.

Appearance (Figure 3.13)
Stem: 0.3–1 m high, erect, branching in the upper part.
Leaves: pale green, sessile, and oblong, with pellucid dots or oil glands.
Flowers: bright cheery yellow in terminal corymb.
In bloom: June–August.

Tradition: Its name has been connected with many ancient superstitions. It was used as aromatic, astringent, resolvent, expectorant and nervine.

Parts used: herb tops, flowers.

Figure 3.13 Hypericum perforatum.

Active ingredients: Aromatic *polycyclic diones* (*pseudohypericin* and *hypericin*).
Documented target cancers: Photodynamic cancer therapy, human cancer cell lines (breast, colon, lung, melanoma), antiretroviral.

Further details

Related compounds

- *Pseudohypericin* and *hypericin*, the major photosensitizing constituents of *Hypericum perforatum*, have been proposed as a photosensitizer for photodynamic cancer therapy (Vandenbogaerde *et al.*, 1988). The presence of foetal calf serum (FCS) or albumin extensively inhibits the photocytotoxic effect of pseudohypericin against A431 tumor cells, and is associated with a large decrease in cellular uptake of the compound. These results suggest that pseudohypericin, in contrast to hypericin, interacts strongly with constituents of FCS, lowering its interaction with cells. Since pseudohypericin is two to three times more abundant in *Hypericum* than *hypericin* and the bioavailabilities of *pseudohypericin* and *hypericin* after oral administration are similar, these results suggest that *hypericin*, and not *pseudohypericin*, is likely to be the constituent responsible for hypericism. Moreover, the dramatic decrease of photosensitizing activity of *pseudohypericin* in the presence of serum may restrict its applicability in clinical situations.

- Hexane extracts of *Hypericum drummondii* showed significant cytotoxic activity on cultured P-388, KB, or human cancer cell lines (breast, colon, lung, melanoma) (Jayasuriya *et al.*, 1989).

Other medical activity

- *Hypericin* and *pseudohypericin* have potent antiretroviral activity and are highly effective in preventing viral-induced manifestations that follow infections with a variety of retroviruses *in vivo* and *in vitro* (Meruelo *et al.*, 1988). Pseudohypericin and hypericin probably interfere with viral infection and/or spread by direct inactivation of the virus or by preventing virus shedding, budding, or assembly at the cell membrane. These compounds have no apparent activity against the transcription, translation, or transport of viral proteins to the cell membrane and also no direct effect on the polymerase. This property distinguishes their mode of action from that of the major antiretro-virus group of nucleoside analogues. Hypericin and pseudohypericin have low *in vitro* cytotoxic activity at concentrations sufficient to produce dramatic antiviral effects in murine tissue culture model systems that use radiation leukemia and Friend viruses. Administration of these compounds to mice at the low doses sufficient to prevent retroviral-induced disease appears devoid of undesirable side effects. This lack of toxicity at therapeutic doses extends to humans, as these compounds have been tested in patients as antidepressants with apparent salutary effects. These observations suggest that pseudohypericin and hypericin could become therapeutic tools against retroviral-induced diseases such as acquired immunodeficiency syndrome (AIDS).

References

Cott, J. (1995) NCDEU update. Natural product formulations available in europe for psychotropic indications. *Psychopharmacol Bull.* 31(4), 745–51.

Jayasuriya, H., McChesney, J.D., Swanson, S.M. and Pezzuto, J.M. (1989) Antimicrobial and cytotoxic activity of rottlerin-type compounds from *Hypericum drummondii. J. Nat. Prod.* 52(2), 325–31.

Muller, W.E. and Rossol, R. (1994) Effects of hypericum extract on the expression of serotonin receptors. *J Geriatr. Psychiatry Neurol.* 71, 63–4.

Meruelo, D., Lavie, G. and Lavie, D. (1988) Therapeutic agents with dramatic antiretroviral activity and little toxicity at effective doses: aromatic polycyclic diones hypericin and pseudohypericin. *Proc. Natl. Acad. Sci. USA* 85(14), 5230–4.

Vandenbogaerde, A.L., Kamuhabwa, A., Delaey, E., Himpens, B.E., Merlevede, W.J. and de Witte, P.A. (1998) Photocytotoxic effect of pseudohypericin versus hypericin. *J Photochem. Photobiol. B.* 45(2–3), 87–94.

***Juniperus virginiana* L. (Juniperus (red cedar)) (Conifereae) — Tumor inhibitor**

Location: North America, Europe, North Africa, and North Asia. It is known as the American Juniper of Bermuda and also as "Pencil Cedar".

Appearance (Figure 3.14)
Stem: 1.5 m high, erect trunk, spreading branches covered with a shreddy bark.
Leaves: straight and rigid, awl-shaped, 0.8–1.5 cm long, with sharp, prickly points.
Flowers: in short cones.

Figure 3.14 Juniperus virginiana.

In bloom: April–May.

Tradition: It is used in the preparation of insecticides, in making liniments and other medicinal preparations and perfumed soaps. The leaves have diuretic properties.

Parts used: ripe, carefully dried fruits, leaves.

Active ingredients: *Podophyllotoxin*.

Further details

Antitumor activity

- *Podophyllotoxin*, the active principle of *Juniperus virginiana* is a tumor inhibitor (Kupchan *et al*., 1965). However, in mice the use of cedar shavings as bedding increased significantly the incidence of spontaneous tumors of the liver and mammary gland, and also reduced the average time at which tumors appeared (Sabine, 1975).
- Both antitumor-promoting and antitumor activities have been attributed to the crude extract from the leaves of *Juniperus chinensis* (Ali *et al*., 1996).

References

Ali, A.M., Mackeen, M.M., Intan-Safinar, I., Hamid, M., Lajis, N.H., el-Sharkawy, S.H. and Murakoshi, M. (1996) Antitumour-promoting and antitumour activities of the crude extract from the leaves of *Juniperus chinensis*. *J. Ethnopharmacol.* 53(3), 165–9.

Kupchan, S.M., Hemingway, J.C. and Knox, J.R. (1965) Tumor inhibitors. VII. Podophyllotoxin, the active principle of *Juniperus virginiana*. *J. Pharm. Sci.* 54(4), 659–60.

Sabine, J.R. (1975) Exposure to an environment containing the aromatic red cedar, *Juniperus virginiana*: procarcinogenic, enzyme-inducing and insecticidal effects. *Toxicology* 5(2), 221–35.

Mallotus philippinensis **(Mallotus (Kamala)) (Euphorbiaceae)** — **Tumor inhibitor**

Location: India, Malay Archipelago, Orissa, Bengal, Bombay, Southern Arabia, China, Australia.

Appearance
Stem: 7–10 m high, 1–1.5 cm in diameter.
Leaves: alternate, articulate petioles 1–2 in long, ovate with two obscure glands at base.
Flowers: dioecious, covered with ferrugineous tomentosum.
In bloom: November–January.

Tradition: The root of the tree is used in dyeing and for cutaneous eruptions. It was used by the Arabs internally for leprosy and in solution to remove freckles and pustules.

Part used: pericarps.

Active ingredients

- Maytansinoid tumor inhibitors: *rottlerin, mallotojaponin, phloroglucinol* derivatives: *mallotolerin, mallotochromanol, mallotophenone, mallotochromene.*
- *ent-kaurane* and *rosane* diterpenoids.

Documented target cancers

- CaM kinase III inhibitor. Cytotoxic (glioblastomas-human, mice).
- Skin tumor (mice), human larynx (HEp-2) and lung (PC-13) carcinoma cells as well as mouse B16 melanoma, leukemia P388, and L5178Y cells.

Further details

Related compounds

- *Mallotus phillippinensis*: pericarps contain rottlerin, a 5,7-dihydroxy-2,2-dimethyl-6-(2,4,6-trihydroxy-3-methyl-5-acetylbenzyl)-8-cinnamoyl-1,2-hromene which has been shown to be an effective CaM kinase III inhibitor. Rottlerin decreased growth and induced cytotoxicity in rat (C6) and two human gliomas (T98G and U138MG) at concentrations that inhibited the activity of CaM kinase III *in vitro* and *in vivo* (Parmer *et al.*, 1997). Far less demonstrable effects were observed on other

Ca^{2++}/CaM-sensitive kinases. Incubation of glial cells with rottlerin produced a block at the G1-S interface and the appearance of a population of cells with a complement of DNA. In addition, rottlerin induced changes in cellular morphology such as cell shrinkage, accumulation of cytoplasmic vacuoles, and packaging of cellular components within membranes.

- The pericarps of *Mallotus japonicus* (Euphorbiaceae) contain *mallotojaponin*, which inhibited the action of tumor promoter *in vitro* and *in vivo* (Satomi *et al.*, 1994); it inhibited tumor promoter-enhanced phospholipid metabolism in cultured cells, and also suppressed the promoting effect of 12-*O*-tetradecanoylphorbol-13-acetate on skin tumor formation in mice initiated with 7,12-dimethylbenz-[a]anthracene (Satomi *et al.*, 1994).
- In addition, pericarps contain a variety of phloroglucinol derivatives which were proved to be significantly cytotoxic in culture against human larynx (HEp-2) and lung (PC-13) carcinoma cells as well as mouse B16 melanoma, leukemia P-388, the KB system and L5178Y cells. These phloroglucinol derivatives are: *mallotolerin* (3-(3-methyl-2-hydroxybut-3-enyl)-5-(3-acetyl-2,4-dihydroxy-5-methyl-6-methoxybenxyl)-phlorbutyrophenone), *mallotochromanol* (8-acetyl-5,7-dihydroxy-6-(3-acetyl-2,4-dihydroxy-5-methyl-6-methoxybenxyl) 2,2-dimethyl-3-hydroxychroman), allotophenone (5-methylene-bis-2, 6-dihydroxy-3-methyl-4-methoxyacetophenone), *mallotochromene* (8-acetyl-5, 7-dihydroxy-6-(3-acetyl-2,4-dihydroxy-5-methyl-6-methoxybenzyl)2,2-dimethylchromene), 3-(3,3-dimethylallyl)-5-(3-acetyl-2,4-dihydroxy-5-methyl-6-methoxybenzyl)-phloracetophenone, and 2,6-dihydroxy-3-methyl-4-methoxyacetophenone (Arisawa *et al.*, 1990).
- *Mallotus anomalus* Meer et Chun contains *ent-kaurane* and *rosane* diterpenoids (Xu, 1991).

References

Arisawa, M., Fujita, A., Morita, N. and Koshimura, S. (1990) Cytotoxic and antitumor constituents in pericarps of *Mallotus japonicus. Planta Med.* **56**(4), 377–9.

Arisawa, M., Fujita, A., Saga, M., Hayashi, T., Morita, N., Kawano, N. and Koshimura, S. (1986) Studies on cytotoxic constituents in pericarps of *Mallotus japonicus*, Part II. *J. Nat. Prod.* **49**(2), 298–302.

Arisawa, M., Fujita, A., Suzuki, R., Hayashi, T., Morita, N., Kawano, N. and Koshimura, S. (1985) Studies on cytotoxic constituents in pericarps of *Mallotus japonicus*, Part I. *J. Nat. Prod.* **48**(3), 455–9.

Mi, J.F., Xu, R.S., Yang, Y.P. and Yang, P.M. (1993) Studies on circular dichroism of diterpenoids from Mallotus anomalus and sesquiterpenoidtussilagone *YaoXueXueBao* **28**(2), 105–9.

Parmer, T.G., Ward, M.D. and Hait, W.N. (1997) Effects of rottlerin, an inhibitor of calmodulin-dependent protein kinase III, on cellular proliferation, viability, and cell cycle distribution in malignant glioma cells. *Cell Growth Differ.* **8**(3), 327–34.

Satomi, Y., Arisawa, M., Nishino, H., Iwashima, A. (1994) Antitumor-promoting activity of mallotojaponin, a major constituent of pericarps of *Mallotus japonicus. Oncology*, **51**(3), 215–9.

Xu, R.S., Tang, Z.J., Feng, S.C., Yang, Y.P., Lin, W.H., Zhong, Q.X. and Zhong, Y. (1991) Studies on bioactive components from Chinese medicinal plants. *Mem. Inst. Oswaldo Cruz* **86**(Suppl 2), 55–9.

Maytenus boaria (Maytenus) (Celastraceae) **Cytotoxic**

Location: Mountains of South America.

Appearance
Stem: 34 m.
Leaves: alternate, simple, narrow, elliptic to lanceolate, tiny teeth, pointed tip.

Active ingredients: *Maytenin*
Ansa macrolide (*maytansine*).

Documented target cancers: basic cellular carcinoma, Kaposi's sarcomatosis, leukemia.

Further details

Related compounds

- Maytenin demonstrates a low irritant action and late antineoplastic properties (Melo *et al.*, 1974).
- Some more species of the same genus appear to have a cytotoxic effect against cancer tumors such as: *Maytenus guangsiensis* Cheng et Sha (anti-leukemic) (Qian *et al.*, 1979), *Maytenus ovatus* (anti-leukemic) (maytansine) (Kupchan *et al.*, 1972), *Maytenus senegalensis* (Tin-Wa *et al.*, 1971).
- *Maytenus wallichiana* Raju et Babu and *Maytenus emarginata* Ding Hou (lymphocytic leukemia).

Biotechnology

- Plant tissue cultures of *Maytenus wallichiana* Raju et Babu and *Maytenus emarginata* Ding Hou were initiated (Dymowski and Furmanowa, 1989) Growth conditions of the callus and the optimum medium composition have been established. Increments of callus wet mass and dynamics of callus growth were determined. Morphological and microscopic observations were also performed. The most efficient growth of the callus, resulting in increments of its wet mass up to 6460%, was obtained on the modified Murashige and Skoog medium. Extracts of the callus were found to be inactive against microorganisms, but proved cytotoxic for lymphocytic leukemia.

References

Dymowski, W. and Furmanowa, M. (1989) The search for cytostatic substances in the tissues of plants of the genus *Maytenus molina in vitro* culture. I. Callas culture and biological studies of its extracts. *Acta Pol. Pharm.* 46(1), 81–9.

Dymowski, W. and Furmanowa, M. (1990) Investigating cytostatic substances in tissue of plants *Maytenus Molina* in *in-vitro* cultures. II. chromatographic test of extracts from callus of *Maytenus wallichiana* R. et B. *Acta Pol. Pharm.* 47(5–6), 51–4.

Dymowski, W. and Furmanowa, M. (1992) Searching for cytostatic substances in plant tissue of Maytenus molina by *in vitro* culture. III. Release of substances from active biological fractions from the callus extract of *Maytenus wallichiana* R. and B. *Acta Pol. Pharm.* 49(1–2), 29–33.

Kuo, Y.H., Chen, C.H., Kuo, L.M., King, M.L., Wu, T.S., Haruna, M. and Lee, K.H. (1990) Antitumor agents, 112. *Emarginatine B*, a novel potent cytotoxic sesquiterpene pyridine alkaloid from *Maytenus emarginata*. *J. Nat. Prod.* 53(2), 422–8.

Kuo, Y.H., King, M.L., Chen, C.F., Chen, H.Y., Chen, C.H., Chen, K. and Lee, K.H. (1994) Two new macrolide sesquiterpene pyridine alkaloids from *Maytenus emarginata: emarginatine G* and the cytotoxic *emarginatine F. J. Nat. Prod.* 57(2), 263–9.

Kupchan, S.M., Komoda, Y., Court, W.A., Thomas, G.J., Smith, R.M., Karim, A., Gilmore, C.J., Haltiwanger, R.C. and Bryan, R.F. (1972) Maytansine, a novel anti-leukemic ansa macrolide from *Maytenus ovatus. J. Am. Chem. Soc.* 94(4), 1354–6.

Melo, A.M., Jardim, M.L., De Santana, C.F., Lacet, Y., Lobo Filho, J., De Lima and Ivan Leoncio, O.G. (1974) First observations on the topical use of Primin, Plumbagin and Maytenin in patients with skin cancer. *Rev. Inst. Antibiot.* (Recife) 14(1–2), 9–16.

Pandey, R.C. (1998) Prospecting for potentially new pharmaceuticals from natural sources. *Med. Res. Rev.* 18(5), 333–46. Review.

Qian, X., Gai, C. and Yao, S. (1979) Studies on the anti-leukemic principle of *Maytenus guangsiensis. Cheng et Sha. Yao Hsueh Hsueh Pao.* 14(3), 182.

Sneden, A.T. and Beemsterboer, G.L. (1980) Normaytansine, a new anti-leukemic ansa macrolide from *Maytenus buchananii. J. Nat. Prod.* 43(5), 637–40.

Tin-Wa, M., Farnsworth, N.R., Fong, H.H., Blomster, R.N., Trojanek, J., Abraham, D.J., Persinos, G.J. and Dokosi, O.B. (1971) Biological and phytochemical evaluation of plants. IX. Antitumor activity of *Maytenus senegalensis* (Celastraceae) and a preliminary phytochemical investigation. *Lloydia*, 34(1), 79–87.

Melia azedarach (Melia) (Meliaceae) — Cytotoxic

Location: Northern India, China, the Himalayas.

Appearance (Figure 3.15)
Stem: 10–17 m high, reddish brown bark.
Leaves: bipinnate, 1–2 in long. The individual leaflets, each about 2 cm long, are pointed at the tips and have toothed edges.
Flowers: large branches of lilac, fragrant, star shaped flowers, that arch or droop in 8 cm panicles.
In bloom: spring – early summer.

Parts used: the bark of the root and trunk, seed.

Figure 3.15 Melia azedarach.

Active ingredients

- Limonoids: *toosendanal, 28-deacetyl sendanin,12-0-methylvolkensin, meliatoxin B1, trichillin H, and toosendanin, 12-deacetyltrichilin I 1-acetyltrichilin H, 3-deacetyltrichilin H, 1-acetyl-3-deacetyltrichilin H, 1-acetyl-2-deacetyltrichilin H, meliatoxin B1, trichilin H, trichilin D and 1,12-diacetyltrichilin B.*
- *Meliavolkinin, melianin C, 3-diacetylvilasinin and melianin B.*

Documented target cancers

- KB cells (meliatoxin B1 and toosendanin).
- P388 cells (limonoids of *Melia azedarach*) (Itokawa *et al.*, 1995).
- Human prostate (PC-3) and pancreatic (PACA-2) cell lines (3, 23,24-diketomelianin B).

Further details

Related compounds

- The root bark of *Melia azedarach*, contains the trichilin-type limonoids 12-deacetyltrichilin I 1-acetyltrichilin H, 3-deacetyltrichilin H, 1-acetyl-3-deacetyltrichilin H, 1-acetyl-2-deacetyltrichilin H, meliatoxin B1, trichilin H, trichilin D and 1,12-diacetyltrichilin B (Takeya *et al.*, 1996).
- The limonoid compound (28-deacetyl sendanin) isolated from the fruit of *Melia toosendan* SIEB. et ZUCC. was evaluated on anticancer activity. It has been proved that 28-deacetyl sendanin has more sensitive and selective inhibitory effects on *in vitro* growth of human cancer cell lines in comparison with adriamycin (Tada *et al.*, 1999).
- The fruits of *Melia toosendan* Sieb. et Zucc. contain the limonoids toosendanal, 12-0-methylvolkensin, meliatoxin B1, trichillin H, toosendanin and 28-deacetyl sendanin (Tada *et al.*, 1999).
- The root bark of *Melia volkensii* contains meliavolkinin, melianin C, 1,3-diacetylvilasinin and melianin B, which all showed marginal cytotoxicities against certain human tumor cell lines (Rogers *et al.*, 1998). Jones oxidation of melianin B4 gave 3, 23,24-diketomelianin B, which showed selective cytotoxicities for the human prostate (PC-3) and pancreatic (PACA-2) cell lines with potencies comparable to those of adriamycin.

References

Itokawa, H., Qiao, Z.S., Hirobe, C. and Takeya, K. (1995) Cytotoxic limonoids and tetranortriterpenoids from *Melia azedarach. Chem Pharm. Bull.* (Tokyo) 43(7), 1171–5.

Rogers, L.L., Zeng, L., Kozlowski, J.F., Shimada, H., Alali, F.Q., Johnson, H.A. and McLaughlin, J.L. (1998) New bioactive triterpenoids from *Melia volkensii. J. Nat. Prod.* 61(1), 64–70.

Tada, K., Takido, M. and Kitanaka, S. (1999) Limonoids from fruit of *Melia toosendan* and their cytotoxic activity. *Phytochemistry* 51(6), 787–91.

Takeya, K., Quio, Z.S., Hirobe, C. and Itokawa, H. (1996) Cytotoxic trichilin-type limonoids from *Melia azedarach. Bioorg. Med. Chem.* 4(8), 1355–9.

Mormodica charantia (Mormodica (Bitter melon)) (Cucurbitaceae) — Anti-leukemic

Location: East India.

Appearance (Figure 3.16)
Stem: thin, crawly.
Leaves: dark, green, and deeply lobed.
Flowers: dioecious, yellow.

Part used: the fruit deprived of the seeds.

Active ingredients: Protein (molecular weight of 11,000 Da).

Documented target cancers: The fruit and seeds of the bitter melon (Momordica charantia) have been reported to have anti-leukemic and antiviral activities:

- Antitumor (mice),
- Antiviral–anti-leukemic (human, selective),
- Immunostimulating (mice).

Further details

Anti-leukemic activity

- This anti-leukemic and antiviral action was associated with an activation of murine lymphocytes. This activity is associated with a single protein component with an apparent molecular weight of 11,000 Da. The factor is not sensitive to boiling or to pretreatments with trypsin, ribonuclease (RNAse), or deoxyribonuclease (DNAse) (Cunnick *et al.*, 1990). As determined by radioactive precursor uptake studies, the purified factor preferentially inhibits RNA synthesis in intact tissue culture cells. Some inhibition of protein synthesis and DNA synthesis also occurs. The factor is preferentially cytostatic

Figure 3.16 Mormodica.

for IM9 human leukemic lymphocytes when compared to normal human peripheral blood lymphocytes. In addition, it preferentially inhibits the soluble guanylate cyclase from leukemic lymphocytes. This inhibition correlates with its preferential cytotoxic effects for these same cells, since cyclic GMP is thought to be involved in lymphocytic cell proliferation and leukemogenesis and, in general, the nucleotide is elevated in leukemic versus normal lymphocytes and changes have been reported to occur during remission and relapse of this disease (Takemoto *et al.*, 1980, 1982).

- At least part of the anti-leukemic activity of the bitter melon extract is due to the activation of NK cells in the host organism (mouse), that is, *in vivo* enhancement of immune functions may contribute to the antitumor effects of the bitter melon extract. In humans, the extract has both cytostatic and cytotoxic activities and can kill leukemic lymphocytes in a dose-dependent manner while not affecting the viability of normal human lymphocyte cells at these same doses (Takemoto *et al.*, 1982). These activities are not due to the presence of the lectins from bitter melon seeds, as these purified proteins had no activity against human lymphocytic cells (Jilka *et al.*, 1983).

References

Cunnick, J.E., Sakamoto, K., Chapes, S.K., Fortner, G.W. and Takemoto, D.J. (1990) Induction of tumor cytotoxic immune cells using a protein from the bitter melon (*Momordica aharantia*). *Cell Immunol.* 126 (2), 278–89.

Jilka, C., Strifler, B., Fortner, G.W., Hays, E.F. and Takemoto, D.J. (1983) *In vivo* antitumor activity of the bitter melon (*Momordica charantia*). *Cancer Res.* 43(11), 5151–5.

Lin, J.Y., Hou, M.J. and Chen, Y.C. (1978) Isolation of toxic and non-toxic lectins from the bitter pear melon *Momordica charantia* Linn. *Toxiconomis* 16(6), 653–60.

Takemoto, D.J., Dunford, C. and McMurray, M.M. (1982) The cytotoxic and cytostatic effects of the bitter melon (*Momordica charantia*) on human lymphocytes. *Toxiconomy* 20(3), 593–9.

Takemoto, D.J., Dunford, C., Vaughn, D., Kramer, K.J., Smith, A. and Powell, R.G. (1982) Guanylate cyclase activity in human leukemic and normal lymphocytes. Enzyme inhibition and cytotoxicity of plant extracts. *Enzyme* 27(3), 179–88.

Takemoto, D.J., Jilka, C. and Kresie, R. (1982) Purification and characterization of a cytostatic factor from the bitter melon *Momordica charantia*. *Prep. Biochem.* 12(4), 355–75.

Takemoto, D.J., Kresie, R. and Vaughn, D. (1980) Partial purification and characterization of a guanylate cyclase inhibitor with cytotoxic properties from the bitter melon (*Momordica charantia*). *Biochem. Biophys. Res. Commun.* 14 94(1), 332–9.

Nigella sativa L. (Nigella (Fennel flower)) (Ranunculaceae) — Cytotoxic

Location: Asia.

Appearance (Figure 3.17)
Stem: stiff, erect, branching.
Leaves: bears deeply cut greyish-green.
Flowers: greyish blue.
In bloom: early summer.

Tradition: In India, the seeds are believed to increase the secretion of milk and are considered as stimulant, diaphoretic. They also use it in tonics. Romans used it in cooking (Roman Coriander). The French used it as a substitute for pepper.

Figure 3.17 Nigella.

Parts used: seed, herb.

Active ingredients: thymoquinone and dithymoquinone, fatty acids.

Documented target cancers: multi-drug resistant (MDR) human tumor cell lines, Ehrlich ascites carcinoma (EAC), Dalton's lymphonia ascites (DLA) and Sarcoma-180 (S-180) cells. Skin cancer (mice).

Further details

Antitumor activity

- *Nigella sativa*: Seeds contain thymoquinone (TQ) and dithymoquinone (DIM), which were cytotoxic *in vitro* against MDR human tumor cell lines (IC50's 78–393 μM). Both the parental cell lines and their corresponding MDR variants, over 10-fold more resistant to the standard antineoplastic agents doxorubicin (DOX) and etoposide (ETP), as compared to their respective parental controls, were equally sensitive to TQ and DIM. The inclusion of the competitive MDR modulator quinine in the assay reversed MDR Dx-5 cell resistance to DOX and ETP by 6- to 16-fold, but had no effect on the cytotoxicity of TQ or DIM. Quinine also increased MDR Dx-5 cell accumulation of the P-glycoprotein substrate 3H-taxol in a dose-dependent manner. However, neither TQ nor DIM significantly altered cellular accumulation of 3H-taxol. The inclusion of 0.5% v/v of the radical scavenger DMSO in the assay reduced the cytotoxicity of DOX by as much as 39%, but did not affect that of TQ or DIM. These studies suggest that TQ and DIM, which are cytotoxic for several types of human tumor cells, may not be MDR substrates, and that radical generation may not be critical to their cytotoxic activity (Salomi *et al.*, 1992).

- *Nigella sativa* seeds also contain certain fatty acids which are cytotoxic *in vitro* against Ehrlich ascites carcinoma (EAC), Dalton's lymphonia ascites (DLA) and Sarcoma-180 (S-180) cells. *In vitro* cytotoxic studies showed 50% cytotoxicity to Ehrlich ascites carcinoma, Dalton's lymphoma ascites and Sarcoma-180 cells at a concentration of 1.5 μg, 3 μg and 1.5 μg respectively with little activity against lymphocytes. The cell growth of KB cells in culture was inhibited by the active principle while K-562 cells resumed near control values on day 2 and day 3. Tritiated thymidine incorporation studies indicated the possible action of an active principle at DNA level. *In vivo* EAC tumor development was completely inhibited by the active principle at the dose of 2 mg per day × 10 for each mouse (Salomi *et al.*, 1992).
- Topical application of *Nigella sativa* inhibited two-stage initiation/promotion [dimethylbenz[a]anthracene (DMBA)/croton oil] skin carcinogenesis in mice. A dose of 100 mg kg^{-1} body wt of these extracts delayed the onset of papilloma formation and reduced the mean number of papillomas per mouse, respectively. The possibility that these extracts could inhibit the action of 20-methylcholanthrene (MCA)-induced soft tissue sarcomas was evaluated by studying the effect of these extracts on MCA-induced soft tissue sarcomas in albino mice. Intraperitoneal administration of Nigella sativa (100 mg $kg^{-1/}$ body wt) and oral administration of *Crocus sativus* (100 mg kg^{-1} body wt) 30 days after subcutaneous administration of MCA (745 nmol × 2 days) restricted tumor incidence to 33.3% and 10%, respectively, compared with 100% in MCA-treated controls (Salomi *et al.*, 1991).

References

Salomi, N.J., Nair, S.C., Jayawardhanan, K.K., Varghese, C.D. and Panikkar, K.R. (1992) Antitumour principles from Nigella sativa seeds. *Cancer Lett.* 63(1), 41–6.

Salomi, M.J., Nair, S.C. and Panikkar, K.R. (1991) Inhibitory effects of *Nigella sativa* and saffron (*Crocus sativus*) on chemical carcinogenesis in mice. *Nutr. Cancer* 16(1), 67–72.

Worthen, D.R., Ghosheh, O.A. and Crooks, P.A. (1998) The *in vitro* anti-tumor activity of some crude and purified components of blackseed, *Nigella sativa* L. *Anticancer Res.* 18(3A), 1527–32.

Origanum vulgare, O. majorana (Oregano (marjoram)) (Lamiaceae) — Anticancer

Location: Mediterranean region of Europe and Asia.

Appearance (Figure 3.18)
Stem: bushy, semi-woody sub-shrub with upright or spreading stems and branches.
Leaves: aromatic, oval-shaped, about 4 cm long and usually pubescent.
Flowers: throughout the summer oregano bears tiny (0.3 cm long) purple tube-shaped flowers that peek out of whorls of purplish-green leafy bracts about an inch long.
In bloom: summer.

Tradition: It was used from very early years for its medicinal properties, as a remedy for narcotic poisons, convulsions and dropsy. The whole plant has a strong fragrant, balsamic odor and an aromatic taste.

Figure 3.18 Origanum majorana.

Parts used: herb, oil, leaves.

Active ingredients: flavonoids, *galangin* and *quercetin*, water-alcoholic extracts and of isolated compounds (*arbutin, methylarbutin* and their *aglycons* – *hydroquinone* and *hydroquinone monomethyl ether*).

Antitumor-promoting activity or *in vitro* cytotoxic effects towards different tumor cell lines were attributed also to *Origanum majorana* extracts or their constituents. When studying cytotoxic activity of *O. majorana* water-alcoholic extracts and of isolated compounds (*arbutin, methylarbutin* and their aglycons – *hydroquinone* and *hydroquinone monomethyl ether*) towards cultured rat hepatoma cells (HTC line), a high dose-dependent HTC cytotoxicity of *hydroquinone.*

Indicative dosage and application: At 300 μM *hydroquinone* caused 40% cellular mortality after 24 h of incubation.

Documented target cancers

- Antitumor-promoting activity or *in vitro* cytotoxic effects towards different tumor cell lines (rat hepatoma cells (HTC line))
- Immunostimulant
- Antimutagenic.

Further details

Other medical activity

- Some studies have shown that oregano extracts or herbal mixtures with *Origanum* spp. possess *in vitro* antiviral activity or have immunostimulating effects both *in vitro* and *in vivo*. However, little knowledge has been attained so far on mechanisms of immunomodulating activity or underlying active compounds. It has been shown that ethanol extracts of *Origanum vulgare* inhibited intracellular propagation of $ECHO_9$ Hill virus and also showed interferon inducing activity *in vitro.* Flavonoid *luteoline*, a constituent of *Origani herba*, has been considered as responsible for the induction of an interferon-like substance. A mixture of herbal preparation containing rosemary, sage, thyme and oregano (*Origanum vulgare*) showed radical scavenging activity and inhibition of the human immunodeficiency virus (HIV) infection at very low concentrations. It was suggested that the main active compounds of herbal preparations were *carnosol, carnosic acid, carvacrol* and *thymol*. Significant inhibitory effects of *Origanum vulgare* extracts against HIV-1 induced cytopathogenicity in MT-4 cells were also observed by Yamasaki *et al.* (1998). According to Krukowski and co-workers, an increase in immunoglobulin (IgG) levels was observed in reared calves, fed with a conventional concentrate supplemented by a mineral–herbal mixture containing *Origanum majorana.*

- A strong and dose-dependent capacity of inactivating dietary mutagen Trp-P-1 in the *Salmonella typhimurium* TA 98 assay was observed in *Origanum vulgare* water extracts, that exhibited significant antimutagenic effects *in vitro* (Ueda *et al.*, 1991). *Origanum majorana* aqueous extracts were also able to suppress the mutagenicity of liver-specific carcinogen Trp-P-2 (Natake *et al.*, 1989). When studying the mechanism of suppressing the mutagenicity of Trp-P-2 in *Origanum vulgare*, it was found that two flavonoids, *galangin* and *quercetin* acted as Trp-P-2 specific desmutagens, which neutralized this mutagen during or before mutating the bacteria (*Salmonella typhimurium* TA 98) (Kanazawa *et al.*, 1995). The amounts of *galangin* and *quercetin* required for 50% inhibition (IC_{50}) against 20 ng of Trp-P-2 were 0.12 μg and 0.81 μg, respectively. It was also found that *quercetin* acted as a mutagen at high concentrations (>10 μg per plate), but was a desmutagen when applied at low ($>0.1<10$ μg per plate) concentrations. Milic and Milic (1998) have found that isolated phenolic compounds from different spice plants, including *Origanum vulgare*, strongly inhibited pyrazine cation free radical formation in the Maillard reaction and the formation of mutagenic and carcinogenic amino-imidazoazarene in creatinine containing model systems.

Antitumor activity

- In a literature survey, referring to the anticancer activity of *Origanum* genus, different approaches, testing systems and cell lines have been used by different authors when assessing the carcinogenic potential of plants or their isolated compounds. However, there are no available data on practical/clinical use of oregano in cancer prevention. In 1966, an international project was performed with the aim of screening the native plants of former Yugoslavia for their potential agricultural use in the USA and Yugoslavia. In the frame of this project 1,466 samples of 754 plant species were analyzed for chemical and antitumor activity. According to the results of the Cancer Chemotherapy National Service Center Screening Laboratories (Washington, DC) a high carvacrol (60–85%) containing *Origanum heracleoticum* (= *O. vulgare* spp. *hirtum* (Link) Ietswaart) was reported to show high antitumor activity. Zheng (1991) has found that essential oil of *Origanum vulgare* fed to mice, induced the activity of glutathione S-transferase (GST) in various tissues. The GST enzyme system is involved in detoxification of chemical carcinogens and plays an important role in prevention of carcinogenesis, which would explain the anticancer potential of *O. vulgare* essential oil. This oil exhibited high levels of cytotoxicity (at dilutions of up to 1:10,000) against four permanent eukaryotic cell lines including two derived from human cancers (epidermoid larynx carcinoma: Hep-2 and epitheloid cervix carcinoma: HeLa). Other studies, that refer to *in vitro* cytotoxic and/or anti-proliferative effects of *Origanum vulgare* extracts or isolated compounds (*carvacrol, thymol*) include those of Bocharova and He, who observed moderate suppressing activities of *O. vulgare* extracts ($CE_{50} = 220\,mg\,ml^{-1}$) on human ovarian carcinoma cells (CaOv), or of isolated *carvacrol* and *thymol* ($IC_{50} = 120\,\mu mol\,l^{-1}$) on Murine B 16(F10) melanoma cells – a tumor cell line with high metastatic potential.

- Antitumor-promoting activity or *in vitro* cytotoxic effects towards different tumor cell lines were attributed also to *Origanum majorana* extracts or their constituents. When studying the cytotoxic activity of *O. majorana* water-alcoholic extracts and of isolated compounds (*arbutin, methylarbutin* and their aglycons – *hydroquinone* and *hydroquinone monomethyl ether*) towards cultured rat hepatoma cells (HTC line), a high dose-dependent HTC cytotoxicity of *hydroquinone* was observed, whilst *arbutin* was not active (Assaf *et al.*, 1987). At 300 μM *hydroquinone* caused 40% cellular mortality after 24 h of incubation, but no cells remained viable after 72 h. It has been established that this well-known antiseptic of the urinary tract was a more potent cytotoxic compound towards rat hepatoma cells than many classic antitumor agents like *azauridin* or *colchicin*, but less than *valtrate*, a monoterpenic ester of *Valeriana* spp.

References

Adam, K., Sivropoulou, A., Kokkini, S., Lanaras, T. and Arsenakis, M. (1998) Antifungal activities of *Origanum vulgare* subsp. *hirtum, Mentha spicata, Lavandula angustifolia*, and *Salvia fruticosa* essential oils against human pathogenic fungi. *J. Agri. Food Chem.* 46(5), 1739–45.

Aruoma, O.I., Spencer, J.P.E., Rossi, R., Aeschbach, R., Khan, A., Mahmood, N., Munoz, A., Murcia, A., Butler, J. and Halliwell, B. (1996) An evaluation of the antioxidant and antiviral action of extracts of rosemary and provencal herbs. *Food Chem. Toxicol.* 34(5), 449–56.

Assaf, M.H., Ali, A.A., Makboul, M.A., Beck, J.P. and Anton, R. (1987) Preliminary study of phenolic glycosides from *Origanum majorana*, quantitative estimation of arbutin, cytotoxic activity of hydroquinone. *Planta Med.* 53(4), 343–5.

Bocharova, O.A., Karpova, R.V., Kasatkina, N.N., Polunina, L.G., Komarova, T. S. and Lygenkova, M.A. (1999) The antiproliferative activity for tumor cells is important to compose the phytomixture for prophylactic oncology. *Farmacevtski Vestnik* 50, 378–379.

Kanazawa, K., Kawasaki, H., Samejima, K., Ashida, H. and Danno, G. (1995) Specific desmutagens (antimutagens) in Oregano against dietary carcinogen, Trp-P-2, are galangin and quercetin. *J. Agri. Food Chem.* 43(2), 404–9.

Mayer, E., Sadar, V. and Spanring, J. (1971) New crops screening of native plants of Yugoslavia of potential use in the agricultures of the USA and SFRJ. University of Ljubljana, Biotechical Faculty, Final Technical Report. Printed by Partizanska knjiga Ljubljana, 210 pp.

Milic, B.L. and Milic, N.B. (1998) Protective effects of spice plants on mutagenesis. *Phytotherapy Res.* 12(Suppl. 1), S3–6.

Sivropoulou, A., Papanikolaou, E., Nikolaou, C., Kokkini, S., Lanaras, T. and Arsenakis, M. (1996) Antimicrobial and cytotoxic activities of *Origanum* essential oils. *J. Agri. Food Chem.* 44(5), 1202–5.

Skwarek, T., Tynecka, Z., Glowniak, K. and Lutostanska, E. (1994) Plant inducers of interferons. *Herba Polonica* 40(1–2), 42–9.

Ueda, S., Kuwabara, Y., Hirai, N., Sasaki, H. and Sugahara, T. (1991) Antimutagenic capacities of different kinds of vegetables and mushrooms. *J. Jap. Soc. Food Sci. Technol.* 38(6), 507–14.

Krukowski, H., Nowakowicz-Debek, B., Saba, L. and Stenzel, R. (1998) Effect of mineral–herbal mixtures on IgG blood serum level in growing calves. *Roczniki Naukowe Zootechniki* 25(4), 97–103.

Yamasaki, K., Nakano, M., Kawahata, T., Mori, H., Otake, T., Ueba, N., Oishi, I., Inami, R., Yamane, M., Nakamura, M., Murata, H. and Nakanishi, T. (1998) Anti-HIV-1 activity of herbs in Labiatae. *Biol. Pharm. Bull.* 21(8), 829–33.

Natake, M., Kanazawa, K., Mizuno, M., Ueno, N., Kobayashi, T., Danno, G. and Minamoto, S. (1989) Herb-water extracts markedly suppress the mutagenicity of Trp-P-2. *Agri. Biol. Chem.* 53(5), 1423–5.

Zheng, S., Wang, X., Gao, L., Shen, X. and Liu, Z. (1997) Studies on the flavonoid compounds of Origanum vulgare L. *Indian J. Chem.* 36, 104–106.

Paeonia officinalis **L. (Paeonia (Paeony)) (Ranunculaceae)** — **Tumor inhibitor**

Location: Only grows wild on an island called the Steep Holmes, in the Severn, Great Britain.

Appearance (Figure 3.19)
Stem: green (red when quite young), about 1 m high.
Root: composed of several roundish, thick knobs of tubers, which hang below each other's, connected by strings.
Leaves: composed of several unequal lobes, which are cut into many segments.
Flowers: deep purple, fragrant.
In bloom: late spring.

Figure 3.19 Paeonia officinalis.

Tradition: The genus is supposed to have been named after the physician Paeos, who cured gods of wounds received during the Trojan War with the aid of this plant. In ancient times it was connected with many superstitions. It was used as antispasmodic and tonic.

Part used: root.

Active ingredients: LHRH antagonist and a weak anti-estrogen on the uterine DNA synthesis in immature rats.

Documented target cancers: Intestinal metaplasia, atypical hyperplasia of the gastric mucosa (*Paeonia lactiflora*), uterine myomas (*Paeonia lactiflora, Paeonia suffruticosa*).

Further details

Related species

- Shi-Quan-Da-Bu-Tang (Ten Significant Tonic Decoction), or SQT (Juzentaihoto, TJ-48) was formulated by Taiping Hui-Min Ju (Public Welfare Pharmacy Bureau) in Chinese Song Dynasty in AD 1200. It is prepared by extracting a mixture of ten medical herbs (*Rehmannia glutinosa, Paeonia lactiflora, Liqusticum wallichii, Angelica sinesis, Glycyrrhiza uralensis, Poria cocos, Atractylodes macrocephala, Panax ginseng. Astragalus membranaceus and Cinnamomum cassia*) that tone the blood and vital energy, and strengthen health and immunity. This potent and popular prescription has traditionally been used against anemia, anorexia, extreme exhaustion, fatigue, kidney and spleen insufficiency and general weakness, particularly after illness (Zee-Cheng, 1992).
- *Paeonia alba* is one of the herbal constituents of Xiao Wei Yan Powder (some of the other constituents are *Smilax glabrae, Hedyotis diffusae, Taraxacum mongolicum, Caesalpinia sappan, Cyperus rotundus, Bletilla striata* and *Glycyrrhiza uralensis*). This preparation has been used for the treatment of intestinal metaplasia and atypical hyperplasia of the gastric mucosa of chronic gastritis, administrated orally at $5\text{–}7\,\mathrm{g\,d.}^{-1}$ After 2–4 months of administration, the total remission rate exceeded 90%. It was 91.3% and that of the AH was 92.16%, while in control group, they were 21.3% and 14.46% respectively. The animal experiments revealed no toxic effect, so safety guarantee was provided for its clinical application (Liu *et al.*, 1992).
- The root of *Paeonia lactiflora* Pall. and the root bark of *Paeonia suffruticosa* Andr. (Paeoniaceae) are components of Kuei-chih-fu-ling-wan (Keishi-bukuryo-gan), a traditional Chinese herbal remedy which contains three components: the bark of *Cinnamomum cassia* Bl. (Lauraceae), seeds of *Prunus persica* Batsch. or *P. persiba* Batsch.var.davidiana Maxim. (Rosaceae) and carpophores of *Poria cocos* Wolf. (Polyporaceae). This prescription has been frequently used in the treatment of gynecological disorders such as hypermenorrhea, dysmenorrhea and sterility. After treatment with the preparation, clinical symptoms of hypermenorrhea and dysmenorrhea were improved in more than 90% of the cases with shrinking of uterine myomas in roughly 60% of the cases (Sakamoto *et al.*, 1992).

References

Aburada, M., Takeda, S., Ito, E., Nakamura, M. and Hosoya, E. (1983) Protective effects of juzentaihoto, dried decoctum of 10 Chinese herbs mixture, upon the adverse effects of mitomycin C in mice. *J. Pharmacobiodyn.* 6(12), 1000–4.

Liu, X.R., Han, W.Q. and Sun, D.R. (1992) Treatment of intestinal metaplasia and atypical hyperplasia of gastric mucosa with xiao wei yan powder. *Chung Kuo Chung Hsi I Chieh Ho Tsa Chih* 12(10), 580, 602–3.

Sakamoto, S., Yoshino, H., Shirahata, Y., Shimodairo, K. and Okamoto, R. (1992) Pharmacotherapeutic effects of kuei-chih-fu-ling-wan (keishi-bukuryo-gan) on human uterine myomas. *Am. J. Chin. Med.* 20(3–4), 313–17.

Zee-Cheng, R.K. (1992) Shi-quan-da-bu-tang (ten significant tonic decoction), SQT. A potent Chinese biological response modifier in cancer immunotherapy, potentiation and detoxification of anticancer drugs. *Methods Find Exp. Clin. Pharmacol.* 14(9), 725–36.

***Panax quinquefolium* (Linn.) (Ginseng) (Araliaceae) — Immunomodulator**

Location: Of Manchuria, China and other parts of eastern Asia origin. It is easy to find it in most of the forests of the countries of Southeast Asia but also in the United States and Canada. It is also cultivated.

Appearance (Figure 3.20)
Stem: simple, erect about 30.5 cm high.
Root: it is 10–25 cm long and 1–2 cm diameter.
Leaves: each divided into five finely toothed leaflets.
Flowers: single terminal umbel, with a few small, yellowish flowers.

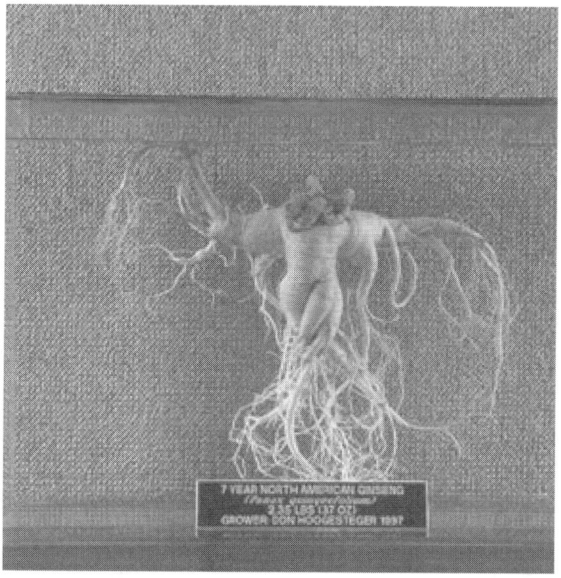

Figure 3.20 Ginseng/Panax.

Tradition: The root has been used for centuries in traditional Chinese medicine. It is believed that it makes those who use it stronger and younger.

Part used: Roots (Ginseng radix).

Active ingredients: *Ginsenosides* (*saponins*), *ginsan* (*acidic polysaccharide*), *panaxytriol* and *panaxydol* (*polyacetylenic alcohols*).

Particular value: It is used particularly for dyspepsia, vomiting and nervous disorders.

Documented target cancers

- *Ginsenosides* appear to have antitumor promoting activity and antimetastatic action in several cancers such as ovarian cancer, breast cancer, stomach cancer and melanoma.
- *Ginsan* has antineoplastic activity. It has been proved that it induces Th1 cell and macrophage cytokines (Kim *et al.*, 1998).
- Antineoplastic activity, cancer chemoprevention, effects on cytochemical components of SGC-823 gastriccarcinoma (in cell culture), Ehrlich ascites tumor cells (mouse), inhibition of autochthonous tumor, effects on adenocarcinoma of the human ovary, stomach cancer, melanoma cells (Xiaoguang *et al.*, 1998).

Further details

Related compounds

- *Panaxytriol* (possible action) is cytotoxic. It is responsible for inhibition of mitochondrial respiration. *Panaxydol* has antiproliferative activity and has affinity for target cell membrane.

Other medical activity

- Red ginseng is a traditional Chinese medicine. Its extracts A and B are the active components of *Panax ginseng*. As it is considered as a tonic many studies have been carried out on ginseng and the immune function of the human body. Some studies refer to the effects of red ginseng extracts on transplantable tumors and proliferation of lymphocyte. It has been proved that in a two-stage model, red ginseng extracts had a significant cancer chemoprevention (Xiaoguang *et al.*, 1998). At the dose of 50–400 $mg\,kg^{-1}$, the extracts could inhibit DMBA/Croton oil-induced skin papilloma in mice and decrease the incidence of papilloma. The red ginseng extract B seems to have a stronger antioxidative effect than that of extract A. Those doses (50 approximately 400 $mg\,kg^{-1}$) could significantly inhibit the growth of transplantable mouse sarcoma S-180 and melanoma B16. In lower doses (extract A 0.5 $mg\,ml^{-1}$ and B 0.1 and 0.25 $mg\,ml^{-1}$) might effectively promote the transformation of T lymphocyte.
- Another study took place in Korea with Korean red ginseng, evaluating the effects of ginseng in inhibition or prevention of carcinogenesis. It was administered orally to ICR new born mice (Yun *et al.*, 1983). Tumors were induced by various chemical carcinogens within 24 h after birth. The newborn mices were injected in the ubscapular

region by 9, 10-dimethyl-1, 2-benzanthracene (DMBA), urethane, and aflatoxin B1. Autopsy was done on the mices immediately following sacrifice and an examination of all their organs was conducted (histopathological examinations, weight, etc.). The decrease in the average diameter and in the weight of lung adenomas was over 23%, while the incidence of diffuse pulmonary infiltration was 63%. The results of the study indicate that Korean red ginseng extract inhibited the incidence and also the proliferation of tumors induced by DMBA, urethane and aflatoxin B1.

Related species

- *Panax ginseng*: most of the compounds come from the methanolic extract of the root. Only the roots are used in medicine. It contains *ginsenosides*, *ginsan*, *panaxytriol* and *panaxydol*. A new chloride is also produced that is cytotoxic. It is used against various human cancers such stomach, breast, ovary, lung, leukemia, hepatoma, adenocarcinomas. Administration is either oral or by injection (shenmai injection).
- *Panax vietnamensis*: The root contains the ginsenosides: *majonoside* −R2, *ginsenoside* −R2, and *ginsenoside* –Rg1. It is used for its inhibitory effects on tumor growth (human ovarian cancer cells) and for its antitumor promoting activity. *Ginsenoside* −Rg1 seems to downregulate glycocorticoid receptors and displays synergistic effects with CAMP (Lee *et al.*, 1997).
- *Panax quinquefolius* L.: It is the American version of the ginseng. The extract of the root was used in studies and the administration was done orally. *Ginsenosides* were contained, also, and the effects showed a decrease of serum gamma globulin and IgG1 isotype (in mice) and ps2 expression in MCF-7 breast cancer cells (Kim *et al.*, 1997).
- *Panax ginseng (red)*: It is the Korean *Panax ginseng*. The extract of the root is used in medicine: A and B which contain *ginsenoside* −Rg3, −Rb2, −Rh2, −Rh4, −20(R), −20(S). Inhibits the tumor metastasis, tumor angiogenesis and improves the cell immune system. In studies related to stomach cancer the shenmai injection is used which is produced from red ginseng extract.

References

Bernart, M.W., Cardellina, II J.H. Balaschak, M.S., Alexander, M.R., Shoemaker, R.H. and Boyd, M.R. (1996) Cytotoxic falcarinol oxylipins from *Dendropanax arboreus*. *J. Nat. Prod.* 59(8), 748–53.

Kim, K.H., Lee, Y.S., Jung, I.S., Park, S.Y., Chung, H.Y., Lee, I.R. and Yun, Y.S. (1998) Acidic polysaccharide from Panax ginseng, ginsan, induces Th1 cell and macrophage cytokines and generates LAK cells in synergy with rIL-2. *Planta Med.* 64(2), 110–15.

Kim, Y.W., Song, D.K., Kim, W.H., Lee, K.M., Wie, M.B., Kim, Y.H., Kee, S.H. and Cho, M.K. (1997) Long-term oral administration of ginseng extract decreases serum gamma-globulin and IgG1 isotype in mice. *J. Ethnopharmacol.* 58(1), 55–8.

Lee, Y.J., Chung, E., Lee, K.Y., Lee, Y.H., Huh, B. and Lee, S.K. (1997) Ginsenoside-Rg1, one of the major active molecules from Panax ginseng, is a functional ligand of glucocorticoid receptor. *Mol. Cell Endocrinol.* 133(2), 135–40.

Li, C.P. (1975) A new medical trend in China. *Am. J. Chin. Med.* 3(3), 213–21.

Lin, S.Y., Liu, L.M. and Wu, L.C. (1995) Effects of Shenmai injection on immune function in stomach cancer patients after chemotherapy *Chung Kuo Chung Hsi I Chieh Ho Tsa Chih* 15(8), 451–3.

Yamamoto, M., Kumagai, A. and Yamamura, Y. (1983) Plasma lipid-lowering and lipogenesis-stimulating actions of ginseng saponins in tumor-bearing rats. *Am. J. Chin. Med.* 11(1–4), 88–95.

Wakabayashi, C., Hasegawa, H., Murata, J. and Saiki, I. (1997) *In vivo* antimetastatic action of ginseng protopanaxadiol saponins is based on their intestinal bacterial metabolites after oral administration. *Oncol. Res.* 9(8), 411–17.

Wakabayashi, C., Murakami, K., Hasegawa, H., Murata, J. and Saiki, I. (1998) An intestinal bacterial metabolite of ginseng protopanaxadiol saponins has the ability to induce apoptosis in tumor cells. *Biochem. Biophys. Res. Commun.* 246(3), 725–30.

Xiaoguang, C., Hongyan, L., Xiaohong, L., Zhaodi, F., Yan, L., Lihua, T. and Rui, H. (1998) Cancer chemopreventive and therapeutic activities of red ginseng. *J Ethnopharmacol.* 60(1), 71–8.

Yun, T.K., Yun, Y.S. and Han, I.W. (1983) Anticarcinogenic effect of long-term oral administration of red ginseng on newborn mice exposed to various chemical carcinogens. *Cancer Detect. Prev.* 6(6), 515–25.

Salikhova, R.A., Umnova, N.V., Fomina, M.M. and Poroshenko, G.G. (1994) An antimutagenicity study of bioginseng. *Izv Akad Nauk Ser Biol* 1, 48–55.

***Phyllanthus niruri* (Phyllanthus) (Euphorbiaceae)** — Antitumor, Cytotoxic

Location: Northern Asia.

Appearance
Stem: 0.5 m high, erect, red.
Leaves: small, green, oblonged and feathered.
Flowers: greenish white.

Parts used: root, fruit, leaf.

Active ingredients: *Glycosides: phyllanthoside, phyllanthostatin* (*Phyllanthus acuminatus*).

Particular value: It is used as antitumor, anti-leukemic (*Phyllanthus acuminatus*), antiviral, cytotoxic, chemopreventive.

Indicative dosage and application: Against the growth of the murine P-388 lymphocytic leukemia cell line in a dose of 0.35 $\mu g\,ml^{-1}$.

Documented target cancers

- Treatment of acute and chronic hepatitis B and healthy carriers of HBV hepatocellular carcinoma (liver cancer) (*Phyllanthus urinaria, Phyllanthus amarus*) (Blumberg *et al.*, 1989, 1990).
- Dalton's lymphoma ascites (DLA) tumor (mice) (*Phyllanthus emblica*) (Suresh and Vasudevan, 1994).
- Phyllanthostatin inhibits the growth of the murine P-388 lymphocytic leukemia cell line (Pettit *et al.*, 1990).

Further details

Other medical activity

- The aqueous extract of *Phyllanthus amarus* contains some components that are able to inhibit *in vitro* HBsAg secretion on a dose-dependent manner. Various hepatoma cell lines, such as the Alexander cell line, human-derived cell line which has the property of secreting HBsAg in the supernatant (Jayaram and Thyagarajan, 1996).

- Extracts of *Phyllanthus amarus* have also been shown to inhibit the DNA polymerase of HBV and woodchuck hepatitis virus (WHV) *in vitro* (Blumberg *et al.*, 1989).

Related compounds

- *Phyllanthus niruri L.*: the MeOH extract of the dried leaf contains niruriside, a potent antiviral compound (Qian-Cutrone *et al.*, 1996).
- The roots of *Phyllanthus acuminatus* contain the glycosides, phyllanthoside (a major antineoplastic constituent) and phyllanthostatin which inhibits ($ED_{50} = 0.35$ $\mu g\ ml^{-1}$) the growth of the murine P-388 lymphocytic leukemia cell line. This species contains also didesacetylphyllanthostatin and descinnamoylphyllanthocindiol (Pettit *et al.*, 1990).
- Aqueous extracts of edible dried fruits of *Phyllanthus emblica* prevented the incidence of carcinogenesis in mice treated with nickel chloride. Ascorbic acid, a major constituent of the fruit, fed for seven consecutive days in equivalent concentration as that present in the fruit, however, could only alleviate the cytotoxic effects induced by low doses of nickel; at the higher doses it was ineffective. The greater efficacy of the fruit extract could be due to the interaction of its various natural components rather than to any single constituent (Dhir *et al.*, 1991).

Related species

- *Phyllanthus emblica* is an excellent source of vitamin C (ascorbate) and, when administered orally, has been found to enhance natural killer (NK) cell activity and antibody-dependent cellular cytotoxicity. Enhanced activity was highly significant on days 3, 5, 7 and 9 after tumor inoculation with respect to the untreated tumor bearing control. The following have been documented: (a) an absolute requirement for a functional NK cell or K-cell population in order that *P. emblica* can exert its effect on tumor-bearing animals, and (b) the antitumor activity of *P. emblica* is mediated primarily through the ability of the drug to augment natural cell-mediated cytotoxicity (Suresh and Vasudevan, 1994).

References

Blumberg, B.S., Millman, I., Venkateswaran, P.S. and Thyagarajan, S.P. (1990) Hepatitis B virus and primary hepatocellular carcinoma: treatment of HBV carriers with *Phyllanthus amarus. Vaccine* 8(Suppl.S), 86–92.

Blumberg, B.S., Millman, I., Venkateswaran, P.S. and Thyagarajan, S.P. (1989) Hepatitis B virus and hepatocellular carcinoma – treatment of HBV carriers with *Phyllanthus amarus. Cancer Detect Prev.* 14(2), 195–201.

Dhir, H., Agarwal, K., Sharma, A. and Talukder, G. (1991) Modifying role of *Phyllanthus emblica* and ascorbic acid against nickel clastogenicity in mice. *Cancer Lett.* 59(1), 9–18.

Jayaram, S. and Thyagarajan, S.P. (1996) Inhibition of HBsAg secretion from Alexander cell line by *Phyllanthus amarus. Indian J. Pathol. Microbiol.* 39(3), 211–15.

Ji, X.H., Qin, Y.Z., Wang, W.Y., Zhu, J.Y. and Liu, X.T. (1993) Effects of extracts from *Phyllanthus urinaria* L. on HBsAg production in PLC/PRF/5 cell line. *Chung Kuo Chung Yao Tsa Chih.* **18**(8), 496–8, 511.

Pettit, G.R., Schaufelberger, D.E., Nieman, R.A., Dufresne, C. and Saenz-Renauld, J.A. (1990) Antineoplastic agents, 177. Isolation and structure of phyllanthostatin 6. *J. Nat. Prod.* **53**(6), 1406–13.

Qian-Cutrone, J., Huang, S., Trimble, J., Li, H., Lin, P.F., Alam, M., Klohr, S.E. and Kadow, K.F. (1996) Niruriside, a new HIV REV/RRE binding inhibitor from *Phyllanthus niruri*. *J. Nat. Prod.* **59**(2), 196–9.

Suresh, K. and Vasudevan, D.M. (1994) Augmentation of murine natural killer cell and antibody dependent cellular cytotoxicity activities by *Phyllanthus emblica*, a new immunomodulator. *J. Ethnopharmacol.* 44(1), 55–60.

Yeh, S.F., Hong, C.Y., Huang, Y.L., Liu, T.Y., Choo, K.B. and Chou, C.K. (1993) Effect of an extract from *Phyllanthus amarus* on hepatitis B surface antigen gene expression in human hepatoma cells. *Antiviral Res.* 20(3), 185–92.

Plumeria sp. (Plumeria) (Apocynaceae) — Cytotoxic

Location: warm tropical areas of the Pacific Islands, Caribbean, South America and Mexico (Plumeria rubra: Indonesia, Thailand).

Appearance (Figure 3.21)
Stem: 10–12 m, widely spaced thick succulent branches.
Leaves: round or pointed, long leather, fleshy leaves in clusters near the branch tips.
Flowers: large, waxy, red, white, yellow, pink and multiple pastels, fragrant.
In bloom: early summer through the early fall months.

Tradition: Traditional medicinal plant of Thailand.

Part used: bark.

Figure 3.21 Plumeria.

Active ingredients

- *Petroleum-ether- and CHCl3-soluble extracts: (1) iridoids: fulvoplumierin, allamcin and allamandin, (2) 2,5-dimethoxy-p-benzoquinone.*
- H_2O-soluble extract: (1) *iridoids: plumericin, isoplumericin, (2) lignan: liriodendrin.*

Documented target cancers: murine lymphocytic leukemia (P-388) and a number of human cancer cell types (breast, colon, fibrosarcoma, lung, melanoma, KB).

Further details

Other medical activity

- The iridoids: plumericin, isoplumericin except their cytotoxic activity, they also have antibacterial activity

Related compounds

- Five additional iridoids, 15-demethylplumieride, plumieride, alpha-allamcidin], beta-allamcidin, and 13-*O*-trans-p-coumaroylplumieride, were obtained as inactive constituents (Hamburger *et al.*, 1991). Compound 15-demethylplumieride was found to be a novel natural product, and its structure was determined by spectroscopic methods and by conversion to plumieride (Kardono *et al.*, 1990).

References

Borchert, R. and Rivera, G. (2001) Photoperiodic control of seasonal development and dormancy in tropical stem-succulent trees. *Tree Physiol.* 21(4), 213–21.

Franca, O.O., Brown, R.T. and Santos, C.A. (2000) Uleine and demethoxyaspidospermine from the bark of *Plumeria lancifolia*. *Fitoterapia* 71(2), 208–10.

Guevara, A.P., Amor, E. and Russell, G. (1996) Antimutagens from *Plumeria acuminata* Ait. *Mutat. Res.* 12, 361, (2–3), 67–72.

Hamburger, M.O., Cordell, G.A. and Ruangrungsi, N. (1991) Traditional medicinal plants of Thailand. XVII. Biologically active constituents of *Plumeria rubra*. *J. Ethnopharmacol.* 33(3), 289–92.

Kardono, L.B., Tsauri, S., Padmawinata, K., Pezzuto, J.M. and Kinghorn, A.D. (1990) Cytotoxic constituents of the bark of *Plumeria rubra* collected in Indonesia. *J. Nat. Prod.* 53(6), 1447–55.

Muir, C.K. and Hoe, K.F. (1982) Pharmacological action of leaves of *Plumeria acuminata*. *Planta Med.* 44(1), 61–3.

Tan, G.T., Pezzuto, J.M., Kinghorn, A.D. and Hughes, S.H. (1991) Evaluation of natural products as inhibitors of human immunodeficiency virus type 1 (HIV-1) reverse transcriptase. *J. Nat. Prod.* 54(1), 143–54.

Radford, D.J., Gillies, A.D., Hinds, J.A. and Duffy, P. (1986) Naturally occurring cardiac glycosides. *Med. J. Aust.* 12, 144(10), 540–4.

Polanisia dodecandra L. (Polanisia) (Capparaceae) — Cytotoxic

Location: plants are found from Quebec and Maryland to southern Saskatchewan and Manitoba south to Arkansas and northern Mexico at elevations under 6,000 feet.

Appearance (Figure 3.22)
Stem: 0.3–1 m, simple, strong dark odor.
Leaves: 5 cm long and bear three leaflets about an inch long.
Flowers: 20 flowers are clustered at the top of the plant. 1 cm long, white with purple basis.
In bloom: May–October.

Active ingredients: *Flavonols*: 5,3′-dihydroxy-3,6,7,8,4′-pentamethoxyflavone [1], 5,4′-dihydroxy-3,6,7,8,3′-pentamethoxyflavone [2].

Documented target cancers: It is used in: cancer of the central nervous system (SF-268, SF-539, SNB-75, U-251), non-small cell lung cancer (HOP-62, NCI-H266, NCI-H460, NCI-H522), small-cell lung cancer (DMS-114), ovarian cancer (OVCAR-3, SK-OV-3), colon cancer (HCT-116), renal cancer (UO-31), melanoma cell line (SK-MEL-5), leukemia cell lines (HL-60 [TB], SR), medulloblastoma (TE-671) tumor cells.

Further details

Other medical activity

- 5,3′-dihydroxy-3,6,7,8,4′-pentamethoxyflavone inhibited tubulin polymerization ($IC_{50} = 0.83 \pm 0.2\,\mu M$) and the binding of radiolabeled colchicine to tubulin with 59% inhibition when present in equimolar concentrations with colchicine. It is the first example of a flavonol that exhibits potent inhibition of tubulin polymerization and, therefore, warrants further investigation as an antimitotic agent (Shi *et al.*, 1995).

Figure 3.22 Polanisia dodecandra.

References

Shi, Q., Chen, K., Li, L., Chang, J.J., Autry, C., Kozuka, M., Konoshima, T., Estes, J.R., Lin, C.M. and Hamel, E. (1995) Antitumor agents, 154. Cytotoxic and antimitotic flavonols from *Polanisia dodecandra*. *J. Nat. Prod.* 58(4), 475–82.

Wang, H.K., Xia, Y., Yang, Z.Y., Natschke, S.L. and Lee, K.H. (1998) Recent advances in the discovery and development of flavonoids and their analogues as antitumor and anti-HIV agents. *Adv. Exp. Med. Biol.* 439, 191–225.

Shi, Q., Chen, K., Fujioka, T., Kashiwada, Y., Chang, J.J., Kozuka, M., Estes, J.R., McPhail, A.T., McPhail, D.R. and Lee, K.H. (1992) Antitumor agents, 135. Structure and stereochemistry of polacandrin, a new cytotoxic triterpene from *Polanisia dodecandra*. *J. Nat. Prod.* 55(10), 1488–97.

Rabdosia rubescens **(Rabdosia) (Lamiaceae)** **Cytotoxic**

Location: West China.

Tradition: It is a traditional medicinal herb of China.

Active ingredients

- *Unsaturated lactone*: 10-epi-olguine (*Rabdosia ternifolia* (D. Don) Hara)
- *Oridonin* (*Rabdosia rubescens*) (cytotoxic + cisplatin, inducing DNA damage)
- *Diterpenoids*: enmein-, oridonin- and trichorabdal-type (*Rabdosia trichocarpa*)
- Antitumor constituent: *rabdophyllin G* (*Rabdosia macrophylla*).

Particular value: Important use in medicine for fighting cancer.

Precautions: Careful use because of its toxicity.

Documented target cancers: Human cancer cell lines, Ehrlich ascites carcinoma (mice), anti-leukemic.

Further details

Antitumor activity

- Trichorabdal-type diterpenoids showed the highest antitumor activity against Ehrlich ascites carcinoma in mice. *In vitro* activity against HeLa cells and *in vivo* activity against P-388 lymphocytic leukemia were also determined, but no synergistic increase in activity due to plural active sites was observed in those cases (Fuji *et al.*, 1989).
- From August 1974 to January 1987, 650 cases of moderate and advanced esophageal carcinoma were treated with a combination of chemotherapy and Rabdosia rubescens and/or traditional Chinese medicinal prescription. After treatment, 40 patients survived for over 5 years (5-year survival rate 6.15%): 32 for over 6 years, 23 for more than 10 years, 5 for more than 15 years and 20 died of tumors (16 cases) or other diseases (4 cases). There were 20 patients who lived or more than 18 years (Wang, 1993). Analyzing the data, it is believed that the age, the state of activity, the length

of illness, the effectiveness of primary treatment, the multi-course extensive therapy, long-term maintenance treatment, etc. are all important factors affecting the results of drug treatment.

- One hundred and fifteen patients with inoperable esophageal carcinoma were treated by either chemotherapy alone or chemotherapy plus *Rabdosia rubescens*. In group A, out of 31 patients treated with pingyangmycin (P) and nitrocaphane (N), 10 (32.3%) responded to the treatment. Among them, 2 showed partial response (greater than 50% tumor regression) and 8 minimal response (greater than 50% tumor regression). In group B, out of 84 patients treated with PN plus *Rabdosia rubescens*, 59 (70.2%) responded. Of them, 10 showed complete response (100% tumor regression), 16 partial response and 33 minimal response. The one-year survival rates of group A and B were 13.6% and 41.3%. Statistical significance was present in these two groups both in the response rate and one-year survival rate. As regards the drug toxicity, there was no significant difference between these two groups. Alopecia, anorexia, nausea and hyperpyrexia occurred in more than 30% of patients. Mild leukopenia and thrombocytopenia and interstitial pneumonia were noted in some patients, and two patients died of toxicity in the lungs (Wang, 1993; Wang *et al.*, 1986).

References

Cheng, P.Y., Xu, M.J., Lin, Y.L. and Shi, J.C. (1982) The structure of rabdophyllin G, an antitumor constituent of *Rabdosia macrophylla. Yao Hsueh Hsueh Pao.* 17(12), 917–21.

Fuji, K., Node, M., Sai, M., Fujita, E., Takeda, S. and Unemi, N. (1989) Terpenoids. LIII. Antitumor activity of trichorabdals and related compounds. *Chem. Pharm. Bull.* (Tokyo) 37(6), 1472–6.

Gao, Z.G., Ye, Q.X. and Zhang, T.M. (1993) See RSynergistic effect of oridonin and cisplatin on cytotoxicity and DNA cross-link against mouse sarcoma S180 cells in culture. *Chung Kuo Yao Li Hsueh Pao* 14(6), 561–4.

Lu, G.H., Wang, F.P., Pezzuto, J.M., Tam, T.C., Williams, I.D. and Che, C.T. (1997) 10-Epi-olguine from *Rabdosia ternifolia. J. Nat. Prod.* 60(4), 425–7.

Nagao, Y., Ito, N., Kohno, T., Kuroda, H. and Fujita, E. (1982) Antitumor activity of *Rabdosia* and *Teucrium diterpenoids* against P 388 lymphocytic leukemia in mice. *Chem. Pharm. Bull.* (Tokyo) 30(2), 727–9.

Wang, R.L. (1993) A report of 40 cases of esophageal carcinoma surviving for more than 5 years after treatment with drugs. *Chung Hua Chung Liu Tsa Chih.* 15(4), 300–2.

Wang, R.L., Gao, B.L., Xiong, M.L., Mei, Q.D., Fan, K.S., Zuo, Z.K., Lang, T.L., Gao, G.Q., Ji, Z.C. and Wie, D.C. (1986) Potentiation by *Rabdosia rubescens* on chemotherapy of advanced esophageal carcinoma. *Chung Hua Chung Liu Tsa Chih* 8(4), 297–9.

Rubia cordifolia L. (Rubia (Bengal madder)) (Rubiaceae) — Antitumor

Location: India

Appearance

Stem: 3 m high, stalks are very weak that they often lie along the ground preventing the plant from rising.

Root: main and side roots, the side roots run under the surface of the ground for some distance sending up shoots.

Leaves: have spines along the midrib on the underside.
Flowers: the flower-shoots spring from the joints in pairs, the loose spikes of yellow
In bloom: June.

Part used: root.

Active ingredients

- The root of *R. cordifolia:* naphthohydroquinones, naphthohydroquinone dimers, naphthohydroquinone, naphthoquinone, anthraquinones, naphthohydroquinone dimer, bicyclic hexapeptides: RA-XI, -XII, XIII, -XIV, -XV and -XVI (P388).
- *Rubia akane, R. cordifolia:* cyclic hexapeptide: RA-700.

Indicative dosage and application: RA-700 was given from 0.2 to 1.4 $mg\,m^{-2}$ in single i.v. dose study, from 0.4 to 2.0 $mg\,m^{-2}$ in 5-day i.v.

Documented target cancers: Various tumors *in vivo* and *in vitro* (such as: P388, L1210, L5178Y, B16 melanoma, Lewis lung carcinoma and sarcoma-180) (Brunet *et al.*, 1997; Larana *et al.*, 1997; Sanz *et al.*, 1998; Urbano-Ispizua *et al.*, 1997, 1998).

Further details

Other medical effects

- RA-700 has been tested in a phase I clinical study conducted by the RA-700 clinical study group consisting of 6 institutions. A single dose administration and 5-day schedule administration were evaluated with 14 patients respectively. RA-700 was given from 0.2 to 1.4 $mg\,m^{-2}$ in single i.v. dose study, from 0.4 to 2.0 $mg\,m^{-2}$ in 5-day i.v. schedule study. Nausea and vomiting, fever, stomachache, mild hypotension and slight abnormality of electric-cardiogram were observed as the toxicities. In a pharmacokinetic study, the elimination half-lives (t1/2) of RA-700 in plasma were 55 min, of alpha-phase and 3.9 h of beta-phase by single dose study, and 23–25 min of alpha-phase and 6–14 h of beta-phase by a 5-day schedule study. Accumulation was not found by 5-day schedule administration, and metabolite were not observed in plasma and urine. It seems that RA-700 is metabolized by the liver and excreted in the feces (Yoshida *et al.*, 1994). In conclusion, the maximum tolerated dose was 1.4 $mg\,m^{-2}$ for 5-day schedule administration.

Other medical activity

- Further studies have shown that: (1) changes in cardiac function were noted in both groups, (2) changes in blood pressure, sigma QRS, ejection fraction, and fractional shortening of the second group tended to be more extreme than those of the first group. Care for continuity is a concern with long-term and high doses of RA-700,

(3) because of the small sample, we could find no relationship between the changes in cardiac function and the injection doses of RA-700, (4) therefore, the cardiac function must be checked by giving anti-neoplastic drugs to neoplastic patients.

Antitumor activity

- The antitumor activity of RA-700 was evaluated in comparison with *deoxy-bouvardin* and *vincristine (VCR)*. As regards the proliferation of L1210 cultured cells, the cytotoxicity of RA-700 was similar to that of VCR but superior to that of deoxy-bouvardin (Yoshida *et al.*, 1994). The IC50 value of RA-700 was 0.05 mcg ml^{-1} under our experimental conditions. RA-700 inhibited the incorporation of ^{14}C-leucine at a concentration at which no effects were observed on the incorporation of 3H-thymidine and 3H-uridine in L1210 culture cells *in vitro*. The antitumor activity of RA-700 was similar to that of deoxy-bouvardin and VCR against P388 leukemia. Daily treatment with RA-700 at an optimal dose resulted in 118% ILS. As with deoxy-bouvardin and VCR, the therapeutic efficacy of RA-700 depends on the time schedule. RA-700 showed marginal activity against L1210 leukemia (50% ILS), similar to that of deoxy-bouvardin but inferior to that of VCR. RA-700 inhibited Lewis tumor growth in the early stage after tumor implantation, whereas deoxy-bouvardin and VCR did not. As regards toxicity, a slight reduction of peripheral WBC counts was observed with the drug, but no reduction of RBC and platelet counts. BUN, creatinine, GPT and GOT levels in plasma did not change with the administration of the drug.

Related compounds

- Another anticancer principle isolated from *Rubia cordifolia* is RC-18, which has been used against a spectrum of experimental murine tumors, namely P388, L1210, L5178Y, B16 melanoma, Lewis lung carcinoma and sarcoma-180. RC-18 exhibited significant increase in the life span of ascites leukaemia P388, L1210, L5178Y and a solid tumor B16 melanoma. However, it failed to show any inhibitory effect on solid tumors, Lewis lung carcinoma and sarcoma-180. Promising results against a spectrum of experimental tumors suggest that RC-18 may lead to the development of a potential anti-cancer agent (Brunet *et al.*, 1997; Larana *et al.*, 1997; Sanz *et al.*, 1998; Urbano-Ispizua *et al.*, 1997, 1998).

Related species

- Madder root, *Rubia tinctorum* L., is a traditional herbal medicine used against kidney stones. This species contains lucidin, a hydroxyanthraquinone derivative present in this plant and is mutagenic in bacteria and mammalian cells. In these respect, the use of madder root for medicinal purposes is associated with a carcinogenic risk (Westendorf *et al.*, 1998).

References

Alegre, A., Diaz-Mediavilla, J., San-Miguel, J., Martinez, R., Garcia Larana, J., Sureda, A., Lahuerta, J.J., Morales, D., Blade, J., Caballero, D., De la Rubia, J., Escudero, A., Diez-Martin, J.L., Hernandez-Navarro, F., Rifon, J., Odriozola, J., Brunet, S., De la Serna, J., Besalduch, J., Vidal, M.J., Solano, C., Leon, A., Sanchez, JJ., Martinez-Chamorro, C. and Fernandez-Ranada, J.M. (1998) Autologous peripheral blood stem cell transplantation for multiple myeloma: a report of 259 cases from the Spanish Registry. Spanish Registry for Transplant in MM (Grupo Espanol de Trasplante Hematopoyetico-GETH) and PETHEMA. *Bone Marrow Transplant.* 21(2), 133–40.

Brunet, S., Urbano-Ispizua, A., Solano, C., Ojeda, E., Caballero, D., Torrabadella, M., de la Rubia, J., Perez-Oteiza, J., Moraleda, J., Espigado, I., de la Serna, J., Petit, J., Bargay, J., Mataix, R., Vivancos, P., Figuera, A., Sierra, J., Domingo-Albos, A., Hernandez, F., Garcia Conde, J. and Rozman, C. (1997) Allogenic transplant of non-manipulated hematopoietic progenitor peripheral blood cells. Spanish experience of 79 cases. Allo-Peripheral Blood Transplantation Group. *Sangre (Barc)* 42(Suppl. 1), 42–3.

Garcia Larana, J., Diaz Mediavilla, J., Martinez, R., Lahuerta, J.J., Alegre, A., Odriozola, J., Sureda, A., San Miguel, J., De la Rubia, J., Escudero, A., Conde, E., Blade, J., Cabrera, R., Gastearena, J., Besalduch, J., Vidal, M.J., Hernandez, F., Rifon, J., Leon, A., Mataix, R., Parody, R., Moraleda, J.M., Solano, C., de Pablos, J.M. and Sanchez, J.J. (1997) Maintenance treatment with interferon-alfa in multiple myeloma after autotransplantation of peripheral blood progenitor cells. Spanish Register of Transplantation in Myeloma. *Sangre (Barc)* 42(Suppl. 1), 38–41.

Lopez, F., Jarque, I., Martin, G., Sanz, G.F., Palau, J., Martinez, J., de la Rubia, J., Larrea, L., Arnao, M., Solves, P., Cervera, J., Martinez, M.L., Peman, J., Gobernado, M. and Sanz, M.A. (1998) Invasive fungal infections in patients with blood disorders. *Med. Clin. (Barc)* 110(11), 401–5.

Lopez, A., de la Rubia, J., Arriaga, F., Jimenez, C., Sanz, G.F., Carpio, N. and Marty, M.L. (1998) Severe hemolytic anemia due to multiple red cell alloantibodies after an ABO-incompatible allogeneic bone marrow transplant. *Transfusion* 38(3), 247–51.

Sanz, M.A., de la Rubia, J., Bonanad, S., Barragan, E., Sempere, A., Martin, G., Martinez, J.A., Jimenez, C., Cervera, J., Bolufer, P. and Sanz, G.F. (1998) Prolonged molecular remission after PML/RAR alpha-positive autologous peripheral blood stem cell transplantation in acute promyelocytic leukemia: is relevant pretransplant minimal residual disease in the graft? *Leukemia* 12(6), 992–5.

Urbano-Ispizua, A., Solano, C., Brunet, S., de la Rubia, J., Odriozola, J., Zuazu, J., Figuera, A., Caballero, D., Martinez, C., Garcia, J., Sanz, G., Torrabadella, M., Alegre, A., Perez-Oteiza, J., Jurado, M., Oyonarte, S., Sierra, J., Garcia-Conde, J. and Rozman, C. (1998) Allogeneic transplantation of selected CD34+ cells from peripheral blood: experience of 62 cases using immunoadsorption or immunomagnetic technique. Spanish Group of Allo-PBT. *Bone Marrow Transplant.* 22(6), 519–25.

Urbano-Ispizua, A., Solano, C., Brunet, S., de la Rubia, J., Odriozola, J., Zuazu, J., Figuera, A., Caballero, D., Martinez, C., Garcia, J., Sanz, G., Torrabadella, M., Alegre, A., Perez-Oteyza, J., Sierra, J., Garcia-Conde, J. and Rozman, C. (1997) Allogeneic transplant of CD34+ peripheral blood cells from HLA-identical donors: Spanish experience of 40 cases. Allo-Peripheral Blood Transplantation Group. *Sangre (Barc)* 42(Suppl. 1), 44–5.

Westendorf, J., Pfau, W. and Schulte, A. (1998) Carcinogenicity and DNA adduct formation observed in ACI rats after long-term treatment with madder root, *Rubia tinctorum* L. *Carcinogenesis* 19(12), 2163–8.

Salvia sclarea (Salvia (Clarry)) (Lamiaceae) — Anti-leukemic

Location: Middle Europe.

Appearance (Figure 3.23)

Stem: It is a biennial plant with square brownish stems 0.5–1 m high, hairy, with few new branches.

Figure 3.23 Salvia sclarea.

Leaves: are arranged in pairs, almost stalkless, almost as large as the hand, heart-shaped and covered with velvety hairs.

Flowers: are interspersed with large colored, membraneous bracts, longer than the spiny calyx. Blue or white.

In bloom: Summer.

Tradition: This herb was first brought into use by the wine merchants of Germany and later was employed as a substitute for sophisticating beer, communicating considerable bitterness and intoxicating property. In ancient times and in the middle ages it was used for its curative properties.

Part used: seeds.

Active ingredients: A specific *lectin* from the seeds of *Salvia sclarea*.

Documented target cancers: Inhibitory activity against human erythroleukemic cell line K562, T leukemia cells Jurkat.

Further details

Related compounds

- From the seeds of *Salvia sclarea* (SSA) was isolated a lectin specific for GalNac-Ser/Thr studied in human erythroleukemic cell line K562. Another study proved that glycoproteins from the human T leukemia cells Jurkat were found to bind to the GalNac-Ser/Thr specific lectin from SSA. Studies show that this specific lectin has an inhibitory activity against the human erythroleukemic cell line K562 and T leukemia cells.

Other medical activity

- Some strong natural antioxidants like carnosol were proved to exhibit anti-inflammatory and inhibitory effects with regard to tumor-initiation activities in mice test systems. Also some sage compounds (ursolic and/or oleanolic acid) that show no antioxidant may turn promising in future research of inflammation and of cancer prevention. A squalene derived triterpenoid ursolic acid and its isomer oleanolic acid (up to 4% in sage leaves, dry weight basis) act anti-inflammatory and inhibit tumorigenesis in mouse skin. Recent data on the anti-inflammatory activity of sage (*S. officinalis* L.) extracts when applied topically ($ID_{50} = 2040\,\mu g\,cm^{-2}$) and evaluated as edema inhibition after Croton oil-induced dermatitis in mouse ear, confirm/suggest ursolic acid to be the main active ingredient, responsible for sage anti-inflammatory effect. The data on the pharmacological effects of these metabolites promise new therapeutic possibilities of sage extracts.

Anti-leukemic activity

- Ursolic acid showed significant cytotoxicity in lymphatic leukemia cells P-388 ($ED_{50} = 3.15\,\mu g\,ml^{-1}$) and L-1210 ($ED_{50} = 4.00\,\mu g\,ml^{-1}$) as well as human lung carcinoma cell A-549 ($ED_{50} = 4.00\,\mu g\,ml^{-1}$) (Lee *et al.*, 1987; Fang and McLaughlin, 1989). Both carnosol and ursolic acid are referred to as being strong inhibitors of 12-*O*-tetradecanoylphorbol-13-acetate (TPA)-induced ornithine decarboxylase activity and of TPA-induced tumor promotion in mouse skin. The tumorigenesis-prevention potential of ursolic acid was comparable to that of retinoic acid (RA) – a known inhibitor of tumor promotion. Both ursolic acid- and oleanolic acid- treatment (41 nmol of each), when applied continuously before each TPA-treatment (4.1 nmol), delayed the formation of papillomas in mouse skin, significantly reduced the rate of papilloma-bearing mice and reduced the number of papillomas per mouse, when compared with the control group (only TPA treatment). Ursolic acid acted more effectively in a single application before initial TPA-treatment when compared to the effect of RA and/or oleanolic acid. So, the mechanism of the inhibitory action of ursolic acid (inhibition of the first critical cellular event in tumor promotion step caused by TPA) may differ slightly from those of RA and/or oleanolic acid, which block a critical second stage process in tumor promotion by TPA (induction of ornithine decarboxylase and polyamine levels).

Antitumor activity

- A possible tumorigenesis preventing effect can be predicted for abietane diterpene galdosol, isolated from *S. canariensis* L., which showed significant cytostatic activity ($ID_{50} = 0.50\,\mu g\,ml^{-1}$) when inhibition of development of single-layer culture of HeLA 229 cells was measured in *in vitro* experiment.
- One of the most dangerous environmental sources of cytogenetic damage is ionizing radiation, which acts either directly or by secondary reactions and induces ionization in tissues. Interaction of ionizing radiation with water and other protoplasmatic constituents in oxidative metabolism causes formation of harmful oxygen radicals. DNA lesions, caused by reactive oxygen species in mammalian cells are the initial event

which may lead to possible mutagenesis and/or carcinogenesis and form the basis of spontaneous cancer incidence. Free radicals play an important role in preventing deleterious alterations in cellular DNA and genotoxic effects caused by ionizing radiation in mammalian tissues. Many drugs and chemicals (for example sulfhydryl compounds) are known to increase the survival rate in animals. Based on animal models studies, *S. miltiorrhiza* and its extracts were shown to have a potential to prevent X-radiation-induced pulmonary injuries and high dosage gamma-irradiation-induced platelet aggregation lesions.

- The antiproliferative activity of tanshinones against five human tumor cells, that is, A-549 (lung), SK-OV-3 (ovary), SK-MEL-2 (melanoma), XF-498 (central nerve system) and HCT-15 (colon), was evaluated by sulfrhodamine-B method. Eighteen isolated tanshinones exhibited significant but presumably nonspecific cytotoxicity against all tested tumor cells, which might be attributed to common naphtoquinone skeleton rather than to substituents attached to it. Methylenetanshiquinone and tanshindiol C exhibited most powerful cytotoxic effects against tested tumor cells, with IC_{50} ranging from $0.4\,\mu g\,ml^{-1}$ in A-549 cells to $2.2\,\mu g\,ml^{-1}$ in SK-MEL-2 cells and IC_{50} from $0.3\,\mu g\,ml^{-1}$ in SK-MEL-2 cells to $0.9\,\mu g\,ml^{-1}$ in SK-OV-3 cancer cell lines respectively.

Related species

- From *S. przewalskii* Maxim. var. *mandarinorum* Stib., a strong bacteriostatic compound, przewaquinone A was isolated (Yang *et al.*, 1981, 1984). Przewaquinone A was reported to possess potential for inhibiting Lewis lung carcinoma and melanoma B-16.

References

Darias, V., Bravo, L., Rabanal, R., Sαnchez-Mateo, C.C. and Martνn-Herrera, D.A. (1990) Cytostatic and antibacterial activity of some compounds isolated from several lamiaceae species from the Canary Islands. *Planta Med.* 56, 70–2.

Du, H., Quian, Z. and Wang, Z. (1990) Prevention of radiation injury of the lungs by *Salvia miltiorrhiza* in mice. *Chinese J. Mod. Develop. Trad. Med.* 10(4), 230–1.

Fang, X.P. and McLaughlin, J.L. (1989) Ursolic acid, a cytotoxic component of the berries of Ilex verticillata. *Fitoterapia* 61(1), 176–7.

Hanawalt, P.C. (1998) Genomic instability: environmental invasion and the enemies within. *Mutation Res.* 400(1–2), 117–25.

Huang, M.T., Ho, C.T., Wang, Z.X., Ferraro, T., Lou, Y.R., Stauber, K., Ma, W., Georgiadis, C., Laskin, J.D. and Conney, A.H. (1994) Inhibition of skin tumorigenesis by Rosemary and its constituents carnosol and ursolic acid. *Cancer Res.* 54, 701–8.

Lee, A.R., Chang, W.L., Lin, H.C. and King, M.L. (1987) Isolation and bioactivity of new tanshinones. *J. Nat. Prod.*, 50, 157–61.

Lutz, W.K. (1998) Dose–response relationships in chemical carcinogenesis: superposition of different mechanisms of action, resulting in linear–nonlinear curves, practical treshholds, J-shapes. *Mutation Res.* 405(2), 117–24.

Ryu, S.Y., Lee, C.O. and Choi, S.U. (1997) *In Vitro* cytotoxicity of tanshinones from *Salvia Miltiorrhiza. Planta Med.* 63, 339–42.

Tokuda, H., Onigashi, H. and Koshimizu, K. (1986) Inhibitory effects of ursolic and oleanolic acid on skin tumor promotion by 12-o-tetradecanoilphorbol-13-acetate. *Cancer Lett.* 33, 279–85.

Wang, H.F., Li, X.D., Chen, Y.M., Yuan, L.B. and Foye, W.O. (1991) Radiation-protective and platelet aggregation inhibitory effects of five traditional Chinese drugs and acetylsalicylic acid following high-dose-gamma-irradiation. *J. Ethnopharmacol.* 34(2–3), 215–19.

Xiao, P.G. (1989) Excerpts of the Chinese pharmacopoeia. In *Herbs, Spices and Medicinal Plants*. (Eds. L.E. Craker and J.E. Simon), *Recent advances in Botany, Horticulture and Pharmacology,* vol. 4, Oryx Press, Arizona, pp. 42–114.

***Sargassum bacciferum* (Sargassum) (Fucaceae) Antimetastatic**

Location in: North Atlantic Ocean.

Appearance (Figure 3.24)

Thallus: coarse, light yellow or brownish-green, erect, 0.5–1 m in height. Attaches itself to the rocks by branched, rootlike, woody extremities, developed from the base of the stalk. The front is almost fan shaped, narrow and trap shaped at the base, the rest is flat and leaf-like in form, wavy, many times divided into two, with erect divisions having a very strong, broad, compressed midrib running to the apex.

Parts used: dried mass of root, stem and leaves.

Active ingredients

- Aqueous extract: *Fucoidan polysaccharides.*
- Methanolic extract: *dihydroxysargaquinone.*

Documented target cancers: Antimetastatic (lung cancer, Ehrlich carcinoma) (mice); Anti-leukemic (dihydroxysargaquinone); Immunostimulatory; Cytotoxic (dihydroxysargaquinone).

Figure 3.24 Sargassum.

Further details

Related species

- *Sargassum thunbergii*, the brown seaweed umitoranoo contains neutral and acidic polysaccharides. Antitumor activity has been attributed to two fractions, GIV-A ($[\alpha]^{25}{}_D$ $-127°$ and mol. wt., 19,000) and GIV-B ($[\alpha]^{25}{}_D$ $-110°$ and mol. wt., 13,500) (Itoh *et al.*, 1993). These compounds were found to be a fucoidan or L-fucan containing approx. 30% sulfate ester groups per fucose residue, about 10% uronic acid, and less than 2% protein.
- *Sargassum fulvellum* contains a polysaccharide fraction (either a sulphated peptidoglycuronoglycan or a sulphated glycuronoglycan) with remarkable tumor-inhibiting effect against sarcoma-180 implanted subcutaneously in mice.
- *Sargassum tortile*: The CC14 partition fractions from methanolic extracts of this species contain dihydroxysargaquinone, which is cytotoxic against cultured P-388 lymphocytic leukemia cells (Numata *et al.*, 1991).
- *Sargassum kjellmanianum* is also effective in the *in vivo* growth inhibition of the implanted Sarcoma-180 cells (Yamamoto *et al.*, 1981).

Related compounds

- GIV-A markedly inhibited the growth of Ehrlich ascites carcinoma at the dose of 20 mg kg^{-1} X10 with no sign of toxicity in mice. It is acting as a so-called activator of the reticuloendothelial system. Fucoidan enhanced the phagocytosis and chemiluminescence of macrophages. By the immunofluorescent method, binding of the third component of complement (C3) cleavage product to macrophages and the proportion of C3 positive cells were increased. These results suggest that the antitumor activity of fucoidan is related to the enhancement of immune responses (Itoh *et al.*, 1995). The present results indicate that fucoidan may open new perspectives in cancer chemotherapy.

References

Amagata, T., Minoura, K. and Numata, A. (1998) Cytotoxic metabolites produced by a fungal strain from a *Sargassum alga. J. Antibiot.* (*Tokyo*) **51**(4), 432–4.

Iizima-Mizui, N., Fujihara, M., Himeno, J., Komiyama, K., Umezawa, I. and Nagumo, T. (1985) Antitumor activity of polysaccharide fractions from the brown seaweed *Sargassum kjellmanianum. Kitasato Arch. Exp. Med.* **58**(3), 59–71.

Itoh, H., Noda, H., Amano, H. and Ito, H. (1995) Immunological analysis of inhibition of lung metastases by fucoidan (GIV-A) prepared from brown seaweed *Sargassum thunbergii. Anticancer Res.* **15**(5B), 1937–47.

Itoh, H., Noda, H., Amano, H., Zhuaug, C., Mizuno, T. and Ito, H. (1993) Antitumor activity and immunological properties of marine algal polysaccharides, especially fucoidan, prepared from *Sargassum thunbergii* of Phaeophyceae. *Anticancer Res.* **13**(6A), 2045–52.

Numata, A., Kanbara, S., Takahashi, C., Fujiki, R., Yoneda, M., Fujita, E. and Nabeshima, Y. (1991) Cytotoxic activity of marine algae and a cytotoxic principle of the brown alga *Sargassum tortile. Chem. Pharm. Bull.* (Tokyo) **39**(8), 2129–31.

Yamamoto, I., Takahashi, M., Suzuki, T., Seino, H. and Mori, H. (1984) Antitumor effect of seaweeds. IV. Enhancement of antitumor activity by sulfation of a crude fucoidan fraction from *Sargassum kjellmanianum*. *Jpn. J Exp. Med.* 54(4), 143–51.

Yamamoto, I., Nagumo, T., Takahashi, M., Fujihara, M., Suzuki, Y., Iizima, N. (1981) Antitumor effect of seaweeds. III. Antitumor effect of an extract from *Sargassum kjellmanianum*. *Jpn. J. Exp. Med.* 51(3), 187–9.

Zhuang, C., Itoh, H., Mizuno, T. and Ito, H. (1995) Antitumor active fucoidan from the brown seaweed, umitoranoo (*Sargassum thunbergii*). *Biosci. Biotechnol. Biochem.* 59(4), 563–7.

Scutellaria baicalensis **Georgii (Scutellaria (Scullcap)) (Labiatae)** — **Antitumor**

Location: USA, Great Britain.

Appearance

Stem: square, 15–45 cm high, somewhat slender, either paniculately branched or in small specimens.

Root: perennial and creeping root-stock.

Leaves: opposite downy leaves, oblong and tapering, heart-shaped at the base, 1–5 cm long, notched and short petioles.

Flowers: in pairs, each growing from the axils of the upper, leaf-like bracts, bright blue with white inside.

In bloom: July–September.

Part used: The whole herb.

Active ingredients

- Flavonoids: *baicalin, baicalein and wogonin.*
- *Flavones: 5,7,2′-trihydroxy- and 5,7,2′,3′-tetrahydroxyflavone.*

Documented target cancers: Hepatoma cell lines, Pliss' lymphosarcoma, Epstein–Barr virus, skin cancer (mice).

Further details

Related compounds

- *Scutellaria baicalensis* Georgi (methanol extract) contains the flavonoids *baicalin, baicalein* and *wogonin* which induce the quinone reductase in the Hepa 1c1c7 murine hepatoma cell line (Park *et al.*, 1998). Baicalin may be the major active principle of QR induction mediated by scutellaria radix extract. In addition, the flavones *5,7,2′-trihydroxy-* and *5,7,2′,3′-tetrahydroxyflavone* exhibit remarkable inhibitory effects on mouse skin tumor promotion in an *in vivo* two-stage carcinogenesis test and on the Epstein–Barr virus early antigen activation.
- Isolation of E-1-(4′-Hydroxyphenyl)-but-1-en-3-one from *Scutellaria barbata* (Ducki *et al.*, 1996).

- Ten known glycosidic compounds, *betulalbuside A* (1), *8-hydroxylinaloyl,3-0-beta-D-glucopyranoside* (2) *(monoterpen glycosides), ipolamide* (3) *(iridoid glycoside), acteoside (verbascoside)* (4), *leucosceptoside A* (5), *martynoside* (6), *forsythoside B* (7), *phlinoside B* (8), *phlinoside C* (9), *and teuerioside* (10) *(phenylpropanoid glycosides)* were isolated from methanolic extracts of *Phlomis armeniaca* and *Scutellaria salviifolia (Labiatae)* (Yamashiki *et al.*, 1997). Structure elucidations were carried out using 1H-, 13C-NMR and FAB-MS spectra, as well as chemical evidence. The cytotoxic and cytostatic activities of isolated compounds were investigated by the 3-[4,5-dimethylthiazol-2-yl]-2,5-diphenyltetrazolium bromide (MTT) method. Among the glycosides obtained here, caffeic acid-containing phenylpropanoid (or phenethyl alcohol, or phenylethanoid) glycosides were found to show activity against several kinds of cancer cells. However, they didn't affect the growth and viability of primary-cultured rat hepatocytes. Study of the structure–activity relationship indicated that ortho-dihydroxy aromatic systems of phenylpropanoid glycosides are necessary for their cytotoxic and cytostatic activities.

Antitumor activity

- The advancement of Pliss' lymphosarcoma in rats was shown to be associated with disorders of platelet-mediated hemostasis presenting with either lowered or increased aggregation activity of platelets. In the latter case, a direct correlation was observed between functional activity of thrombocytes, on the one hand, and degree of tumor advancement and its metastatic activity, on the other. The extract of *Scutellaria baicalensis Georgi* was shown to produce a normalizing effect on platelet-mediated hemostasis whatever the pattern of alteration that points to the adaptogenic activity of the drug (Gol'dberg *et al.*, 1997). This activity is thought to be responsible for the drug's antitumor and, particularly, metastasis-preventing effect.
- In experiments with murine and rat transplantable tumors, *Scutellaria baicalensis Georgii* extract treatment was shown to ameliorate cyclophosphamide and 5-fluorouracil-induced myelotoxicity and to decrease tumor cell viability. This was partly attributed to a pronounced antistressor action of the extract and its normalizing effect on some homeostatic parameters.
- As a supplement to conventional chemotherapy: cytostatic therapy of patients with lung cancer is attended with decrease in the relative number of T-lymphocytes and their theophylline-resistant population. Patients who were given *Scutellaria barbata* (SB) showed a tendency towards increase of these parameters during antitumor chemotherapy. The immunoregulation index (IRI) in this case was approximately twice the background values during the whole period of investigation. The inclusion of SB in the therapeutic complex promotes increase in the number of immunoglobulins A at a stable level of *immunoglobulin G* (Smol'ianinov *et al.*, 1997).

Other medical activity

- Glial cells have a role in maintaining the function of neural cells. A study was undertaken to clarify the effects of baicalin and baicalein, flavonoids isolated from an important medicinal plant *Scutellariae Radix* (the root of *Scutellaria baicalensis Georgi*), on glial cell function using C6 rat glioma cells (Kyo *et al.*, 1998). *Baicalin* and *baicalein* caused concentration-dependent inhibition of a histamine-induced increase in intracellular Ca^{2+} concentrations ($[Ca^{2+}]_i$). The potency of baicalein was significantly greater than that of baicalin. The noradrenaline- and carbachol-induced increase in $[Ca^{2+}]_i$ was also inhibited by baicalein and both drugs inhibited histamine-induced accumulation of total [3H]inositol phosphates, consistent with their inhibition of the increase in $[Ca^{2+}]_i$. These results suggest that baicalin and baicalein inhibit $[Ca^{2+}]_i$ elevation by reducing phospholipase C activity. The inhibitory effects of baicalin and baicalein on $[Ca^{2+}]_i$ elevation might be important in the interpretation of their pharmacological action on glial cells, such as inhibition of Ca^{2+}-required enzyme phospholipase A2.

- Hemopoiesis was studied in 88 patients with lung cancer during antitumor chemotherapy and its combination with a dry SB extract. Administration of the plant preparation was accompanied with hemopoiesis stimulation, intensification of bone-marrow erythro- and granulocytopoiesis and increase in the content of circulating precursors of the type of erythroid and granulomonocytic colony-forming units.

Related species

- *Oldenlandia diffusa* (OD) and *Scutellaria barbata* (SB) have been used in traditional Chinese medicine for treating liver, lung and rectal tumors while *Astragalus membranaceus* (AM) and *Ligustrum lucidum* (LL) are often used as adjuncts in cancer therapy. The effects of aqueous extracts of these four herbs on aflatoxin B1 (AFB1)-induced mutagenesis were investigated using *Salmonella typhimurium* TA100 as the bacterial tester strain and rat liver 9000 x*g* supernatant as the activation system. The effects of these herbs on [3H]AFB1 binding to calf-thymus DNA were assessed. Organosoluble and water-soluble metabolites of AFB1 were extracted and analyzed by high-performance liquid chromatography (HPLC). Mutagenesis assays revealed that all of these herbs produced a concentration-dependent inhibition of histidine-independent revertant (His+) colonies induced by AFB1. At a concentration of 1.5 mg per plate, SB and OD in combination exhibited an additive effect. The trend of inhibition of these four herbs on AFB1-induced mutagenesis was: SB greater than LL greater than AM. LL, OD and SB significantly inhibited AFB1 binding to DNA, reduced AFB1–DNA adduct formation, and also significantly decreased the formation of organosoluble metabolites of AFB1. This data suggest that these Chinese medicinal herbs possess cancer chemopreventive properties (Yamashiki *et al.*, 1997).

References

Ducki, S., Hadfield, J.A., Lawrence, N.J., Liu, C.Y., McGown, A.T. and Zhang, X. (1996) Isolation of E-1-(4′-Hydroxyphenyl)-but-1-en-3-one from *Scutellaria barbata. Planta Med.* 62(2), 185–6.

Gol'dberg, V.E., Ryzhakov, V.M., Matiash, M.G., Stepovaia, E.A., Boldyshev, D.A., Litvinenko, V.I. and Dygai, A.M. (1997) Dry extract of *Scutellaria baicalensis* as a hemostimulant in antineoplastic chemotherapy in patents with lung cancer. *Eksp. Klin. Farmakol.* 60(6), 28–30.

Kyo, R., Nakahata, N., Sakakibara, I., Kubo, M. and Ohizumi. Y. (1998) Effects of Sho-saiko-to, San'o-shashin-to and Scutellariae Radix on intracellular Ca2+ mobilization in C6 rat glioma cells. *Biol. Pharm. Bull.* 21(10), 1067–71.

Kyo, R., Nakahata, N., Sakakibara, I., Kubo, M. and Ohizumi, Y. (1998) Baicalin and baicalein, constituents of an important medicinal plant, inhibit intracellular Ca2+ elevation by reducing phospholipase C activity in C6 rat glioma cells. *J. Pharm. Pharmacol.* 50(10), 1179–82.

Park, H.J., Lee, Y.W., Park, H.H., Lee, Y.S., Kwon, I.B. and Yu, J.H. (1998) Induction of quinone reductase by a methanol extract of *Scutellaria baicalensis* and its flavonoids in murine Hepa 1c1c7 cells. *Eur. J. Cancer Prev.* 7(6), 465–71.

Razina, T.G., Zueva, E.P., Litvinenko, V.I. and Kovalev, I.P. (1998) A semisynthetic flavonoid from the Baikal skullcap (*Scutellaria baicalensis*) as an agent to enhance the efficacy of chemotherapy in experimental tumors. *Eksp. Klin. Farmakol.* 61(2), 54–6.

Saracoglu, I., Inoue, M., Calis, I. and Ogihara, Y. (1995) Studies on constituents with cytotoxic and cytostatic activity of two Turkish medicinal plants *Phlomis armeniaca* and *Scutellaria salviifolia. Biol. Pharm. Bull.* 18(10), 1396–400.

Smol'ianinov, E.S., Gol'dberg, V.E., Matiash, M.G., Ryzhakov, V.M., Boldyshev, D.A., Litvinenko, V.I. and Dygai, A.M. (1997) Effect of *Scutellaria baicalensis* extract on the immunologic status of patients with lung cancer receiving antineoplastic chemotherapy. *Eksp. Klin. Farmakol.* 60(6), 49–51.

Yamashiki, M., Nishimura, A., Suzuki, H., Sakaguchi, S. and Kosaka, Y. (1997) Effects of the Japanese herbal medicine "Sho-saiko-to" (TJ-9) *on in vitro* interleukin-10 production by peripheral blood mononuclear cells of patients with chronic hepatitis C. *Hepatology* 25(6), 1390–7.

***Stellera chamaejasme* (Stellera) (Thymelaceae) Cytotoxic**

Location: China.

Active ingredients: *Diterpene: gnidimacrin.*

Indicative dosage and application: Gnidimacrin has been used at the dosages of 0.02–0.03 mg kg^{-1}. intraperitoneally against mouse leukemia P-388 and L-1210 *in vivo* and showed significant antitumor activities.

Documented target cancers: Human leukemias, stomach cancers and non-small cell lung cancers *in vitro*.

Further details

Related species

- *Stellera chamaejasme* L.: The root (methanolic extract) contains the daphnane-type diterpene gnidimacrin. Gnidimacrin acts as a protein kinase C activator for tumor cells.

Antitumor activity

- *Gnidimacrin* was found to strongly inhibit cell growth of human leukemias, stomach cancers and non-small-cell lung cancers *in vitro* at concentrations of 10^{-9} to 10^{-10} M (Feng *et al.*, 1995). On the other hand, even at 10^{-6} to 10^{-5} M, the small-cell lung cancer cell line H69 and the hepatoma cell line HLE were refractory to gnidimacrin. The agent showed significant antitumor activity against murine leukemias and solid tumors in an *in vivo* system. In K562, a sensitive human leukemia cell line, gnidimacrin induced blebbing of the cell surface, which was completely inhibited by staurosporine at concentrations above 10^{-8} M, and arrested the cell cycle transiently to G2 and finally the G1 phase at growth-inhibitory concentrations. It inhibited phorbol-12,13-dibutyrate(PDBu) binding to K562 cells and directly stimulated protein kinase C (PKC) activity in the cells in a dose-dependent manner (3–100 nM). Although activation of PKC isolated from refractory H69 cells was observed only with 100 nM gnidimacrin, the degree of activation was lower than that produced by 3 nM in K562 cells.

- *Gnidimacrin* showed significant antitumor activities against mouse leukemia P-388 and L-1210 *in vivo* (Yoshida *et al.*, 1996). At the dosages of 0.02–0.03 mg kg^{-1} i.p., the increase in life span (ILS) was 70% and 80%, respectively. *Gnidimacrin* was also active against murine solid tumors *in vivo*, such as Lewis lung carcinoma, B-16 melanoma and colon cancer 26. It showed ILSs of 40%, 49% and 41% at the dosages of 0.01–0.02 mg kg^{-1} i.p., respectively. Gnidimacrin strongly inhibited cell proliferation of human cancer cell lines such as leukemia K562, stomach cancers Kato-III, MKN-28, MKN-45, and mouse L-1210 by the MTT assay and colony forming assay *in vitro*. The IC50 of gnidimacrin was 0.007–0.00012 $\mu g\,ml^{-1}$.
- Inhibitory effects of *Stellera chamaejasme* on the growth of a transplantable tumor in mice (Yang, 1986).

References

Feng, W., Tetsuro, I. and Mitsuzi, Y. (1995) The antitumor activities of gnidimacrin isolated from *Stellera chamaejasme* L. *Chung Hua Chung Liu Tsa Chih* 17(1), 24–6.

Yang, B.Y. (1986) Inhibitory effects of Stellera chamaejasme on the growth of a transplantable tumor in mice. *Chung Yao Tung Pao.* 11(1), 58–9.

Yoshida, M., Feng, W., Saijo, N. and Ikekawa, T. (1996) Antitumor activity of daphnane-type diterpene gnidimacrin isolated from *Stellera chamaejasme* L. *Int. J. Cancer* 66(2), 268–73.

***Trifolium pratense* L. (Clover, Red) (Leguminosae) — Chemopreventive**

Synonyms: Trefoil, purple clover.

Location: It can be found throughout Europe, central and northern Asia from the Mediterranean to the Arctic Circle and high up in the mountains.

Appearance

Stem: several stems 0.3–0.6 m high.

Root: one root, slightly hairy.

Leaves: ternate, leaflets ovate, nearly smooth.

Tradition: Fomentations and poultices of the herb have been used as local treatment.

Parts used: leaves, flowers.

Active ingredients: isoflavone biochanin A.

Particular value: The fluid extract is used as an alterative and antispasmodic.

Documented target cancers: The ability of the isoflavone biochanin A to inhibit carcinogen activation in cells in culture suggests that *in vivo* studies of this compound as a potential chemopreventive agent are warranted (Cassady *et al.*, 1988).

Further details

Chemopreventive activity

- Based on the epidemiological evidence for a relationship between consumption of certain foods and decreased cancer incidence in humans, an assay was developed to screen and fractionate plant extracts for chemopreventive potential. This assay measures effects on the metabolism of [3H]benzo(a)pyrene [B(a)P] in hamster embryo cell cultures. Screening of several plant extracts has generated a number of activity leads. The 95% ethyl alcohol extract of one of these actives, *Trifolium pratense* L. Leguminosae, red clover, significantly inhibited the metabolism of B(a)P and decreased the level of binding of B(a)P to DNA by 30–40%. Using activity-directed fractionation by solvent partitioning and then silica gel chromatography, a major active compound was isolated and identified as the isoflavone, biochanin A. The pure compound decreased the metabolism of B(a)P by 54% in comparison to control cultures and decreased B(a)P-DNA binding by 37–50% at a dose of $25\,\mu g\,ml^{-1}$. These studies demonstrate that the hydrocarbon metabolism assay can detect and guide the fractionation of potential anticarcinogens from plants (Cassady *et al.*, 1988).

Related compounds

- The tannins, *delphinidin* and *procyanidin* were isolated from flowers of white clover (*Trifolium repens*) and the leaves of Arnot Bristly Locust (*Robina fertilis*) respectively, and tested for mutagenic properties in a range of systems. There was no evidence for either compound causing significant levels of frameshift or base-pair mutagenesis in bacterial mutagenicity assays, although both were weakly positive in a bacterial DNA-repair test. Both compounds very slightly increased the frequency of petite mutagenesis in Saccharomyces cerevisiae strain D5. In V79 Chinese hamster cells, both were efficient inducers of micronuclei. In each of these test systems, increasing the potential of the compound for metabolic activation by addition of "S9" mix had little effect on toxicity or mutagenicity of either tannin. It would seem that potential chromosome-breaking activity of condensed tannins could represent a carcinogenic hazard for animals grazing on pastures of white clover in flower. It may also have wider implications for human carcinogenesis by some, if not all, condensed tannins (Ferguson *et al.*, 1985).

References

Cassady, J.M., Zennie, T.M., Chae, Y.H., Ferin, M.A., Portuondo, N.E. and Baird, W.M. (1988) Use of a mammalian cell culture benzo(a)pyrene metabolism assay for the detection of potential anticarcinogens from natural products: inhibition of metabolism by biochanin A, an isoflavone from *Trifolium pratense* L. *Cancer Res.* 48(22), 6257–61.

Ferguson, L.R., van Zijl, P., Holloway, W.D. and Jones, W.T. (1985) Condensed tannins induce micronuclei in cultured V79 Chinese hamster cells. *Mutat. Res.* 158(1–2), 89–95.

Liu, J., Burdette, J.E., Xy, H., Gu, C.,Van Breemen, R.B., Bhat, K.P., Booth, N., Constantinou, A.I., Pezzuto, J.M., Fong, H.H., Farnsworth, N.R. and Bolton, J.L. (2001) Evaluation of estrogenic activity of plant extracts for the potential treatment of menopausal symptoms. *J. Agric. Food Chem.* 49(5), 2472–9.

Moyad, M.A. (2002) Complementary/alternative therapies for reducing hot flashes in prostate cancer patients: reevaluating the existing indirect data from studies of breast cancer and postmenopausal women. *Urology* 59(4 Suppl. 1), 20–33. Review.

Viola odorata (Violet sweet) (Violaceae) — Cytotoxic

Other names: Sweet-scented violet.

Location: It is found in tropical and temperate regions of the world, in deciduous woods and hedges.

Appearance
Stem: slightly hairy, up to 10 cm high.
Root: stolon, up to 20 cm long.
Leaves: Rounded, sagittate to heart-shaped, slightly hairy, alternate, up to 6 cm long. The two halves of the young leaves are rolled in two coils.
Flowers: Deep purple (occasionally white or pink), fragrant, with yellow stamens, 0.5–1.5 cm.
Fruit: 3-valved capsule.
In bloom: February–April. Flowers produced in autumn are very small, with no apparent flower-like structure and not fragrant (*cleistogamous*) but are highly seed-setting.

Biology: A perennial plant, violet is propagated either by seed or cuttings (scions). The flowers are great attractors of bees and other insects, due to their high honey content. It is recommended to avoid cultivation near air-polluted areas, because the hairy parts can become accumulating points for smog.

Tradition: The species is supposed to have derived its name from *Viola*, the Latin form of the Greek name *Ione* or *Io*, who was turned into a plant by her beloved Jupiter, the flowers emerging right above the earth so that she could use them as food. Another Greek myth claims that the violet emerged on the spot where a resting Orpheus laid his lyre. Homer and Virgil have mentioned the calming and sedative properties of the plant. It was exactly the same properties that made the species be associated with death, as referred to by Shakespeare in Hamlet.

Parts used: whole plant fresh, flowers and leaves dried, rhizomes.

Active ingredients: *Cyclopentenyl cytosine.*

Particular value: Violet flowers possess slightly laxative properties, well known in the form of syrup. It is also used in ague, epilepsy inflammatation of the eyes, sleeplessness.

Precautions: rhizomes are strongly emetic and purgative.

Indicative dosage and application: It has not yet been standardized as a dose, for example on human glioblastoma cells the levels of the drug range from 0.01 to 1 μM.

Documented target cancers: *Cyclopentenyl cytosine* (*CPEC*) exerts an antiproliferative effect against a wide variety of human and murine tumor lines.

Further details

Antitumor activity

- *CPEC* inhibits the proliferation of tumor cell lines, including a panel of human gliosarcoma and astrocytoma lines (Agdaria *et al.*, 1997). This effect is produced primarily by the 5′-triphosphate metabolite CPEC-TP, an inhibitor of cytidine-5′-triphosphate (CTP) synthase (EC 6.3.4.2). This has been demonstrated, for example, on human glioblastoma cells obtained at surgery and exposed to the drug at levels ranging from 0.01 to 1 μM for 24 h. Dose-dependent accumulation of CPEC-TP was accompanied by a concomitant decrease in CTP pools, with 50% depletion of the latter being achieved at a CPEC level of *c*.0.1 μM. Human glioma cell proliferation was inhibited 50% by 24-h exposure to 0.07 μM CPEC. Post-exposure decay of CPEC-TP was slow, with a half time of 30 h. DNA cytometry showed a dose-dependent shift in cell cycle distribution, with an accumulation of cells in S-phase (Agdaria *et al.*, 1997). The pharmacological effects of CPEC on freshly excised glioblastoma cells are quantitatively similar to those seen in a range of established tissue culture lines, including human glioma, colon carcinoma, and MOLT-4 lymphoblasts, supporting the recommendation that the drug may be advantageous for the treatment of human glioblastoma.

References

Agbaria, R., Kelley, J.A., Jackman, J., Viola, J.J., Ram, Z. and Oldfield, E.J. (1997) Antiproliferative effects of cyclopentenyl cytosine (NSC 75575) in human glioblastoma cells. *DG Oncoles* 9(3), 111–8.

Crucitti, F., Doglietto, G., Frontera, D., Viola, G. and Buononato, M. (1995) Carcinoma of the pancreatic head area. Therapy: resectability and surgical management of resectable tumors. *Rays* 20(3), 304–15. Review.

De Berardis, B., Torresini, G., Viola, V., Imondi, G., Marinelli, S. and Di Pietrantonio, F. (2000) Recurrent giant retroperitoneal leiomyosarcoma. Report of a clinical case. *G. Chir.* 21(5), 239–41.

Fazio, V., Messina, V., Marino, A., Di Trapani, F. and Viola, V. (2002) Treatment with self-expanding metallic enteral stents in occlusion caused by neoplastic stenosis of the sigmoid and rectum. *Chir Ital.* 54(2), 233–9.

Ferrara, F., Annunziata, M., Schiavone, E.M., Copia, C., De Simone, M., Pollio, F., Palmieri, S., Viola, A., Russo, C. and Mele, G. (2001) High-dose idarubicin and busulphan as conditioning for autologous stem cell transplantation in acute myeloid leukemia: a feasibility study. *Hematol. J.* 2(4), 214–9.

Ferrara, F., Palmieri, S., Pocali, B., Pollio, F., Viola, A., Annunziata, S., Sebastio, L., Schiavone, E.M., Mele, G., Gianfaldoni, G. and Leoni, F. (2002) De novo acute myeloid leukemia with multilineage dysplasia: treatment results and prognostic evaluation from a series of 44 patients treated with fludarabine, cytarabine and G-CSF (FLAG). *Eur. J. Haematol.* 68(4), 203–9.

Frontera, D., Doglietto, G., Viola, G. and Crucitti, F. (1995) Carcinoma of the pancreatic head area. Epidemiology, natural history and clinical findings. *Rays* 20(3), 226–36. Review.

Krygier, G., Lombardo, K., Vargas, C., Alvez, I., Costa, R., Ros, M., Echenique, M., Navarro, V., Delgado, L., Viola, A. and Muse, A. (2001) Familial uveal melanoma: report on three sibling cases. *Br. J. Ophthalmol.* 85(8), 1007–8.

Morini, A., Manera, V., Boninsegna, C., Viola, L. and Orrico, D. (2000) Mandibular drop resulting from bilateral metastatic trigeminal neuropathy as the presenting symptom of lung cancer. *J. Neurol.* 247(8), 647–9.

Longo, V.D., Viola, K.L., Klein, W.L. and Finch, C.E. (2000) Reversible inactivation of superoxide-sensitive aconitase in Abeta1–42-treated neuronal cell lines. *J. Neurochem.* 75(5), 1977–85.

Madajewicz, S., Hentschel, P., Burns, P., Caruso, R., Fiore, J., Fried, M., Malhotra, H., Ostrow, S., Sugarman, S. and Viola, M. (2000) Phase I chemotherapy study of biochemical modulation of folinic acid and fluorouracil by gemcitabine in patients with solid tumor malignancies. *J. Clin. Oncol.* 18(20), 3553–7.

Morini, A., Manera, V., Boninsegna, C., Viola, L. and Orrico, D. (2000) Mandibular drop resulting from bilateral metastatic trigeminal neuropathy as the presenting symptom of lung cancer. *J. Neurol.* 247(8), 647–9.

Palmieri, S., Sebastio, L., Mele, G., Annunziata, M., Annunziata, S., Copia, C., Viola, A., De Simone, M., Pocali, B., Schiavone, E.M. and Ferrara, F. (2002) High-dose cytarabine as consolidation treatment for patients with acute myeloid leukemia with t(8;21). *Leuk. Res.* 26(6), 539–43.

Villani, F., Viola, G., Vismara, C., Laffranchi, A., Di Russo, A., Viviani, S. and Bonfante, V. (2002) Lung function and serum concentrations of different cytokines in patients submitted to radiotherapy and intermediate/high dose chemotherapy for Hodgkin's disease. *Anticancer Res.* 22(4), 2403–8.

Wikstroemia indica **(Wikstroemia) (Thymelaeaceae)** — **Anti-leukemic**

Location: Guam and Micronesia.

Appearance (Figure 3.25)
Stem: shrub with smooth, reddish bark.
Leaves: opposite, light green that is rounded at both ends.
Flowers: small, yellowish green, grow in racemes from the leaf axils.

Figure 3.25 Wikstroemia indica.

Active ingredients: *Daphnoretin, tricin, kaempferol-3-O-β-D-glucopyranoside,* and (+)-*nortrachelogenin, wikstroelides.*

Documented target cancers: It is used against Ehrlich ascites carcinoma (mice) (daphnoretin), as anti-leukemic (tricin, kaempferol-3-*O*-β-D-glucopyranoside, and nortrachelogenin), and against P-388 lymphocytic leukemia.

Further details

Related compounds

- *Wikstroemia indica* (Thymelaeaceae): The bark contains kaempferol-3-*O*-β-D-glucopyranoside, huratoxin, pimelea factor P2, wikstroelides A-G, daphnane-type diterpenoids (wikstroelides H-O), tricin, kaempferol-3-O-β-D-glucopyranoside, (+)-nortrachelogenin, daphnoretin, tricin, kaempferol-3-O-β-D-glucopyranoside, and (+)-nortrachelogenin (Wang *et al.*, 1998).
- The ethanol extracts of *Wikstroemia foetida* var. oahuensis and *Wikstroemia uvaursi* showed antitumor activity against the P-388 lymphocytic leukemia (3PS) test system. One PS-active constituent of both plants was the lignan wikstromol (Torrance, 1979).

References

Abe, F., Iwase, Y., Yamauchi, T., Kinjo, K., Yaga, S., Ishii, M. and Iwahana, M. (1998) Minor daphnane-type diterpenoids from *Wikstroemia retusa. Phytochemistry* 47(5), 833–7.

Torrance, S.J., Hoffmann, J.J. and Cole, J.R. (1979) Wikstromol, antitumor lignan from *Wikstroemia foetida* var. oahuensis Gray and *Wikstroemia uvaursi* Gray *Thymelaeaceae. J. Pharm. Sci.* 68(5), 664–5.

Wang, H.K., Xia, Y., Yang, Z.Y., Natschke, S.L. and Lee, K.H. (1998) Recent advances in the discovery and development of flavonoids and their analogues as antitumor and anti-HIV agents. *Adv. Exp. Med. Biol.* 439, 191–225.

3.2.3. The fable: where tradition fails to meet reality

Aconitum napellus L. (Aconite) (Ranunculaceae) — **Poisonous**

Location: It is found in lower mountain slopes of north portion of Eastern Hemisphere, from Himalayas through Europe to Great Britain.

Appearance (Figure 3.26)
Stem: 3 ft. high.
Root: fleshy, spindle-shaped, pale-colored when young, dark brown skin when mature.
Leaves: dark, green glossy, deeply divided in palmate manner.
Flowers: in erect clusters of a dark blue or white color.
In bloom: late spring–early summer.

Figure 3.26 Aconitumfischeri.

Tradition: One of the most useful drugs. It was used for many years as an anodyne, diuretic and diaphoretic. It was used, also, for poisoning the arrows. It is mentioned by Dioscorides that arrows tipped with the juice would kill wolves.

Parts used: The whole plant, but especially the root (Aconiti tuber).

Active ingredients: alkaloids: *aconitine*, *aconine*, *benzaconine* (*picraconitine*).

Particular value: It produces highly toxic alkaloids, so all procedures must be done carefully.

Precautions: Keep away from children, even in gardens. In the dose of 3–6 mg it can cause death.

Documented target cancers: It is tested as a possible anticancer drug. The results of the tests have not been announced yet.

Further details

Related compounds

- The whole plant contains diterpene alkaloids (N-deethylaconotine) (*Aconitum napellus* and *Aconitum napellus* ssp. *neomontanum*) (Grieve, 1994).

References

Ameri, A. and Simmet, T. (1999) Interaction of the structurally related aconitum alkaloids, aconitine and 6-benzyolheteratisine, in the rat hippocampus. *Eur. J. Pharmacol.* **386**(2–3), 187–94.

Been, A. (1992) Aconitum: genus of powerful and sensational plants. *Pharm. Hist.* **34**(1), 35–7.

Cole, C.T. and Kuchenreuther, M.A. (2001) Molecular markers reveal little genetic differentiation among *Aconitum noveboracense* and *A. columbianum* (*Ranunculaceae*) populations. *Am. J. Bot.* **88**(2), 337–47.

Colombo, M.L, Bravin, M. and Tome, F. (1988) A study of the diterpene alkaloids of *Aconitum napellus* ssp. neomontanum during its onthogenetic cycle. *Pharmacol. Res. Commun* **15**(Supp.), 123–8.

Fico, G., Braca, A., De Tommasi, N., Tome, F. and Morelli, I. (2001) Flavonoids from *Aconitum napellus* subsp. neomontanum. *Phytochemistry* 57(4), 543–6.

Grieve, M. (1994) *A Modern Herbal.* Edited and introduced by Mrs C.F. Leyel Tiger books international, London.

Imazio, M., Belli, R., Pomari, F., Cecchi., E., Chinaglia, A., Gaschino, G., Ghisio, A., Trinchero, R. and Brusca, A. (2000) Malignant ventricular arrhythmias due to *Aconitum napellus* seeds. *Circulation* **102**(23), 2907–8.

Kim, D.K., Kwon, H.Y., Lee, K.R., Rhee, D.K. and Zee, O.P. (1998) Isolation of a multidrug resistance inhibitor from *Aconitum* pseudo-laeve var. erectum. *Arch Pharm Res.* **21**(3), 344–7.

Li, Z.B. and Wang, F.P. (1998) Two new diterpenoid alkaloids, beiwusines A and B, from *Aconitum kusnezoffii. J. Asian Nat. Prod. Res.* 1(2), 87–92.

Marchenko, M.M., Kopylchuk, H.P. and Hrygorieva, O.V. (2000) Activity of cytoplasmic proteinases from rat liver in Heren's carcinoma during tumor growth and treatment with medicinal herbs. *Ukr. Biokhim. Zh.* 72(3), 91–4.

Peng, C.S., Wang, F.P. and Jian, X.X. (2000) New norditerpenoid alkaloids from *Aconitum hemsley anum var. pengzhouense. J. Asian Nat Prod Res.* 2(4), 245–9.

Ulubelen, A., Mericli, A.H., Mericli, F., Kilincer, N., Ferizli, A.G., Emekci, M. and Pelletier, S.W. (2001) Insect repellent activity of diterpenoid alkaloids. *Phytother. Res.* 15(2), 170–1.

Wang, F.P., Peng, C.S., Jian, X.X. and Chen, D.L. (2001) Five new norditerpenoid alkaloids from *Aconitum sinomontanum. J. Asian Nat. Prod. Res.* 3(1), 15–22.

Yamanaka, H., Doi, A., Ishibashi, H. and Akaike, N. (2002) Aconitine facilitates spontaneous transmitter release at rat ventromedial hypothalamic neurons. *Br. J. Pharmacol.* 135(3), 816–22.

Strychnos Nux-vomica (Strychnos) (Loganiaceae) — Cytotoxic, Poisonous

Location: India, in the Malay Archipelago.

Appearance
Stem: medium-sized tree with short, thick trunk.
Root: very bitter.
Leaves: opposite.
Flowers: small, greeny-white.

Tradition: The powdered seeds are employed in atonic dyspepsia. The tincture of *Nux Vomica* is often used in mixtures, for its stimulant action on the gastro-intestinal track.

Parts used: seeds.

Active ingredients: strychnopentamine (a dimeric indole alkaloid) from *Strychnos usambarensis.*

Particular value: *Strychnine* is the chief alkaloid constituent of the seeds and acts as a bitter. It improves the pulse and raises blood pressure, acts as a tonic to the circulatory system in cardiac failure, but in small doses, because it can be poisonous.

Precautions: Application of the drug can cause partial haemolysis and liver damage.
Indicative dosage and application

- Four subcutaneous injections of 1.5 mg *strychnopentamine* (one per day) induce a significant decrease of the number of Ehrlich ascites tumor cells.
- *Strychnopentamine* at a relatively low concentration (less than 1 μg) after 72 h of treatment on B16 melanoma cells and on non-cancer human fibroblasts cultured *in vitro*.

Documented target cancers

- Against Ehrlich ascites tumor cells with a significant increase of the survival of the treated mice.
- *Strychnopentamine* applied on B16 melanoma cells and on non-cancer human fibroblasts cultured *in vitro* strongly inhibits cell proliferation and induces cell death.

Further details

Related compounds

- Strychnopentamine (SP) is an alkaloid isolated from *Strychnos usambarensis* Gilg, is a potential anticancer agent, which strongly inhibits cell proliferation and induces cell death on B16 melanoma cells and on non-cancer human fibroblasts cultured *in vitro* and induce a significant decrease of the number of Ehrlich ascites tumor cells (Quetin-Leclercq *et al.*, 1993).
- *Strychnopentamine*, in a low concentration (less than 1 μg) after 72 h showed that incorporation of *thymidine* and *leucine* by B16 cells significantly decreases after only 1 h of treatment. SP induces the formation of dense lamellar bodies and vacuolization in the cytoplasm intense blebbing at the cell surface and various cytological alterations leading to cell death (Quetin-Leclercq *et al.*, 1991 and 1993).
- Three more alkaloids isolated from *Strychnos usambarensis* on cancer cells in culture (Bassleer *et al.*, 1992).

References

Bassleer, R., Depauw-Gillet, M.C., Massart, B., Marnette, J.M., Wiliquet, P., Caprasse, M. and Angenot, L. (1982) Effects of three alkaloids isolated from *Strychnos usambarensis* on cancer cells in culture. *Planta Med.* 45(2), 123–6.

Cai, B.C., Hattori, M. and Namba, T. (1990) Processing of nux vomica. II. Changes in alkaloid composition of the seeds of *Strychnos nux-vomica* on traditional drug-processing. *Chem. Pharm. Bull.* (Tokyo) 38(5), 1295–8.

Chin, V.T., Hue, P.G., Zwaving, J.H. and Hendriks, H. (1987) Some adaptations in the method of the "pharmacopoeia helvetica editio sixta" for the determination of total alkaloids in *Strychnos* seeds. *Pharm. Weekbl.* [Sci]. 9(6), 324–5.

Gao, H., Sun, W. and Sha, Z. (1990) Quantitative determination of strychnine and brucine in semen *Strychni* and its preparations by gas chromatography. *Chung Kuo Chung Yao Tsa Chih* 15(11), 670–1, 703.

Gong, L.G. (1984) Studies on the processing method of *Strychnos nux-vomica*. *Chung Yao Tung Pao* 9(2), 67–9.

Jacob, M., Soediro-Soetarno, Puech, A., Casadebaig-Lafon, J., Duru, C. and Pellecuer, J. (1985) Comparative study of availability of various extracts of *Strychnos Ligustrina* B. L. *Pharm. Acta Helv.* 60(1), 13–6.

Melo, M.F., Santos, C.A., Chiappeta, A.A., de Mello, J.F. and Mukherjee, R. (1987) Chemistry and pharmacology of a tertiary alkaloid from *Strychnos trinervis* root bark. *J. Ethnopharmacol.* 19(3), 319–25.

Nikoletti, M., Goulart, M.O., de Lima, R.A., Goulart, A.E., Delle Monache, F. and Marini Bettolo, G.B. (1984) Flavonoids and alkaloids from *Strychnos pseudoquina*. *J. Nat. Prod.* 47(6), 953–7.

Ogeto, J.O., Juma, F.D. and Muriuki, G. (1984) Practical therapeutics: some investigations of the toxic effects of the alkaloids extracted from *Strychnos henningsii* (Gilg) "muteta." *East Afr. Med. J.* 61(5), 427–32.

Quetin-Leclercq, J., Angenot, L. and Bisset, N.G. (1990) South American *Strychnos* species. Ethnobotany (except curare) and alkaloid screening. *J. Ethnopharmacol.* 8(1), 1–52. Review.

Quetin-Leclercq, J., Bouzahzah, B., Pons, A., Greimers, R., Angenot, L., Bassleer, R. and Barbason, H. (1993) Strychnopentamine, a potential anticancer agent. *Planta Med.* 59(1), 59–62.

Quetin-Leclercq, J., De Pauw-Gillet, M.C., Angenot, L. and Bassleer, R. (1991) Effects of strychnopentamine on cells cultured *in vitro*. *Chem. Biol. Interact.* 80(2), 203–16.

Tits, M., Damas, J., Quetin-Leclercq, J. and Angenot, L. (1991) From ethnobotanical uses of *Strychnos henningsii* to antiinflammatories, analgesics and antispasmodics. *J. Ethnopharmacol.* 34(2–3), 261–7.

Verpoorte, R., Aadewiel, J., Strombom, J. and Baerheim Svendsen, A. (1984) Alkaloids from *Strychnos chrysophylla*. *J. Ethnopharmacol.* 10(2), 243–7.

Weeratunga, G., Goonetileke, A., Rolfsen, W., Bohlin, L. and Sandberg, F. (1984) Alkaloids in *Strychnos aculeata*. *Acta Pharm. Suec.* 21(2), 135–40.

Wright, C.W., Bray D.H., O'Neill, M.J., Warhust, D.C., Phillipson, J.D., Quetin-Leclercq, J. and Angenot, L. (1991) Antiamoebic and antiplasmodial activities of alkaloids isolated from *Strychnos usambarensis*. *Planta Med.* 57(4), 337–40.

Yuno, K., Yamada, H., Oguri, K. and Yoshimura, H. (1990) Substrate specificity of guinea pig liver flavin-containing monooxygenase for morphine, tropane and strychnos alkaloids. *Biochem Pharmacol.* 15, 40(10), 2380–2.

Symphytum officinale L. (Comfrey) (Boraginaceae) — Antimitotic, Carcinogenic

Probably nature's most famous wound-healing species, comfrey has been often referred to as a cancer-fighting drug. Quite ironically, its use may actually increase the possibility of contracting the disease.

Location: It is found in Europe and temperate Asia, usually in watery places.

Appearance
Stem: leafy, angular, covered with bristly hairs, 60–90 cm high.
Root: fibrous, fleshy, and spindle-shaped.
Leaves: radical leaves are very large (they decrease in size), shape ovate, covered with rough hairs.
Flowers: yellow or purple, growing on short stalks, scorpoid in form.
In bloom: May–July.

Tradition: A green vegetable (roots and leaves). Decoction used as herbal tea.

Parts used: Leaves, root.

Active ingredients: *pyrrolizidine alkaloid-N-oxides: 7-acetyl intermedine, 7-acetyl lycopsamine, lycopsamine, intermedine, symphytine.*

Documented carcinogenic properties

- Its crude watery extract and its protein fraction stimulate the *in vivo* proliferation of neoplastic cells and exert an antimitotic effect on human T lymphocytes (Olinescu *et al.*, 1993).
- When digested, it may cause hepatocellular adenomas (at least in rats!).
- Contains hepatotoxic *pyrrolizidine alkaloids*.
- Alkaloid fractions obtained from the roots demonstrate antimitotic and mutagenic activities against both animal and plant cells.

Further details

Related compounds

- The crude watery extract of *Symphytum officinale* and certain protein and carbohydrate components had remarkable effects on the respiratory burst of human PMN granulocytes stimulated via Fc receptors.
- Pyrrolizidine alkaloids have been linked to liver and lung cancers and a range of other deleterious effects. Some comfrey-containing products were found to contain measurable quantities of one or more of the hepatotoxic pyrrolizidine alkaloids, in ranges from 0.1 to 400.0 ppm. Products containing comfrey leaf in combination with one or more other related compounds were found to contain the lowest alkaloid levels (Couet *et al.*, 1996). Highest levels were found in bulk comfrey root, followed by bulk comfrey leaf.

Carcinogenic activity

- The carcinogenicity of *Symphytum officinale* L. was studied in inbred ACI rats. Three groups of 19–28 rats each were fed comfrey leaves for 480–600 days; four additional groups of 15–24 rats were fed comfrey roots for varying lengths of time. A control group was given a normal diet were induced in all experimental groups that received the diets containing comfrey roots and leaves (Hirono *et al.*, 1978). Hemangioendothelial sarcoma of the liver was infrequently induced.

Other medical activity

- Mutagenic and antimitotic effects have been attributed to aqueous solutions of alkaloid fractions obtained from infusions of *Symphytum officinale* L. (Furmanowaa *et al.*, 1983).

References

Aftab, K., Shaheen, F., Mohammad, F.V., Noorwala, M. and Ahmad, V.U. (1996) Phyto-pharmacology of saponins from *Symphytum officinale L. Adv. Exp. Med. Biol.* 404, 429–42.

Ahmad, V.U., Noorwala, M., Mohammad, F.V. and Sener, B. (1993) A new triterpene glycoside from the roots of *Symphytum officinale. J. Nat. Prod.* 56(3), 329–34.

Ahmad, V.U., Noorwala, M., Mohammad, F.V., Sener, B., Gilani, A.H. and Aftab, K. (1993) Symphytoxide A, a triterpenoid saponin from the roots of *Symphytum officinale. Phytochemistry* 32(4), 1003–6.

Barbakadze, V.V., Kemertelidze, E.P., Targamadze, I.L., Shashkov, A.S. and Usov, A.I. (2002) Novel biologically active polymer of 3-(3,4-dihydroxyphenyl)glyceric acid from two types of the comphrey *Symphytum asperum* and *S. caucasicvum (Boraginoceae). Bioorg Khim.* 28(4), 362–6.

Barthomeuf, C.M., Debiton, E., Barbakadze, V.V. and Kemertelidze, E.P. (2001) Evaluation of the dietetic and therapeutic potential of a high molecular weight hydroxycinnamate-derived polymer from *Symphytum asperum Lepech*. Regarding its antioxidant, antilipoperoxidant, antiinflammatory, and cytotoxic properties. *J. Agric. Food Chem.* 49(8), 3942–6.

Behninger, C., Abel, G., Roder, E., Neuberger, V. and Goggelmann, W. (1989) Studies on the effect of an alkaloid extract of *Symphytum officinale* on human lymphocyte cultures. *Planta Med.* 55(6), 518–22.

Betz, J.M., Eppley, R.M., Taylor, W.C. and Andrzejewski, D. (1994) Determination of pyrrolizidine alkaloids in commercial comfrey products (*Symphytum* sp.). *J. Pharm. Sci.* 83(5), 649–53.

Couet, C.E., Crews, C. and Hanley, A.B. (1996) Analysis, separation, and bioassay of pyrrolizidine alkaloids from comfrey (*Symphytum officinale*). *Nat. Toxins* 4(4), 163–7.

Furmanowa, M., Guzewska, J. and Beldowska, B. (1983) Mutagenic effects of aqueous extracts of *Symphytum officinale L.* and of its alkaloidal fractions. *J. Appl. Toxicol.* 3(3), 127–30.

Hirono, I., Mori, H. and Haga, M. (1978) Carcinogenic activity of *Symphytum officinale. J. Natl. Cancer Inst.* 61(3), 865–9.

Johnson, B.M., Bolton, J.L. and Van Breemen, R.B. (2001) Screening botanical extracts or quinoid metabolites. *Chem. Res. Toxicol.* 14(11), 1546–51.

Kim, N.C., Oberlies, N.H., Brine, D.R., Handy, R.W., Wani, M.C. and Wall, M.E. (2001) Isolation of symlandine from the roots of common comfrey (*Symphytum officinale*) using countercurrent chromatography. *J. Nat. Prod.* 64(2), 251–3.

Lenghel, V., Radu, D.L., Chirila, P. and Olinescu, A. (1995) The influence of some vegetable extracts on the *in vitro* adherence of mouse and human lymphocytes to nylon fibers. *Roum Arch. Microbiol. Immunol.* 54(1–2), 15–30.

Mohammad, F.V., Noorwala, M., Ahmad, V.U. and Sener, B. (1995a) A bidesmosidic hederagenin hexasaccharide from the roots of *Symphytum officinale. Phytochemistry* 40(1), 213–8.

Mohammad, F.V., Noorwala, M., Ahmad, V.U. and Sener, B. (1995b) Bidesmosidic triterpenoidal saponins from the roots of *Symphytum officinale. Planta Med.* 61(1), 94.

Mroczek, T., Glowniak, K. and Wlaszczyk, A. (2002) Simultaneous determination of N-oxides and free bases of pyrrolizidine alkaloids by cation-exchange solid-phase extraction and ion-pair high-performance liquid chromatography. *J. Chromatogr. A.* 949(1–2), 249–62.

Noorwala, M., Mohammad, F.V., Ahmad, V.U. and Sener, B. (1994) A bidesmosidic triterpene glycoside from the roots of *Symphytum officinale. Phytochemistry* 36(2), 439–43.

Olinescu, A., Manda, G., Neagu, M., Hristescu, S. and Dasanu, C. (1993) Action of some proteic and carbohydrate components of *Symphytum officinale* upon normal and neoplastic cells. *Roum Arch. Microbiol. Immunol.* 52(2), 73–80.

Stickel, F. and Seitz, H.K. (2000) The efficacy and safety of comfrey. *Public Health Nutr.* 3(4A), 501–8. Review.

3.2.4. Other species with documented anticancer activity

Acacia catechu (Willd.) (Catechu) (Leguminosae) **Antitumor**

Synonyms: *Catechu nigrum* (*Leguminosae*), catechu black, cutch.

Location: It is found in Burma and India.

Appearance
Stem: handsome trees.

Leaves: compoundly pinnate.
Flowers: are arranged in rounded or elongated clusters.

Tradition: Is sold under the name of Catechu. It occurs in commerce in black, shining pieces or cakes.

Parts used: leaves, young shoots.

Active ingredients: *Proteins: Concanavalin A, abrin B* chain and *trypsin* inhibitor (ACTI) (*Acacia confusa*).

Particular value: It is used as an astringent to overcome relaxation of mucous membranes in general. An infusion can be employed to stop nose-bleeding, and is also employed as an injection for uterine hemorrhage leucorrhoea and gonorrhoea. Externally, it is applied in the form of powder, to boils, ulcers and cutaneous eruptions.

Documented target cancers: sarcoma-180 cells and Hela cell culture (mice).

Further details

Related compounds

- Synthetic chimeric protein (ANB-ACTI) of *abrin B* chain and trypsin inhibitor
- Synthetic chimeric protein (Con A-ACTI) of *Concanavalin A* and trypsin inhibitor.

Mode of action: *Abrin B* chain of chimeric protein may act as a vector to carry ACTI into the tumor cells. ACTI in the chimeric protein potentiates its antitumor activity as well as its resistance to tryptic digestion (Lin *et al.*, 1989).

References

Agrawal, S. and Agarwal, S.S. (1990) Preliminary observations on leukaemia specific agglutinins from seeds. *Indian J. Med. Res.* 92, 38–42.

Hanausek, M., Ganesh, P., Walaszek, Z., Arntzen, C.J., Slaga, T.J. and Gutterman, J.U. (2001) Avicins, a family of triterpenoid saponins from *Acacia victoriae* (Bentham), suppress H-ras mutations and aneuploidy in a murine skin carcinogenesis model. *Proc. Natl. Acad. Sci. USA.* 98(20), 11551–6.

Kaur, K., Arora, S., Hawthorne, M.E., Kaur, S., Kumar, S. and Mentha, R.G. (2002) A correlativestudyon antimutagenic and chemopreventive activity of *Acacia auriculiformis* A. Cunn. and *Acacia nilotica* (L.) Willd. *ExDel. Drug Chem. Toxicol.* Feb., 25(1), 39–64.

Lin, J.Y., Hsieh, Y.S. and Chu, S.C. (1989) Chimeric protein: abrin B chain-trypsin inhibitor conjugate as a new antitumor agent. *Biochem. Int.* 19(2), 313–23.

Lin, J.Y. and Lin, L.L. (1985) Antitumor lectin-trypsin inhibitor conjugate. *J. Natl. Cancer. Inst.*, 74(5), 1031–6.

Lo, Y.L., Hsu, C.Y. and Huang, J.D. (1998) Comparison of effects of surfactants with other MDR reversing agents on intracellular uptake of epirubicin in Caco-2 cell line. *Anticancer Res.* 18(4C), 3005–9.

Pico, J.L., Choquet, C., Rosenfeld, C., Sharif, A. and Bourillon, R. (1976) Quantitative variations of three different lectin receptors as a function of establishment and metabolism of normal and leukaemic human cell lines. *Differentiation* 4, 5,(2–3), 115–7.

Popoca, J., Aguilar, A., Alonso, D. and Villarreal, M.L. (1998) Cytotoxic activity of selected plants used as antitumorals in Mexican traditional medicine. *J. Ethnopharmacol.* 59(3), 173–7.

Aristolochia elegans (Aristolochia) (Aristolochiaceae)

Location: South America, with Brazil being its home territory.

Appearance (Figure 3.27)
Stem: slender woody stems twine gracefully in tight coils around fence wire and other supports to lift the vine to heights of 10 or 12 feet.
Root: short horizontal rhizome with numerous long, slender roots below.
Leaves: rich glossy green, about 3 in long by 2 in wide and grow closely, creating a dense mass of foliage.
Flowers: light green and covered with purple brown spots on the flared lips of the blossom in a pattern reminiscent of calico fabric.
In bloom: Summer.

Parts used: dried rhizome and roots.

Active ingredients: *sesquiterpene lactone versicolactone A*.

Indicative dosage and application: it is still under tests.

Documented target cancers: mutagenic activity in the Ames test.

Further details

Other species

- Aristolochia versicolar: Roots contain the sesquiterpene lactone *versicolactone A*.
- *Aristolochia tagala, Aristolochia rigida*. Two aristolochia acids and a flavonol glycoside have been isolated from *A. rigida*. Only *Aristolochic acid IV* has shown a weak direct mutagenic activity in the Ames test.

Figure 3.27 Chamaecyparis.

References

Konigsbauer, H. (1968) On the usability of *Aristolochia tagala* Cham. in dermatology. *Z. Haut. Geschlechtskr.* 43(4), 159–63.

Pistelli, L., Nieri, E., Bilia, A.R., Marsili, A. and Scarpato, R. (1993) Chemical constituents of *Aristolochia rigida* and mutagenic activity of aristolochic acid IV. *J. Nat. Prod.* 56(9), 1605–8.

Zhang, J., He, L.X., Xue, H.Z., Feng, R. and Pu, Q.L. (1991) The structure of versicolactone A from *Aristolochia versicolar S.M. Hwang. Yao Hsueh Hsueh Pao* 26(11), 846–51.

***Chamaecyparis lawsonianna* (Cypress hinoki) (Cupressaceae)** — Anti-leukemic

Location: Of southern Japan and the island of Taiwan origin, it is found in eastern Asia and North America. The typical form of the Hinoki false cypress is rarely cultivated, and most gardeners are more familiar with one or more of the many dwarf cultivars selected for size, form and foliage color.

Appearance (Figure 3.27)
Stem: reddish evergreen conifer with attractive soft and stringy brown bark, cypress can grow over 3 m tall with a trunk diameter of 12 cm.
Leaves: drooping flat frondlike branchlets bearing small scalelike leaves. Has two kinds of leaves: adult leaves are like closely adpressed overlapping scales; leaves on juvenile branchlets and young plants don't overlap and are shaped more like tiny awls or broad needles. The scalelike leaves are borne in pairs of two unequal sizes and shapes.

Tradition: is used as specimens and for hedging, screening and windbreaks.

Active ingredients: *Alkaloids; hinokitiol, tropolone.*

Documented target cancers: high potency in the P-388 leukemia assay.

Further details

Anti-leukemic activity

- Tropolone derivatives prepared from hinokitiol, which naturally occurs in the plants of *Chamaecyparis* species, show high potency in the P-388 leukemia assay. It preferentially inhibits the soluble guanylate cyclase from leukemic lymphocytes (Yamato *et al.*, 1986). This inhibition correlates with its preferential cytotoxic effects for these same cells, since cyclic GMP is thought to be involved in lymphocytic cell proliferation and leukemogenesis and, in general, the nucleotide is elevated in leukemic versus normal lymphocytes and changes have been reported to occur during remission and relapse of this disease.

References

Debiaggi, M., Pagani, L., Cereda, P.M., Landini, P. and Romero, E., (1988) Antiviral activity of Chamaecyparis lawsoniana extract: study with herpes simplex virus type 2. *Microbiologica* 11(1), 55–61.

Hiroi, T., Miyazaki, Y., Kobayashi, Y., Imaoka, S. and Funae, Y. (1995) Induction of hepatic P450s in rat by essential wood and leaf oils. *Xenobiotica* 25(5), 457–67.

Ito, H., Nishimura, J., Suzuki, M., Mamiya, S., Sato, K., Takagi, I. and Baba, S. (1995) Specific IgE to Japanese cypress (*Chamaecyparis obtusa*) in patients with nasal allergy. *Ann. Allergy Asthma Immunol.* 74(4), 299–303.

Kingetsu, I., Ohno, N., Hayashi, N., Sakaguchi, M., Inouye, S. and Saito, S. (2000) Common antigenicity between Japanese cedar (*Cryptomeria japonica*) pollen and Japanese cypress (*Chamaecyparis obtusa*) pollen, I. H-2 complex affects cross responsiveness to Cry j 1 and Cha o 1 at the T- and B-cell level in mice. *Immunology* 99(4), 625–9.

Koyama, S., Yamaguchi, Y., Tanaka, S. and Motoyoshiya, J. (1997) A new substance (Yoshixol) with an interesting antibiotic mechanism from wood oil of Japanese traditional tree (Kiso-Hinoki), *Chamaecyparis obtusa. Gen. Pharmacol.* 28(5), 797–804.

Kuo, Y.H., Chen, C.H. and Huang, S.L. (1998) New diterpenes from the heartwood of *Chamaecyparis obtusa var. formosana. J. Nat. Prod.* 26, 61(6), 829–31.

Miura, H. (1967) The isolation of isocryptomerin from the leaves of *Chamaecyparis obtusa Endlicher. Yakugaku Zasshi* 87(7), 871–4.

Muto, N., Dota, A., Tanaka, T., Itoh, N., Okabe, M., Inada, A., Nakanishi, T. and Tanaka, K. (1995) Hinokitiol induces differentiation of teratocarcinoma F9 cells. *Biol. Pharm. Bull.* 18(11), 1576–9.

Okano, M., Nishioka, K., Nagano, T., Ohta, N. and Masuda, Y. (1994) Clinical characterization of allergic patients sensitized to *Chamaecyparis obtusa* – using AlaSTAT system. *Arerugi* 43(9), 1179–84.

Panella, N.A., Karchesy, J., Maupin, G.O., Malan, J.C. and Piesman, J. (1997) Susceptibility of immature Ixodes scapularis (Acari: Ixodidae) to plant-derived acaricides. *J. Med. Entomol.* 34(3), 340–5.

Suzuki, M., Ito, M., Ito, H., Baba, S., Takagi, I., Yasueda, H. and Ohta, N. (1996) Antigenic analysis of *Cryptomeria japonica* and *Chamaecyparis obtusa* using anti-Cry j 1 monoclonal antibodies. *Acta Otolaryngol.* 525(Suppl.), 85–9.

Takemoto, D.J., Dunford, C., Vaughn, D., Kramer, K.J., Smith, A. and Powell, R.G. (1982) Guanylate cyclase activity in human leukemic and normal lymphocytes. Enzyme inhibition and cytotoxicity of plant extracts. *Enzyme* 27(3), 179–88.

Toda, T., Chong, Y.S. and Nozoe, T. (1967) New constituents of Chamaecyparis formosensis Matsum. *Chem. Pharm. Bull. (Tokyo)* 15(6), 903–5.

Yamato, M., Hashigaki, K., Kokubu, N., Tashiro, T. and Tsuruo, T. (1986) Synthesis and antitumor activity of tropolone derivatives. 3. *J. Med. Chem.* 29(7), 1202–5.

Crinum asiaticum **(Crinum) (var. toxicarium (Hubert)) (Liliaceae)** **Inhibitor**

Location: wild in low, humid spots in various parts of India and on the coast of Ceylon. It is cultivated in Indian gardens.

Appearance
Stem: large plant.
Root: fibrous.
Leaves: showy.
Flowers: handsome, white.
In bloom: April and May.

Tradition: It was used in India for many years.

Part used: bulbs, leaves.

Active ingredients: *alkaloid: lycorine.*

Particular value: the bulb was admitted to the Pharmacopoeia of India as a valuable emetic.

Documented target cancers

- Lycorine inhibits not only induction of MM46 cell death by calprotectin but also inhibits the suppressive effect of calprotectin on target DNA synthesis at a half effective concentration of 0.1–0.5 $\mu g\,ml^{-1}$.
- Lycorine has been reported to posses inhibitory activity against protein translation.

Further details

Antitumor activity

It has been demostrated that calprotectin, an abundant calcium-binding protein complex in polymorphonuclear leukocytes (PMNs), has the capacity to induce growth inhibition and apoptotic cell death against a variety of tumor cell lines and normal cells such as fibroblasts. Therefore, calprotectin which is released to extracellular spaces, might cause tissue destruction in severe inflammatory conditions. Using MM46 mouse mammary carcinoma cells as targets, hot water extracts of *Crinum asiaticum* (lycorine, is the active inhibitory molecule) showed strong inhibition of calprotectin-induced cytotoxicity *in vitro*. The dose–response relationship between the inhibitory effects of lycorine on calprotectin action and target protein synthesis shows that lycorine inhibition for calprotectin cytotoxicity is not solely due to its inhibitory effect on protein synthesis (Yui *et al.*, 1998).

References

Elgorashi, E.E., Drewes, S.E. and van Staden, J. (2001) Alkaloids from *Crinum moorei*. *Phytochemistry* 56(6), 637–40.

Fennell, C.W. and van Staden, J. (2001) *Crinum* species in traditional and modern medicine. *J. Ethnopharmacol.* 78(1), 15–26. Review.

Kapu, S.D., Ngwai, Y.B., Kayode, O., Akah, P.A., Wambebe, C. and Gamaniel, K. (2001) Anti-inflammatory, analgesic and anti-lymphocytic activities of the aqueous extract of Crinum giganteum. *J. Ethnopharmacol.* 78(1), 7–13.

Min, B.S., Gao, J.J., Nakamura, N., Kim, Y.H. and Hattori, M. (2001) Cytotoxic alkaloids and a flavan from the bulbs of *Crinum asiaticum* var. *japonicum*. *Chem. Pharm. Bull.* (Tokyo) 49(9), 1217–9.

Min, B.S., Kim, Y.H., Tomiyama, M., Nakamura, N., Miyashiro, H., Otake, T. and Hattori, M. (2001) Inhibitory effects of Korean plants on HIV-1 activities. *Phytother. Res.* 15(6), 481–6.

Okpo, S.O., Fatokun, F. and Adeyemi, O.O. (2001) Analgesic and anti-inflammatory activity of *Crinum glaucum* aqueous extract. *J. Ethnopharmacol.* 78(2–3), 207–11.

Samud, A.M., Asmawi, M.Z., Sharma, J.N. and Yusof, A.P. (1999) Anti-inflammatory activity of *Crinum asiaticum* plant and its effect on bradykinin-induced contractions on isolated uterus. *Immunopharmacology* 43(2–3), 311–6.

Yui, S., Mikami, M., Kitahara, M. and Yamazaki, M. (1998) The inhibitory effect of lycorine on tumor cell apoptosis induced by polymorphonuclear leukocyte-derived calprotectin. *Immunopharmacology* 40(2), 151–62.

Zvetkova, E., Wirleitner, B., Tram, N.T., Schennach, H. and Fuchs, D. (2001) Aqueous extracts of *Crinum latifolium* (L.) and *Camellia sinensis* show immunomodulatory properties in human peripheral blood mononuclear cells. *Int. Immunopharmacol.* 1(12), 2143–50.

Casearia sylvestris Sw. (Casearia)(Flacourtiaceae) — Antitumor

Parts used: leaves.

Active ingredients: *Clerodane diterpenes: casearins A-F.*

Further details

The structures, of the Active ingredients mentioned before, have been completely elucidated by two dimensional nuclear magnetic resonance, circular dichroism spectroscopy, X-ray analysis, and chemical evidences (Itokawa *et al.*, 1990).

References

De Carvalho, P.R., Furlan, M., Young, M.C., Kingston, D.G. and Bolzani, V.S. (1998) Acetylated DNA-damaging clerodane diterpenes from *Casearia sylvestris. Phytochemistry* 49(6), 1659–62.

Itokawa, H., Totsuka, N., Morita, H., Takeya, K., Iitaka, Y. Schenkel, E.P. and Motidome (1990) New antitumor principles, casearins A-F, for *Casearia sylvestris* Sw. (Flacourtiaceae). *Chem. Pharm. Bull.* (Tokyo) 38(12), 3384–8.

Oberlies, N.H., Burgess, J.P., Navarro, H.A., Pinos, R.E., Fairchild, C.R., Peterson R.W., Soejarto, D.D., Farnsworth, N.R., Kinghorn, A.D., Wani, M.C. and Wall, M.E. (2002) Novel bioactive clerodane diterpenoids from the leaves and twigs of *Casearia sylvestris. J. Nat. Prod.* 65(2), 95–9.

Simonsen, H.T., Nordskjold, J.B., Smitt, U.W., Nyman, U., Palpu, P., Joshi, P. and Varughese, G. (2001) *In vitro* screening of Indian medicinal plants for antiplasmodial activity. *J. Ethnopharmacol.* 74(2), 195–204.

Eurycoma longifolia (Simaroubaceae) — Cytotoxic

Location: Indonesia.

Part used: roots.

Active ingredients

- Four canthin-6-one alkaloids: *9-methoxycanthin-6-one, 9-methoxycanthin-6-one-N-oxide, 9-hydroxycanthin-6-one, and 9-hydroxycanthin-6-one-N-oxide*, and
- one quassinoid: *eurycomanone.*

Documented target cancers

- Canthin-6-ones 1–4 were found to be active with all cell lines tested: breast, colon, fibrosarcoma, lung, melanoma, KB and murine lymphocytic leukemia (P-388).
- Eurycomanone was significantly active against the human cell lines tested [breast, colon, fibrosarcoma, lung, melanoma, KB and KB-V1 (a multi-drug resistant cell line derived from KB)] but was inactive against murine lymphocytic leukemia (P-388).

Further details

Related compounds

- Two additional isolates from the roots of *Eurycoma longifolia*, the *beta-carboline* alkaloids *beta-carboline-1-propionic acid* and *7-methoxy-beta-carboline-1-propionic acid*, were not significantly active with these cultured cells (Kardono *et al.*, 1991). However, they were found to demonstrate significant antimalarial activity as judged by studies conducted with cultured *Plasmodium falciparum* strains.

References

Kanchanapoom, T., Kasai, R., Chumsri, P., Hiraga, Y. and Yamasah, K. (2001) Canthin-6-one and β-carboline alkaloids from *Eurycoma harmadiana*. *Phytochemistry* 56(4), 383–6.

Kardono, L.B., Angerhofer, C.K., Tsauri, S., Padmawinata, K., Pezzuto, J.M. and Kinghorn, A.D. (1991) Cytotoxic and antimalarial constituents of the roots of *Eurycoma longifolia*. *J. Nat. Prod.* 54(5), 1360–7.

***Glyptopetalum sclerocarpum* (Celastraceae) Cytotoxic**

Active ingredients: *22-hydroxytingenone*.

Documented target cancers: Has been tested against P-388 lymphocytic leukemia, KB carcinoma of the nasopharynx, and a number of human cancer cell types, that is, HT-1080 fibrosarcoma, LU-1 lung cancer, COL-2 colon cancer, MEL-2 melanoma, and BC-1 breast cancer.

Further details

Antitumor activity

- *22-Hydroxytingenone* was isolated from *Glyptopetalum sclerocarpum* M. Laws and its unambiguous ^{13}C-NMR assignments were accomplished through the use of APT, HETCOR, and selective INEPT spectroscopy. Intense, but nonspecific cytotoxic activity was observed when this substance was evaluated with a battery of cell lines comprised of the P-388 lymphocytic leukemia, KB carcinoma of the nasopharynx, and a number of human cancer cell types, that is, HT-1080 fibrosarcoma, LU-1 lung cancer, COL-2 colon cancer, MEL-2 melanoma and BC-1 breast cancer (Bavovada *et al.*, 1990).

References

Bavovada, R., Blasko, G., Shieh, H.L., Pezzuto, J.M. and Cordell, G.A. (1990) Spectral assignment and cytotoxicity of 22-hydroxytingenone from *Glyptopetalum sclerocarpum*. *Planta Med.* 56(4), 380–2.

Sotanaphun, U., Suttisri, R., Lipipum, V. and Bavovada, R. (1998) Quinone-methide triterpenoids from *Glyptopetalum sclerocarpum*. *Phytochemistry* 49(6), 1749–55.

Kigelia pinnata (Kigelia) (Bigoniaceae) **Tumor inhibitor**

Parts used: stembark, fruits.

Active ingredients: *Lapachol*.

Documented target cancers: Effects against four melanoma cell lines and a renal cell carcinoma line (Caki-2).

Further details

Inhibitory activity

- Significant inhibitory activity was shown by the *dichloromethane* extract of the *stembark* and *lapachol* (continuous exposure). Moreover, activity was dose-dependent, the extract being less active after 1 h exposure. Chemosensitivity of the melanoma cell lines to the *stembark* was greater than that seen for the renal adenocarcinoma line. In marked contrast, sensitivity to *lapachol* was similar amongst the five cell lines (Houghton *et al*., 1994). *Lapachol* was not detected in the *stembark* extract.

References

Houghton, P.J., Photiou, A., Uddin, S., Shah, P., Browning, M., Jackson, S.J. and Retsas, S. (1994) Activity of extracts of *Kigelia pinnata* against melanoma and renal carcinoma cell lines. *Planta Med.* 60(5), 430–3.

Jackson, S.J., Houghton, P.J., Retsas, S. and Photiou, A. (2000) *In vitro* cytotoxicity of norviburtinal and isopinnatal from *Kigelia pinnata* against cancer cell lines. *Planta Med.* 66(8), 758–61.

Weiss, C.R., Moideen, S.V., Croft, S.L. and Houghton, P.J. (2000) Activity of extracts and isolated naphthoquinones from *Kigelia pinnata* against *Plasmodium falciparum*. *J. Nat. Prod.* 63(9), 1306–9.

Moideen, S.V., Houghton, P.J., Rock, P., Croft, S.L. and Aboagye-Nyame, F. (1999) Activity of extracts and naphthoquinones from *Kigelia pinnata* against *Trypanosoma brucei brucei* and *Trypanosoma brucei rhodesiense*. *Planta Med.* 65(6), 536–40.

Binutu, O.A., Adesogan, K.E. and Okogun, J.I. (1996) Antibacterial and antifungal compounds from *Kigelia pinnata*. *Planta Med.* 62(4), 352–3.

Akunyili, D.N., Houghton, P.J. and Raman, A. (1991) Antimicrobial activities of the stembark of *Kigelia pinnata*. *J. Ethnopharmacol.* 35(2), 173–7.

Kela, S.L., Ogunsusi, R.A., Ogbogu, V.C. and Nwude, N. (1989) Screening of some Nigerian plants for molluscicidal activity. *Rev. Elev. Med. Vet. Pays Trop.* 42(2), 195–202.

Prakash, A.O., Saxena, V., Shukla, S., Tewari, R.K., Mathur, S., Gupta, A., Sharma, S. and Mathur, R. (1985) Anti-implantation activity of some indigenous plants in rats. *Acta Eur Fertil.* 16(6), 441–8.

Koelreuteria henryi (Sapindaceae) **Tumor inhibitor**

Synonyms: varnish tree.

Location: Of China and Korea origin, it is found in eastern Asia. It can be cultivated?

Appearance (Figure 3.28)
Stem: fast-growing, deciduous tree reaching about 7.5 m in height. At maturity, it has a rounded crown, with a spread equal to or greater than the height.

Figure 3.28 Koelreuteria.

Leaves: compound leaves that give it an overall lacy appearance. The leaves turn yellow before falling.
Flowers: large clusters of showy yellow flowers.

Active ingredients: *Protein-tyrosine kinase inhibitors: anthraquinone, stilbene* and *flavonoid.*

Particular value: In cooler zones, used as a free-standing tree where it can be seen in all its glory! It is also good as a small shade tree where space is limited. Golden rain tree should be used more often as a street and park tree.

Documented target cancers: anthraquinone inhibitor, emodin, displayed highly selective activities against src-Her-2/neu and ras-oncogenes.

Further details

Related compounds

- Protein kinases encoded or modulated by oncogenes were used to prescreen the potential antitumor activity of medicinal plants (Chang *et al.*, 1996). Protein-tyrosine kinase-directed fractionation and separation of the crude extracts of Polygonum cuspidatum and *Koelreuteria henryi* have led to the isolation of three different classes of protein-tyrosine kinase inhibitors, anthraquinone, stilbene and flavonoid.

References

Chang, C.J., Ashendel, C.L., Geahlen, R.L., McLaughlin, J.L. and Waters, D.J. (1996) Oncogene signal transduction inhibitors from medicinal plants. *In Vivo* 10(2), 185–90.

Bonap, U.S. (1998) Checklist, Provided by TAMU-BWG, Texas A&M Bioinformatics Working Group, Based on, Biota of North America Program.

Landsburgia quercifolia **(Cystoseiraceae, Phaeophyta)** **Cytotoxic**

Synonyms: brown algae.

Location: New Zealand.

Active ingredients: *Deoxylapachol, 1,4-Dimethoxy-2-(3-methyl-2-butenyl)-naphthalene,2-(3-methyl-2-butenyl)-2,3-epoxy-1,4-naphthalenedione 4,4-dimethoxy ketal.*

Documented target cancers

- *Deoxylapachol* active against P-388 leukemia cells (IC50 0.6 $\mu g\,ml^{-1}$).

Further details

Related compounds

- *1,4-Dimethoxy-2-(3-methyl-2-butenyl)-naphthalene* was the major low polarity component of extracts of this seaweed, which also contained *2,3-dihydro-2,2-bis(3-methyl-2-butenyl)-1,4-naphthalenedione* and *2-(3-methyl-2-butenyl)-2,3-epoxy-1,4-naphthalenedione 4,4-dimethoxy ketal*. Compound *2-(3-methyl-2-butenyl)-2,3-epoxy-1,4-naphthalenedione 4,4-dimethoxy ketal* was converted to the *2,3-epoxide* of *deoxylapachol*, which had biological activities similar to those of *deoxylapachol* (Perry *et al.*, 1991).

References

Nelson, W.A. (1999) *Landsburgia ilicifolia* (*Cystoseiraceae, Phaeophyta*), a new deep-water species endemic to the Three Kings Islands, New Zealand. *New Zealand Journal of Botany,* 37(1).

Perry, N.B., Blunt, J.W. and Munro, M.H. (1991) A cytotoxic and antifungal 1,4-naphthoquinone and related compounds from a New Zealand brown algae, *Landsburgia quercifolia. J. Nat. Prod.* 54(4), 978–85.

Villouta, E., Chadderton, W.L., Pugsley, C.W., Hay, C.H. (2001) Effects of sea urchin (*Evechinus chloroticus*) grazing in Dusky Sound, Fiordland, New Zealand *New Zealand J. Mar Freshwater Res.* 35, M00006.

Magnolia virginiana **L. (Magnolia) (Magnoliaceae)** **Tumor inhibitor**

Location: North America.

Appearance (Figure 3.29)
Stem: 8 or more ft in height, 3–5 ft diameter, smooth gray trunk.
Leaves: simple, oval, 6 in long by 3 in wide, broad, silvery and slightly hairy underneath.
Flowers: large, white.
In bloom: Spring.

Tradition: It is used in rheumatism and malaria and is contra-indicated in inflammatory symptoms.

Figure 3.29 Magnoliavirginiana.

Parts used: bark of stem and root.

Active ingredients
Neolignans: *magnolol, honokiol and monoterpenylmagnolol*
Parthenolide.

Indicative dosage and application: Is still being tested.

Documented target cancers: Epstein–Barr virus, skin tumor (mice).

Further details

Related species

- *Magnolia officinalis*: The bark contains the neolignans *magnolol*, *honokiol* and *monoterpenylmagnolol*. The MeOH extract of this plant and *magnolol* exhibited remarkable inhibitory effects on mouse skin tumor promotion in an *in vivo* two stage carcinogenesis test (Konoshima *et al.*, 1991).

Related compounds

- Another tumor inhibitory agent, *parthenolide*, has been isolated from *Magnolia grandiflora* I.P (Wiedhopf *et al.*, 1973).

References

Celle, G., Savarino, V., Picciotto, A., Magnolia, M.R., Scalabrini, P. and Dodero, M. (1988) Is hepatic ultrasonography a valid alternative tool to liver biopsy? Report on 507 cases studied with both techniques. *Dig. Dis. Sci.* 33(4), 467–71.

Konoshima, T., Kozuka, M., Tokuda, H., Nishino, H., Iwashima, A., Haruna, M., Ito, K. and Tanabe, L M. (1991) Studies on inhibitors of skin tumor promotion, IX. Neolignans from *Magnolia officinalis*. *J. Nat. Prod.* 54(3), 816–22.

Wiedhopf, R.M., Young, M., Bianchi, E. and Cole, J.R. (1973) Tumor inhibitory agent from *Magnolia grandiflora* (Magnoliaceae). I. Parthenolide. *J. Pharm. Sci.* 62(2), 345.

Naucleo orientalis (Rubiaceae) — Antiproliferative

Part used: leaves.

Active ingredients: Nine angustine-type alkaloids were isolated from ammoniacal extracts of *Nauclea orientalis* (*10-hydroxyangustine, two diastereoisomeric 3,14-dihydroangustolines*).

Documented target cancers: The compounds have been found to exhibit *in vitro* anti-proliferative activity against the human bladder carcinoma T-24 cell line and against EGF (epidermal growth factor)-dependent mouse epidermal keratinocytes.

Further details

Related compounds

- The structures of the isolates were determined with spectroscopic methods, mainly 1D- and 2D-NMR spectroscopy. By using overpressure layer chromatography, it was shown that minor quantities of these alkaloids occur in dried *Nauclea orientalis* leaves. The use of ammonia in the extraction process results in a significant increase in the formation of *angustine*-type alkaloids from *strictosamide*-type precursors (Erdelmeier *et al.*, 1992).

References

Erdelmeier, C.A., Regenass, U., Rali, T. and Sticher, O. (1992) Indole alkaloids with *in vitro* antiproliferative activity from the ammoniacal extract of *Nauclea orientalis*. *Planta Med.* 58(1), 43–8.

Hotellier, F., Delaveau, P. and Pousset, J.L. (1979) Alkaloids and glyco-alkaloids from leaves of *Nauclea latifolia* SM. *Planta Med.* 35(3), 242–6.

Fujita, E., Fujita, T. and Suzuki T. (1967) On the constituents of *Nauclea orientalis* L. I. *Noreugenin* and *naucleoside*, a new glycoside (*Terpenoids V*). *Chem. Pharm. Bull.* (Tokyo) 5(11), 1682–6.

Neurolaena lobata (Neurolaena) (Asteraceae) — Cytotoxic

Location: Guatemala.

Active ingredients: *sesquiterpene lactones*: of the *germacranolide* and *furanoheliangolide* type.

Further details

Antitumour activity

- Aqueous and lipophilic extracts of *Neurolaena lobata* were tested against human carcinoma cell lines with cytotoxic effects (Francois *et al.*, 1996). In addition to that, they were tested, also, against *Plasmodium falciparum in vitro*. Sesquiterpene lactones, isolated from *N. lobata*, were shown to be active against *P. falciparum in vitro* (antiplasmodial activity).

References

Francois, G., Passreiter, C.M., Woerdenbag, H.J. and Van Looveren, M. (1996) Antiplasmodial activities and cytotoxic effects of aqueous extracts and sesquiterpene lactones from *Neurolaena lobata*. *Planta Med.* 62(2), 126–9.

Passreiter, M.C., Stoeber, B.S., Ortega, A., Maldonado, E. and Toscano, A.R. (1999) Gemacranolide type sesquiterpene lactones from *Neurolaena macrocephala*. *Phytochemistry* 50(7), 1153–7.

Passiflora tetrandra **(Passifloraceae)** **Cytotoxic**

Parts used: leaves.

Active ingredients: *4-Hydroxy-2-cyclopentenone*.

Documented target cancers: *4-Hydroxy-2-cyclopentenone* is cytotoxic to P-388 murine leukemia cells (IC50 of less than 1 $\mu g\, ml^{-1}$).

Further details

Other medical activity

- *4-Hydroxy-2-cyclopentenone* is also responsible for the anti-bacterial activity of an extract of leaves from *Passiflora tetrandra* with minimum inhibitory doses (MID) of *c.*10 μg per disk against *Escherichia coli, Bacillus subtilis*, and *Pseudomonas aeruginosa* (Perry *et al.*, 1991).

References

Bergner, P. (1995) Passionflower *Med. Herbalism* 7(1–2).

Blumenthal, M. (ed.) (1998) The Complete German Commission E Monographs: *Therapeutic Guide to Herbal Medicines*. Integrative Medicine Communications, Massachusetts.

Bruneton, J. (1995) *Pharmacognosy, Phytochemistry, Medicinal Plants.*, Hampshire, England, Intercept, Ltd.

Crellin, J.K. and Philpott, J. (1990) *Herbal Medicine Past and Present*. Duke Uni. Press, North Carolina.

Duke, J.A. (1985) CRC *Handbook of Medicinal Herbs*, Ed. CRC Press Boca Raton, FL.

Duke, J. and Vasquez, R. (1994) *Amazonian Ethnobotanical Dictionary*, CRC Press Inc., Boca Raton, FL.

HerbClip (1996) *Passion Flower. "An Herbalist's View of Passion Flower."* American Botanical Council, Austin, TX.

Lung, A. and Foster, S. (1996) *Encyclopedia of Common Natural Ingredients*, Wiley & Sons, New York, NY.

Mowrey, Daniel. (1986) *The Scientific Validation of Herbal Medicine*, Keats Publishing, Inc. New Canaan, CT.

Perry, N.B., Albertson, G.D., Blunt, J.W., Cole, A.L., Munro, M.H. and Walker, (1991) JR4-Hydroxy-2-cyclopentenone: an anti-Pseudomonas and cytotoxic component from *Passiflora tetrandra. Planta Med.* 57(2), 129–31.

***Polyalthia barnesii* (Polyalthia) (Annonaceae) Cytotoxic**

Part used: stem bark.

Active ingredients

- *clerodane diterpenes* (cytotoxic): *16 alpha-hydroxycleroda-3,13(14)Z-dien-15,16-olide.*
- *3 beta, 16 alpha-dihydroxycleroda-4(18),13(14)Z-dien-15,16-olide and 4 beta, 16 alpha-dihydroxyclerod-13(14)Z-en-15,16-olide.*

Documented target cancers: The above compounds are found to exhibit broad cytotoxicity against a panel of human cancer cell lines.

Further details

- The (three) cytotoxic clerodane diterpenes were purified from an ethyl acetate-soluble extract of the stem bark of *Polyalthia barnesii*, namely, *16 alpha-hydroxycleroda-3,13(14)Z-dien-15,16-olide* (Ma *et al.*, 1994).

References

Ma, X., Lee, I.S., Chai, H.B., Zaw, K., Farnsworth, N.R., Soejarto, D.D., Cordell, G.A., Pezzuto, J.M. and Kinghorn, A.D. (1994) Cytotoxic clerodane diterpenes from *Polyalthia barnesii. Phytochemistry* 37(6), 1659–62.

Tuchinda, P., Pohmakotr, M., Reutrakul, V., Thanyachareon, W., Sophasan, S., Yoosook, C., Santisuk, T. and Pezzuto, J.M. (2001) 2-substituted furans from *Polyalthia suberosa. Planta Med.* 67(6), 572–5.

Chen, C.Y., Chang, F.R., Shih, Y.C., Hsieh, T.J., Chia, Y.C., Tseng, H.Y., Chen, H.C., Chen, S.J., Hsu, M.C. and Wu, Y.C. (2000) Cytotoxic constituents of *Polyalthia longifolia var. pendula. J. Nat. Prod.* 63(11), 1475–8.

Li, H.Y., Sun, N.J., Kashiwada, Y., Sun, L., Snider, J.V., Cosentino, L.M. and Lee, K.H. Anti-AIDS agents, 9. Suberosol, a new C31 lanostane-type triterpene and anti-HIV principle from *Polyalthia suberosa. J. Nat. Prod.* 56(7), 1130–3.

Zhao, G.X., Jung, J.H., Smith, D.L., Wood, K.V. and McLaughlin, J.L. (1991) Cytotoxic clerodane diterpenes from *Polyalthia longifolia. Planta Med.* 57(4), 380–3.

Wu, Y.C., Duh, C.Y., Wang, S.K., Chen, K.S. and Yang, T.H. (1990) Two new natural azafluorene alkaloids and a cytotoxic aporphine alkaloid from Polyalthia longifolia. *J. Nat. Prod.* 53(5), 1327–31.

Quevauviller, A. and Hamonniere, M. (1977) Activity of the principal alkaloids of *Polyalthia oliveri Engler* (*Annonaceae*) on the central nervous system and the cardiovascular system. *CR Acad. Sci. Hebd. Seances Acad. Sci. D.* 284(1), 93–6.

Pseudolarix kaempferi (Pseudoradix) (Pinaceae) **Cytotoxic**

Part used: seeds.

Active ingredients

- triterpene lactones *pseudolarolides A, B, C* and *D* and;
- diterpene acids *pseudolaric acid-A and -B*.

Documented target cancers: Against

- Human cancer cell lines: KB (nasopharyngeal), A-549 (lung), and HCT-8 (colon) (*pseudolarolide B, pseudolaric acid-A and -B*).
- Murine leukemia cell line (P-388) (*pseudolarolide B, pseudolaric acid-A and -B*).

Further details

- The seeds contain the triterpene lactones *pseudolarolides A, B, C* and *D* and the diterpene acids *pseudolaric acid-A and -B* (Chen *et al.*, 1993).

References

Chen, G.F., Li, Z.L., Pan, D.J., Tang, C.M., He, X., Xu, G.Y., Chen, K. and Lee, K.H. (1993) The isolation and structural elucidation of four novel triterpene lactones, pseudolarolides A, B, C, and D, from *Pseudolarix kaempferi*. *J. Nat. Prod.*, 56(7), 1114–22.

Chen, G.F., Li, Z.L., Pan, D.J., Jiang, S.H. and Zhu, D.Y. (2001) A novel eleven-membered-ring triterpene dilactone, pseudolarolide F and A related compound, pseudolarolide E, from *Pseudolarix kaempferi*. *J. Asian Nat. Prod. Res.* 3(4), 321–33.

Chen, K., Shi, Q., Li, Z.L., Poon, C.D., Tang, R.J. and Lee K.H. (1999) Structures and stereochemistry of pseudolarolides K and L, novel triterpene dilactones from *pseudolarix kaempferi*. *J. Asian Nat. Prod. Res.* 1(3), 207–14.

Chen, K., Zhang, Y.L., Li, Z.L., Shi, Q., Poon, C.D., Tang, R.J., McPhail, A.T. and Lee, K.H. (1996) Structure and stereochemistry of pseudolarolide J, a novel nortriterpene lactone from *Pseudolarix kaempferi*. *J. Nat. Prod.* 59(12), 1200–2.

Pan, D.J., Li, Z.L., Hu, C.Q., Chen, K., Chang, J.J. and Lee, K.H. (1990) The cytotoxic principles of *Pseudolarix kaempferi*: pseudolaric acid-A and -B and related derivatives. *Planta Med.* 56(4), 383–5.

Yang, S.P. and Yue, J.M. (2001) Two novel cytotoxic and antimicrobial triterpenoids from *Pseudolarix kaempferi*. *Bioorg. Med. Chem. Lett.* 11(24), 3119–22.

Yang, S.P., Wu, Y. and Yue, J.M. (2002) Five new diterpenoids from *Pseudolarix kaempferi*. *J. Nat. Prod.* 65(7), 1041–4.

Zhang, Y.L., Lu, R.Z. and Yan, A.L. (1990) Inhibition of ova fertilizability by pseudolaric acid B in hamster. *Zhongguo Yao Li Xue Bao* 11(1), 60–2.

Psychotria sp. (Psychotria) (Psychotrieae) **Cytotoxic**

Location: Pacific Islands.

Appearance
Stem: slender, which grows partly underground.

Root: fibrous rootlets.
In bloom: January–February.

Parts used: aerial parts and stem bark.

Active ingredients: Alkaloids.

Documented target cancers

- All members of the series exhibited readily detected cytotoxic activity against proliferating and non-proliferating Vero (African green monkey kidney) cells in culture.
- hodgkinsine A exhibited substantial antiviral activity against a DNA virus, herpes simplex type 1, and an RNA virus, vesicular stomatitis virus.

Further details

Related compounds

- *Calycodendron milnei*, a species endemic to the Vate Islands (New Hebrides) synthesize a series of Nb-methyltryptamine-derived alkaloids made by linking together 2 to 8 pyrrolidinoindoline units. Nine alkaloids of this class have been isolated from the aerial parts and stem bark of *Calycodendron milnei*, and examined for potential application as anti-cancer and anti-infective agents (Saad *et al.*, 1995). All members of the series showed readily detected anti-bacterial, anti-fungal, and anti-candidal activities using both tube dilution and disc diffusion assay methods. The most potent antimicrobial alkaloids were *hodgkinsine A* and *quadrigemine C*, which exhibited minimum inhibitory concentration (MIC) values as low as $5\,\mu g\,ml^{-1}$.

References

Adjibade, Y., Kuballa, B., Cabalion, P., Jung, M.L., Beck, J.P. and Anton, R. (1989) Cytotoxicity on human leukemic and rat hepatoma cell lines of alkaloid extracts of *Psychotria forsteriana. Planta Med.* 55(6), 567–8.

Hayashi, T., Smith, F.T. and Lee, K.H. (1987) Antitumor agents. 89. Psychorubrin, a new cytotoxic naphthoquinone from *Psychotria rubra* and its structure–activity relationships. *J. Med. Chem.* 30(11), 2005–8.

Roth, A., Kuballa, B., Bounthanh, C., Cabalion, P., Sevenet, T., Beck, J.P. and Anton, R. (1986) Cytotoxic activity of polyindoline alkaloids of *Psychotria forsteriana* (*Rubiaceae*) (1). *Planta Med.* Dec (6), 450–3.

Saad, H.E., El-Sharkawy, S.H. and Shier, W.T. (1995) Biological activities of pyrrolidinoindoline alkaloids from *Calycodendron milnei. Planta Med.* 61(4), 313–6.

***Rhus succedanea* (Sumach) (Anacardiaceae) — Tumor inhibitor cytotoxic**

Location: Japan.

Appearance
Stem: 1.2 m high.
Leaves: pinnate.

Tradition: As the bark is rich in tannin, it is used in candle-making, for adulterating white beeswax and in making pomades. Japan Wax is obtained in Japan by expression and heat, or by the action of solvents from the fruit of sumach.

Parts used: bark, root, fruit.

Active ingredients

- Tyrosinase inhibitor : 2-hydroxy-4-methoxybenzaldehyde.
- Hinokiflavone (cytotoxic).

Particular value: The root-bark is astringent and diuretic. Used in diabetes.

Further details

Related species

- The root of *Rhus vulgaris* contains *2-hydroxy-4-methoxybenzaldehyde*, which is also found in two other East African medicinal plants the root of *Mondia whitei* (Hook) Skeels (Asclepiaceae), and the bark of *Sclerocarya caffra* Sond (Anacardiaceae) (Kubo, 1999).
- The fruit of *Rhus succedanea* consists almost entirely of *palmitin* and *free palmitic acid*, and is not a true wax.

References

Kubo, I. (1999) Kinst-Hori I 2-Hydroxy-4-methoxybenzaldehyde: a potent tyrosinase inhibitor from African medicinal plants. *Planta Med.* 65(1), 19–22.

Wang, H.K., Xia, Y., Yang, Z.Y., Natschke, S.L. and Lee, K.H. (1998) Recent advances in the discovery and development of flavonoids and their analogues as antitumor and anti-HIV agents. *Adv. Exp. Med. Biol.* 439, 191–225.

Seseli mairei (Apiaceae) (Figure 3.30) — Antitumor

Location: China.

Tradition: leaves are used for making salads.

Part used: roots.

Active ingredients: Cytotoxic *polyacetylene: seselidiol.*

Documented target cancers: cytotoxicity against KB, P-388, and L-1210 tumor cells.

Further details

- *Seselidiol* is a new polyacetylene, that has been isolated from the roots of *Seseli mairei*. On the basis of chemical and spectroscopic evidence, its structure has been established as heptadeca-*1,8(Z)-diene-4,6-diyne-3,10-diol. Seselidiol* and its acetate have been demonstrated to show moderate cytotoxicity against KB, P-388, and L-1210 tumor cells (Hu *et al*., 1990).

Figure 3.30 Seseli.

References

Hu, C.Q., Chang, J.J. and Lee, K.H. (1990) Antitumor agents, 115. Seselidiol, a new cytotoxic polyacetylene from *Seseli mairei*. *J. Nat. Prod.*, 53(4), 932–5.

Nielsen, B.E., Larsen, P.K. and Lemmich, J. (1971) Constituents of umbelliferous plants. XVII. Coumarins from *Seseli gummiferum* Pall. The structure of two new coumarins. *Acta Chem Scand.* 25(2), 529–33.

Nielsen, B.E., Larsen, P.K., Lemmich, J. Constituents of umbelliferous plants. 13. Coumarins from *Seseli gummiferum* Pall. The structure of three new coumarins. *Acta Chem Scand.* 24(8), 2863–7.

Xiao, Y., Yang, L., Cui, S., Liu, X., Liu, D., Baba, K. and Taniguchi, M. (1995) Chemical components of *Seseli yunnanense* Franch. *Zhongguo Zhong Yao Za Zhi* 20(5), 294–5, 319.

Tamarindus indica **(Tamarinds) (Leguminosae) — Immunomodulator**

Synonyms: Implee. *Tamarinus officinalis* (Hook).

Location: It is found in India and tropical Africa, it is cultivated in West Indies.

Appearance

Stem: large handsome tree with spreading branches and a thick straight trunk, 12 m high.
Leaves: alternate, abruptly pinnated.
Flowers: fragrant, yellow-veined, red and purple filaments.

Tradition: In Mauritious the Creoles mix salt with the pulp and use it as a liniment for rheumatism and make a decoction of the bark for asthma. The Bengalese employ tamarind pulp in dysentery, and in times of scarcity use it as food. The natives of India consider that it is unsafe to sleep under the tree owing to the acid they exhale during the moisture of the night.

Parts used: fruits freed from brittle outer part of pericarp.

Active ingredients: polysaccharide.

Particular value: It is used as a cathartic, astrigent, febrifuge, antiseptic, refrigerant. It is useful in correcting bilious disorders. A tamarind pulp is made which is considered a useful drink in febrile conditions and a good diet in convalescence to maintain a slightly laxative action of the bowels. The pulp is said to weaken the action of resinous cathartics, but is frequently prescribed with them as a vehicle for jalap (Grieve, 1994).

Documented target cancers: Immunomodulatory activities such as phagocytic enhancement, leukocyte migration inhibition and inhibition of cell proliferation.

Further details

- A polysaccharide isolated and purified from *Tamarindus indica* shows immunomodulatory activities such as phagocytic enhancement, leukocyte migration inhibition and inhibition of cell proliferation (Sreelekha *et al.*, 1993). These properties suggest that this polysaccharide from *T. indica* may have some biological applications.

References

Coutino-Rodriguez, R., Hernandez-Cruz, P. and Giles-Rios, H. (2001) Lectins in fruits having gastrointestinal activity. Their participation in the hemagglutinating property of *Escherichia coli* 0157:H7. *Arch. Med. Res.* 32(4), 251–7.

Morton, J. (1987) Tamarind. In: *Fruits of warm climates* (Ed. F. Julia) Morton, Miami, FL, pp. 115–21.

Sreelekha, T.T., Vijayakumar, T., Ankanthil, R., Vijayan, K.K. and Nair, M.K. (1993) Immunomodulatory effects of a polysaccharide from *Tamarindus indica. Anticancer Drugs* 4(2), 209–12.

Terminalia arjuna (Combretaceae) — Anticancer

Location: Mauritius medicinal plant.

Parts used: bark, stem and leaves.

Active ingredients: *ellagitannin arjunin* along with *gallic acid, ethyl gallate*, the *flavone luteolin* and *tannins.*

Documented target cancers: Luteolin has a well established record of inhibiting various cancer cell lines.

Further details

- *Luteolin* has a well-established record of inhibiting various cancer cell lines and may account for most of the rationale underlying the use of *T. arjuna* in traditional cancer treatments (Pettit *et al.*, 1996). Luteolin was also found to exhibit specific activity against the pathogenic bacterium *Neisseria gonorrhoeae.*

References

Kandil, F.E. and Nassar, M.I. (1998) A tannin anti-cancer promotor from *Terminalia arjuna. Phytochemistry* 47(8), 1567–8.

Pettit, G.R., Hoard, M.S., Doubek, D.L., Schmidt, J.M., Pettit, R.K., Tackett, L.P. and Chapuis, J. C. (1996) Antineoplastic agents 338. The cancer cell growth inhibitory. Constituents of *Terminalia arjuna* (Combretaceae). *J. Ethnopharmacol.* 53(2), 57–63.

Tropaeolum majus (Nasturtium) (Tropaeolaceae) — Antitumor

Synonyms: garden nasturtium, Indian cress.

Location:It is found in the South American Andes from Bolivia to Columbia.

Appearance (Figure 3.31)
Leaves: rounded or kidney shaped, with wavy margins. Are pale green, about 0.5–1.25 cm across, and are borne on long petioles like an umbrella.
Flowers: bright and happy little flowers, they typically have five petals, although there are double and semi-double varieties. The flowers are about 0.25–0.5 cm in diameter and come in a kaleidoscope of colors including russet, pink, yellow, orange, scarlet and crimson.

Parts used: flowers, leaves and immature seed.

Active ingredients: *benzyl glucosinolate* which, through enzymatic hydrolysis, results in the production of *benzyl isothiocyanate* (*BITC*).

Particular value: The dwarf, bushy nasturtiums add rainbows of cheerful color in annual beds and borders. Used as trailing forms on low fences or trellises, on a gravelly or sandy slope, or in a hanging container. Many gardeners include nasturtiums in the salad garden.

Indicative dosage and application

- Appears promising cytotoxicity in the low μMolar range (0.86–9.4 μM)
- Toxic effects at a dose of $200\,mg\,kg^{-1}$ (within 24 h of drug administration) but no reduction in tumor mass.

Figure 3.31 Tropaeolum.

Documented target cancers: BITC has shows *in vitro* anticancer properties against a variety of human and murine tumor cell lines: human ovarian carcinoma cell lines (SKOV-3, 41-M, CHl, CHlcisR), a human lung tumor (H-69), a murine leukemia (L-1210), and a murine plasmacytoma (PC6/sens).

Further details

Antitumor activity

- Cultured cells of *Tropaeolum majus* produce significant amounts of benzyl glucosinolate. The *in vitro* anticancer properties of BITC against a variety of human and murine tumor cell lines have been studied by four independent methods; SRB, MTT, cell counting, and clonogenic assays. Regardless of the assay used, BITC showed promising cytotoxicity in the low μMolar range (0.86–9.4 μM) against four human ovarian carcinoma cell lines (SKOV-3, 41-M, CHl, CHlcisR), a human lung tumor (H-69), a murine leukemia (L-1210), and a murine plasmacytoma (PC6/sens). The L-1210 cells were most sensitive. BITC administered to mice bearing the ADJ/PC6 plasmacytoma subcutaneous tumor showed toxic effects at a dose of 200 mg kg^{-1} (within 24 h of drug administration) but no reduction in tumor mass (Pintao *et al.*, 1995). However, the growth inhibitory properties of BITC against a range of tumor cell types warrant further *in vivo* antitumor evaluation as well as its biotechnological production.

References

Baran, R., Sulova, Z., Stratilova, E. and Farkas, V. (2000) Ping-pong character of nasturtium-seed xyloglucan endotransglycosylase (XET) reaction. *Gen. Physiol. Biophys.* 19(4), 427–40.

Bettger, W.J., McCorquodale, M.L. and Blackadar, C.B. (2001) The effect of a Tropaeolum speciosum oil supplement on the nervonic acid content of sphingomyelin in rat tissues. *J. Nutr. Biochem.* 12(8), 492–6.

Crombie, H.J., Chengappa, S., Jarman, C., Sidebottom, C. and Reid, J.S. (2002) Molecular characterisation of a xyloglucan oligosaccharide-acting alpha-D- xylosidase from nasturtium (*Tropaeolum majus* L.) cotyledons that resembles plant "apoplastic" alpha-D-glucosidases. *Planta* 214(3), 406–13.

De Medeiros, J.M., Macedo, M., Contancia, J.P., Nguyen, C., Cunningham, G. and Miles, D.H. (2000) Antithrombin activity of medicinal plants of the Azores. *J. Ethnopharmacol.* 72(1–2), 157–65.

Faik, A., Desveaux, D. and MacLachlan, G. (2000) Sugar-nucleotide-binding and autoglycosylating polypeptide(s) from nasturtium fruit: biochemical capacities and potential functions. *Biochem. J.* 347 (Pt 3), 857–64.

Fanutti, C., Gidley, M.J. and Reid, J.S. (1996) Substrate subsite recognition of the xyloglucan endo-transglycosylase or xyloglucan-specific endo-(1–>4)-beta-D-glucanase from the cotyledons of germinated nasturtium (*Tropaeolum majus* L.) seeds. *Planta* 200(2), 221–8.

Ludwig-Muller, J. and Cohen, J.D. (2002) Identification and quantification of three active auxins in different tissues of *Tropaeolum majus*. *Physiol. Plant* 115(2), 320–9.

Lykkesfeldt, J. and Moller, B.L. (1993) Synthesis of Benzylglucosinolate in *Tropaeolum majus* L. (Isothiocyanates as Potent Enzyme Inhibitors). *Plant Physiol.* 102(2), 609–13.

Pintao, A.M., Pais, M.S., Coley, H., Kelland, L.R. and Judson, I.R. (1995) *in vitro* and *in vivo* antitumor activity of benzyl isothiocyanate: a natural product from *Tropaeolum majus*. *Planta Med.* 61(3), 233–6.

Rose, J.K., Brummell, D.A. and Bennett, A.B. (1996) Two divergent xyloglucan endotransglycosylases exhibit mutually exclusive patterns of expression in nasturtium. *Plant Physiol.* 110(2), 493–9.

Valeriana officinalis (Valerian) (Valerianaceae) Cytotoxic

Synonyms: Amantilla, Setwall, All-Heal.

Location: Throughout, mainly in Europe and Northern Asia, in meadows, borders of rivers and open woods on moist soil.

Appearance (Figure 3.32)
Stem: erect, up to 1.5–2 m high.
Root: conical root-stock or rhizome.
Leaves: opposite, pinnate, up to 20 cm long.
Flowers: Pink and small, in umbel-like clusters, 5–6 mm long, with a stinking odor (as the whole plant).
Fruit: capsule.
In bloom: May–September.

Tradition: The term *Phu*, a synonym of the root of valerian indicates its stinking scent. The species has probably derived its name from *Valerius*, who first used it in medicine or the Latin word *valere* ("to be in health"). Valerian is referred to as a calminative in medical texts of the Middle Age.

Biology: The rhizome develops underground for several years before a flowering stem emerges (only one shoot per root). The plant can be propagated either by runners or by seed. For cultivation, adequate fertilization is recommended.

Part used: root.

Active ingredients: *valerianic acid, borneol*, a-pirene, *camphene, valtrate, choline, valerianates* (valerianic acid combines with various bases), *chatarine* and *valerianine* (alkaloids from the root).

Figure 3.32 Valeriana.

Particular value: *Valerian* is a powerful nervine, stimulant, carminative and antispasmodic. It allays pain and promotes sleep. Oil of valerian is used as a remedy for cholera (in a form of cholera drops). The juice of the fresh root (*Energetene* of *valerian*) has been recommended as a narcotic in insomnia and as anti-convulsant in epilepsy.

Precautions: Toxic in high doses. It can cause central paralysis, giddiness, headache, agitation, decrease sensibility, motility and reflex excitability, nausea.

Indicative dosage and application: Still testing. A proposal dose is 300 and 500 $mg\,kg^{-1}$ per day (in rats) but not yet confirmed.

Documented target cancers: Still testing.

Further details

Cytotoxic activity

- Reiterated administration of *Valeriana officinalis* to laboratory animals has been associated with toxic effects. Rats receiving 300 and 600 $mg\,kg^{-1}$ per day of the drug for 30 days. During the period of the treatment, the animals' weight and blood pressure were measured. At the end of the treatment the animals were sacrificed. The principal organs were weighed and hematological and biochemical parameters were determined in blood samples collected. This work is concerned with pharmacological properties which are related to the two plants. The influence of the drugs on the behavior, the pain, the intestinal peristalsis and *strychnine* convulsions are reported (Febri *et al.*, 1991).

Related compounds

- *Colchicine*-treated suspension cultures of *Valeriana wallichii* produce higher amounts of *valepotriates* than did the respective untreated cultures. The ability to produce *valepotriates* in the treated culture remains in the absence of *colchicine* even if the chromosome status returns to normal. When the *colchicine* treatment is repeated, a further increase in *valepotriate* production can be obtained. Besides the known *valepotriates*, a series of fourteen new compounds, hitherto not described for the parent plant, were isolated from the cell suspension culture. Eight of them are also found in plant parts in minor amounts, but six seem to be present only in tissue cultures of *V. wallichii* (Becker and Chavadej 1985).
- Different *in vitro* cultures of *Valerianaceae* were analyzed for valepotriate content [*(iso)valtrate, acevaltrate, didrovaltrate*] in a study on properties of production *in vitro* (plant species, growth conditions, differentiation level, *valepotriate* content of the medium after growth). The *in vitro* cultures were: callus cultures of *Valeriana officinalis* L., *Valerianella locusta* L. and *Centranthus ruber* L.DC.; a suspension culture of *Valeriana officinalis* L. and a root organ culture of *Centranthus ruber* L.DC. All of the cultures produced *valepotriates in vitro* in different amounts. None of the media that had served for growth contained any *valepotriates*. In order to characterize the *in vitro* growth more precisely different parameters (such as fresh and dry weight, lipid and

nitrogen content and (*iso*)*valtrate* content) were analyzed at different time intervals during a growth period in one of the cultures (callus culture of *Valeriana officinalis* L.) (Becker *et al.*, 1977).

- It is possible directly to separate and analyze, quantitatively and qualitatively, the valepotriates from *Valeriana* crude extracts or from commercial *Valeriana* preparations by high-performance liquid chromatography. The separations are achieved on 4 or 8 mm I.D. columns packed with silica gel (particle size 10 μmicron) with n-hexane-ethyl acetate mixtures as eluent. A refractive index detection system is necessary for determining all of the valepotriates. If the concentration differences between didrovaltratum and valtratum are very great, an ultraviolet (UV) detector must be used and the determination must be conducted in two steps. For valtratum drugs UV detection alone will suffice. As internal standards p-dimethylaminobenzaldehyde should be used for extracts and preparations from valtratum races, and benzaldehyde in the presence of didrovaltratum races. This determination is superior to the combined thin-layer chromatographic-hydroxamic acid method used hitherto with respect to time consumption, precision, and sensitivity (Tittel and Wagner, 1978; Suomi *et al.*, 2001).

References

Albrecht, M. and Berger, W. (1995) Psychopharmaceuticals and safety in traffic. *Zeits Allegmeinmed*, 71, 1215–21.

Becker, H. and Chavadej, S. (1985) Valepotriate production of normal and colchicine- treated cell suspension cultures of *Valeriana wallichii*. *J. Nat. Prod.* 48(1), 17–21.

Becker, H., Schrall, R. and Hartmann, W. (1977) Callus cultures of a Valerian species. 1. Installation of a callus culture of *Valeriana Wallichii* DC and 1st analytical studies *Arch Pharm* (*Weinheim*) 310(6), 481–4.

Brown, D.J. (1996) *Herbal Prescriptions for Better Health*. Prima Publishing, Rocklin, CA. pp. 173–8.

Buckova, A., Grznar, K., Haladova, M. and Eisenreichova, E. (1977) Active substances in *Valeriana officinalis L. Cesk Farm* 26(7), 308–9.

Cavadas, C., Araujo, I., Cotrim, M.D., Amaral, T., Cunha, A.P., Macedo, T. and Ribeiro, C.F. (1995) *In vitro* study on the interaction of *Valeriana officinalis* L. extracts and their amino acids on GABAA receptor in rat brain. *Arzneimittelforschung* 45(7), 753–5.

Czabajska, W., Jaruzelski, M. and Ubysz, D. (1976) New methods in the cultivation of *Valeriana officinalis. Planta Med.* 30(1), 9–13.

Della Loggia, R., Tubaro, A. and Redaelli, C. (1981) Evaluation of the activity on the mouse CNS of several plant extracts and a combination of them. *Riv Neurol.* 51(5), 297–310. Review.

Dressing, H., Köhler, S. and Müller, W.E. (1996) Improvement of sleep quality with a high-dose valerian/lemon balm preparation: a placebo-controlled double-blind study. *Psychopharmakotherapie* 6, 32–40.

Fehri, B., Aiache, J.M., Boukef, K., Memmi, A. and Hizaoui, B. (1991) *Valeriana officinalis* and *Crataegus oxyacantha:* toxicity from repeated administration and pharmacologic investigations. *J. Pharm. Belg.* 46(3), 165–76.

Fursa, M.S. (1980) Composition of the flavonoids of *Valeriana officinalis* from the Asiatic part of the USSR. *Farm Zh.* (3), 72–3.

Hendriks, H., Bos, R., Allersma, D.P., Malingre, T.M. and Koster, A.S. (1981) Pharmacological screening of valerenal and some other components of essential oil of *Valeriana officinalis. Planta Med.* 42(1), 62–8.

Hromadkova, Z., Ebringerova, A. and Valachovic, P. (2002) Ultrasound-assisted extraction of water-soluble polysaccharides from the roots of valerian (*Valeriana officinalis. Ultrason Sonochem* (1), 37–44.

Janot, M.M., Guilhem, J., Contz, O., Venera, G. and Cionga, E. (1979) Contribution to the study of valerian alcaloids (*Valeriana officinalis,* L.): actinidine and naphthyridylmethylketone, a new alkaloid. *Ann. Pharm. Fr.* 37(9–10), 413–20.

Kohnen, R. and Oswald, W.D. (1988) The effects of valerian, propranolol and their combination on activation performance and mood of healthy volunteers under social stress conditions. *Pharmacopsychiatry* 21, 447–8.

Kornilievs'kyi, I., Fursa, M.S., Rybal'chenko, A.S. and Koreshchuk, Kie. (1979) Flavonoid makeup of *Valeriana officinalis* from the southern and central provinces of the Ukraine. *Farm Zh.* (4), 71–2.

Leathwood, P.D., Chauffard, F., Heck, E. and Munoz-Box, R. (1982) Aqueous extract of valerian root (*Valeriana officinalis* L.) improves sleep quality in man. *Pharmacol. Biochem. Behav.* 17, 65–71.

Leathwood, P.D. and Chauffard, F. (1982–83) Quantifying the effects of mild sedatives. *J. Psychiatr. Res.* 17(2), 115–22. Review.

Leathwood, P.D. and Chauffard, F. (1985) Aqueous extract of valerian reduces latency to fall asleep in man. *Planta Med.* 51, 144–8.

Mennini, T. and Bernasconi, P.(1993) *In vitro* study on the interaction of extracts and pure compounds from *Valeriana officinalis* roots with GABA, benzodiazepine and barbiturate receptors. *Fitoterapia* 64, 291–300.

Nikul'shina, N.I., Talan, V.A., Bukharov, V.G. and Ivanova, V.M. (1969) Valeroside A – a glycoside from valerian (*Valeriana officinalis* L.). *Farmatsiia* 18(6), 44–7.

Pank, F., Hannig, H.J., Hauschild, J. and Zygmunt, B. (1980) Chemical weed control in the cropping of medicinal plants. Part 1: Valerian (*Valeriana officinalis* L.). *Pharmazie* 35(2), 115–9.

Paris, R., Besson, P. and Herisset, A. (1966) Tests of "industrial lyophilization" of medicinal plants. 3. *Valeriana officinalis* L. Influence of lyophilization on the quality of the drug. *Ann. Pharm. Fr.* 24(11), 669–74.

Perebeinos, V.S. (1974) Permissible content of stalk residues in crude *Valeriana officinalis. Farmatsiia* 23(3), 72–6.

Santos, M.S., Ferreira, F., Cunha, A.P., Carvalho, A.P., Ribeiro, C.F. and Macedo, T. (1994) Synaptosomal GABA release as influenced by valerian root extract – involvement of the GABA carrier. *Arch. Int. Pharmacodyn Ther.* 327(2), 220–31.

Suomi, J., Wiedmer, S.K., Jussila, M. and Riekkola, M.L. (2001) Determination of iridoid glycosides by micellar electrokinetic capillary chromatography-mass spectrometry with use of the partial filling technique. *Electrophoresis* 22(12), 2580–7.

Tamamura, K., Kakimoto, M., Kawaguchi, M. and Iwasaki, T. (1973) Pharmacological studies on the constituents of crude drugs and plants. 1. Pharmacological actions of *Valeriana officinalis* Linne. var. latifolia Miquel. *Yakugaku Zasshi* 93(5), 599–606.

Tittel, G. and Wagner, H. (1978) High-performance liquid chromatographic separation and quantitative determination of valepotriates in valeriana drugs and preparations. *J. Chromatogr.* 1, 148(2), 459–68.

Torssell, K. and Wahlberg, K. (1967) Isolation, structure and synthesis of alkaloids from *Valeriana officinalis* L. *Acta Chem. Scand.* 21(1), 53–62.

Tucakov, J. (1965) Comparative ethnomedical study of *Valeriana officinalis* L. *Glas Srp Akad Nauka {Med.}* (18), 131–50.

Tufik, S., Fujita, K., Seabra, Mde, L. and Lobo, L.L. (1994) Effects of a prolonged administration of valepotriates in rats on the mothers and their offspring. *J. Ethnopharmacol.* 41(1–2), 39–44.

Verzarne Petri, G. (1974) Biosynthesis of alkaloids, valtrates and volatile oils in the roots of *Valeriana officinalis* L. from radioactive precursors. *Acta Pharm. Hung.* 0(0 Suppl. 1), 54–65.

Violon, C., Van Cauwenbergh, N. and Vercruysse, A. (1983) Valepotriate content in different *in vitro* cultures of Valerianaceae and characterization of *Valeriana officinalis L.* callus during a growth period. *Pharm. Weekbl. Sci.* 5(5), 205–9.

Wagner, H., Schaette, R., Horhammer, L. and Holzl, J. (1972) Dependence of the valepotriate and essential oil content in *Valeriana officinalis* L.s.l. on various exogenous and endogenous factors. *Arzneimittelforschung* 22(7), 1204–9.

Yang, G.Y. and Wang, W. (1994) Clinical studies on the treatment of coronary heart disease with *Valeriana officinalis var latifolia. Zhongguo Zhong Xi Yi Jie He Za Zhi* 14(9), 540–2.

Zhang, B.H., Meng, H.P., Wang, T., Dai, Y.C., Shen, J., Tao, C., Wen, S.R., Qi, Z., Ma, L. and Yuan, S.H. (1982) Effects of *Valeriana officinalis* L. extract on cardiovascular system. *Yao Xue Xue Bao* 17(5), 382–4.

Xanthium strumarium (Cocklebur) (Compositae) — Cytotoxic

Location: South Europe, in America near sea-coast, central Asia northwards to the Baltic.

Appearance
Stem: coarse, erect, annual, 0.3–0.6 m high.
Leaves: on long stalks, large broad, heart-shaped, coarsely toothed or angular in both sides.
Flowers: heads, greenish yellow, terminal clusters on short racemes, upper ones male, lower female.

Parts used: the whole plant.

Active ingredients: xanthatin.

Particular value: A valuable and sure specific in the treatment of hydrophobia.

Precautions: Intoxication.

Indicative dosage and application: under investigation.

Documented target cancers: serofibrinous ascites, edema of the gallbladder wall, and lobular accentuation of the liver.

Further details

Cytotoxic activity

- Cocklebur (*Xanthium strumarium*) fed to feeder pigs was associated with acute to subacute hepatotoxicosis. Cotyledonary seedings fed at 0.75–3% of body weight or ground bur fed at 20–30% of the ration caused acute depression, convulsions, and death (Stuart *et al.*, 1981). Principle gross lesions were marked serofibrinous ascites, edema of the gallbladder wall, and lobular accentuation of the liver. Acute to subacute centrilobular hepatic necrosis was present microscopically. The previously reported toxic principle, hydroquinone, was not recovered from the plant or bur of *X. strumarium*. Authentic hydroquinone administered orally failed to produce lesions typical of cocklebur intoxication but did produce marked hyperglycemia. Carboxyatractyloside recovered from the aqueous extract of *X. strumarium* and authentic carboxyatractyloside, when fed to pigs, caused signs and lesions typical of cocklebur intoxication. Marked hypoglycemia and elevated serum glutamic oxaloacetic transaminase and serum isocitric dehydrogenase concentrations occurred in pigs with acute hepatic necrosis that had received either cocklebur seedlings, ground bur or carboxyatractyloside (Stuart *et al.*, 1981).

References

Battle, R.W., Gaunt, J.K. and Laidman, D.L. (1976) The effect of photoperiod on endogenous gamma-tocopherol and plastochromanol in leaves of *Xanthium strumarium* L. (cocklebur). *Biochem. Soc. Trans.* 4(3), 484–6.

Chu, T.R. and Wei, Y.C. (1965) Studies on the principal unsaturated fatty acids of the seed oil of *Xanthium strumarium* L. *Yao Xue Xue Bao* 12(11), 709–12.

Cole, R.J., Stuart, B.P., Lansden, J.A. and Cox, R.H. (1980) Isolation and redefinition of the toxic agent from cocklebur (*Xanthium strumarium*). *J. Agric. Food Chem.* 28(6), 1330–2.

Hatch, R.C., Jain, A.V., Weiss, R. and Clark, J.D. (1982) Toxicologic study of carboxyatractyloside (active principle in cocklebur – *Xanthium strumarium*) in rats treated with enzyme inducers and inhibitors and glutathione precursor and depletor. *Am. J. Vet. Res.* 43(1), 111–6.

Jain, S.R. (1968) Investigations on antileucodermic activity of *Xanthium strumarium. Planta Med.* 16(4), 467–8.

Kapoor, V.K., Chawla, A.S., Gupta, A.K. and Bedi, K.L. (1976) Studies on the oil of *Xanthium strumarium. J. Am. Oil. Chem. Soc.* 53(8).

Khafagy, S.M., Sabry, N.N., Metwally, A.M. and el-Naggar, S.F. (1974) Phytochemical investigation of *Xanthium strumarium. Planta Med.* 26(1), 75–8.

Kuo, Y.C., Sun, C.M., Tsai, W.J., Ou, J.C., Chen, W.P. and Lin, C.Y. (1998) Chinese herbs as modulators of human mesangial cell proliferation: preliminary studies. *J. Lab. Clin. Med.* 132(1), 76–85.

Kupiecki, F.P., Ogzewalla, C.D. and Schell, F.M. (1974) Isolation and characterization of a hypoglycemic agent from *Xanthium strumarium. J. Pharm. Sci.* 63(7), 1166–7.

McMillan, C. (1973) Partial fertility of artificial hybrids between Asiatic and American cockleburs (*Xanthium strumarium* L.). *Nat. New Biol.* 246(153), 151–3.

Pashchenko, M.M. and Pivnenko, G.P. (1970) Polyphenol substances in *Xanthium riparium* and *Xanthium strumarium. Farm Zh.* 25(6), 41–3.

Roussakis, C., Chinou, I., Vayas, C., Harvala, C. and Verbist, J.F. (1994) Cytotoxic activity of xanthatin and the crude extracts of *Xanthium strumarium. Planta Med.* 60(5), 473–4.

Stuart, B.P., Cole, R.J. and Gosser, H.S. (1981) Cocklebur (*Xanthium strumarium,* L. var. *strumarium*) intoxication in swine: review and redefinition of the toxic principle. *Vet. Pathol.* 18(3), 368–83.

Sila, V.I. and Lisenko, L.V. (1971) A pharmacological study of the sum of *Xanthium strumarium alkaloids. Farm Zh.* 26(2), 71–3.

Xylopia aromatica (Annonaceae) — Cytotoxic

Part used: bark.

Active ingredients: *Annonaceous acetogenins: asimicin, venezenin, xylopien, xylomaterin, xylopianin, xylopiacin, xylomaticin, annomontacin, gigantetronenin, gigantetrocin A, and annonacin.*

Documented target cancers: acetogenins showed cytotoxicity, comparable or superior to adriamycin, against three human solid tumor cell lines.

Further details

- *Xylopia aromatica*: the bark (EtOH extract) contains the acetogenins we have already mentioned. These acetogenins showed reduction of the 10-keto of 1 to the racemic OH-10 derivative enhanced the bioactivity, as did the conversion of 1 to 6 and 7. Venezenin like other Annonaceous acetogenins, showed inhibition of oxygen uptake by rat liver mitochondria and demonstrated that the THF ring may not be essential to this mode of action (Colman-Saizarbitoria *et al.*, 1994).

References

Ahammadsahib, K.I., Hollingworth, R.M., McGovren, J.P., Hui, Y.H. and McLaughlin, J.L. (1993) Mode of action of bullatacin: a potent antitumor and par pesticidal annonaceous acetogenin. *Life Sciences* 53, 1113–20.

Colman-Saizarbitoria, T., Zambrano, J., Ferrigni, N.R., Gu, Z.M., Ng, J.H., Smith, D.L. and McLaughlin, J.L. (1994) Bioactive annonaceous acetogenins from the bark of *Xylopia aromatica*. *J. Nat. Prod.* 57(4), 486–93.
Moerman, D.E. (1986) Medicinal plants of native America. U. Mich. Mus. Anthop. Tech. Rept. No. 19. 2 vols. Ann Arbor, Michigan.

Zieridium pseudobtusifolium (Rutaceae) — Tumor inhibitor cytotoxic

Active ingredients: *flavonols*: *5,3′-dihydroxy-3,6,7,8,4′-pentamethoxyflavone, digicitrin, 5-hydroxy-3,6,7,8,3′,4′-hexamethoxyflavone, 3-O-demethyldigicitrin, 3,5,3′-trihydroxy-6,7,8,4′-tetramethoxyflavone, and 3,5-dihydroxy-6,7,8,3′,4′-pentamethoxyflavone.*

Indicative dosage and application

- IC50 0.04 $\mu g\, ml^{-1}$ against (KB) human nasopharyngeal carcinoma cells
- IC50 12 μM inhibited tubulin.

Documented target cancers

- cytotoxic activity against KB cells
- human nasopharyngeal carcinoma cells
- inhibits tubulin assembly into microtubules.

Further details

- Bioassay-guided fractionation of the extracts of *Zieridium pseudobtusifolium* and *Acronychia porteri* led to the isolation of *5,3′-dihydroxy-3,6,7,8,4′-pentamethoxyflavone* which showed activity against (KB) human nasopharyngeal carcinoma cells (IC50 0.04 $\mu g\, ml^{-1}$) and inhibited tubulin assembly into microtubules (IC50 12 μM). Of all these mentioned (in the Active ingredients) *flavonols* showed cytotoxic activity against KB cells (Lichius *et al.*, 1994).

References

Jaffré, T., Reeves, R., Becquer, Th. (eds), 1997. Ecologie des milieux sur roches ultramafiques et sur sols métallifères. Actes de la deuxième Conférence Internationale sur les Milieux Serpentiniques. Nouméa, ORSTOM, (Documents scientifiques et techniques, III : 2), 306 p.
Jaffré, T., Morat, Ph., Veillon, J.M., Rigault, F., Dagostini, G., 2001. Composition et caractéristiques de la flore de la Nouvelle-Calédonie/Composition and Characteristics of the native flora of New Caledonia. Nouméa, IRD (Documents scientifiques et techniques, II :4), 121 p + 16 planches photos.
Le Pierres, D., 1999. Les apports des recherches en génétique sur l'avenir de la culture du café en Nouvelle-Calédonie. *La Calédonie Agricole*, 76, 34–37; 77, 36–8.
Lichius, J.J., Thoison, O., Montagnac, A., Pais, M., Gueritte-Voegelein, F., Sevenet, T., Cosson, J.P., Hadi, A.H. (1994) Antimitotic and cytotoxic flavonols from *Zieridium pseudobtusifolium* and *Acronychia porteri*. *J. Nat. Prod.* 57(7), 1012–6.
Verotta, L., Dell'Agli, M., Giolito, A., Guerrini, M., Cabalion, P., Bosisio, E. 200. In vitro antiplasmodial activity of extracts of Tristaniopsis species and identification of the active constituents : Ellagic acid and 3,4,5-trimethoxyphenyl-(6′-O-galloyl)-O-β-D-glucopyranoside. *Journal of Natural Products* 64(5), 603–7.

Chapter 4

Cytotoxic metabolites from marine algae

Vassilios Roussis, Costas Vagias and Leto A. Tziveleka

4.1 Cytotoxic metabolites from marine algae

The pharmacological importance of hundreds of plants has been known since ancient times and there are documents on their properties dating as early as 2000 BC. The vast majority of bioactive metabolites though has only been discovered and studied scientifically the last 50 years. At this point it is estimated that more than 120 pure chemical substances extracted from higher plants are used in medicine throughout the world. The influence of natural products upon anticancer drug discovery and design cannot be overestimated. Approximately 60% of all drugs now in clinical trials for the multiplicity of cancers are either natural products, compounds derived from natural products, containing pharmacophores derived from active natural products or are "old drugs in new clothes," where modified natural products are attached to targeting systems (Cragg and Newman, 2000).

Most of the efforts towards the discovery of new bioactive metabolites have focused for many years on the easily accessible higher plants. Though in the last few decades, obscure and rare organisms became accessible because of the scientific advancement in the areas of chromatography, spectroscopy and marine technology.

Prior to the development of reliable scuba diving techniques some 40 years ago, the collection of marine organisms was limited to those obtainable by free diving. Subsequently, depths from approximately 3–40 m became routinely attainable and the marine environment has been increasingly explored as a source of novel bioactive agents. Deep water collections can be made by dredging or trawling, but these methods suffer from disadvantages, such as environmental damage and non-selective sampling. These disadvantages can be partially overcome by the use of manned submersibles or remotely operated vehicles. However, the high cost of these means of collecting, precludes their extensive use in routine collection operations. However, the expansion of rebreather techniques in the last few years has begun to open up depths of 100 m to relatively routine collections and one-man flexible suits such as the "Nyut suit" will extend the limit to close to 330 m in due course.

Although the traditional sources of secondary metabolites were terrestrial higher plants, animals and microorganisms, marine organisms have become major targets for natural products research in the past decade.

If the novelty and complexity of compounds discovered from marine sources were the only criteria, then the success of research in this area would be assured for there are many marine natural products that have no counterpart in the terrestrial world. For example the structures assigned to maitotoxin represents perhaps the most complex secondary metabolite described to date. The surprisingly large proportion of marine natural products with interesting pharmacological properties has coined the term "Drugs from the Sea."

Marine organisms have exhibited an impressive spectrum of biological properties and several representatives have been investigated in depth as potential new biotechnological agents with activities including: cytotoxicity; antibiotic activity; anti-inflammatory and antispasmodic activity; antiviral activity; cardiotonic and cardiovascular activity; antioxidant activity; enzyme inhibition activity and many others.

Macroscopic seaweeds and unicellular or colonial phytoplankton, collectively called algae and sea grasses are the primary producers in the sea. With the effect of solar light, they are involved in the fixation of carbon dioxide resulting in evolution of oxygen. Strictly speaking, the distinction between algae and vascular plants is very weak. Even though the cell walls of seaweeds lack lignins, a vascular system similar to that of the higher plants is apparent in many algae.

Economics determine the direction of all industries today and the algal products industry is no exception. Where non-biological sources of compounds traditionally obtained from algae have been found, economics frequently dictate that these be exploited resulting in the decline of the algal based industry, for example, the soda ash industry. The algal products industry of today may be divided into two main areas; the farming of edible seaweeds and the production of fine chemicals and polysaccharide phycocolloids.

Pharmaceutical compounds constitute one of the largest potential markets for algal products. Prior to the 1950s, the use of seaweed extracts and microalgae as drugs or drug sources was restricted to folk medicine. Use of algae in this context was recorded as long ago as 2700 BC in Chinese *Materia Medica*. To date there has been little commercial development of algal products as pharmaceutical agents. The vermifuge α-kainic acid from the red algae *Digenea simplex* was marketed in the past but is no longer available in Western countries. However, there is a tremendous potential for the development of algae as sources of pharmaceutical compounds since in the recent years researchers have ascribed a wide range of biological activities to metabolites produced by algae.

Isolation of pharmacologically active compounds from marine algae has been a subject of many intensive investigations and comprehensive account of such work in this field is given by Baslow (1969), Hoppe (1969), Guven *et al.* (1990), Pietra (1990), Lincoln *et al.* (1991), McConnell *et al.* (1994), Riguera (1997), Tringali (1997), Mayer (1998), Munro *et al.* (1999), Kerr and Kerr (1999), Cragg and Newman (2000), Mayer and Lehmann (2001), Faulkner (2001).

In vivo screens for the detection of antineoplastic activity and *in vitro* cytotoxicity assays have been used in the detection of antineoplastic and cytotoxic metabolites (Margiolis and Wilson, 1977; Hodgson, 1987; Noda *et al.*, 1989; Boyd, 1997). Initial *in vivo* screens followed by *in vitro* cytotoxicity testing to monitor purification of the active compound constitute the most common method of investigation. Many compounds, such as polysaccharides isolated from brown algae, act *via* stimulation/activation of the immune system.

Marine microalgae compose the majority of living species found in the oceans. There is no definite estimate of the total number of the existing species. New species are being discovered constantly, and the number is ever increasing. Currently, more than 10,000 known species are divided into five major divisions of marine microalgae: Chlorophyta (green algae), Chrysophyta (golden-brown, yellow algae and diatoms) Pyrrhophyta (dinoflagellates), Euglenophyta, and Cyanophyta (blue-green algae) (Shimizu, 1993). The phylogenic positions and physiologic characteristics of the organisms are important to consider in studying their metabolism and biochemistry. However, the taxonomy and phylogenic relationship of microalgae are the subjects on which taxonomists have never agreed (Sieburth, 1979) (Figure 4.1).

One important issue is the handling of Cyanophyta, "Blue-green algae." Strict disciplinarians place them in bacteria (cyanobacteria) and refuse to include them in the category of algae,

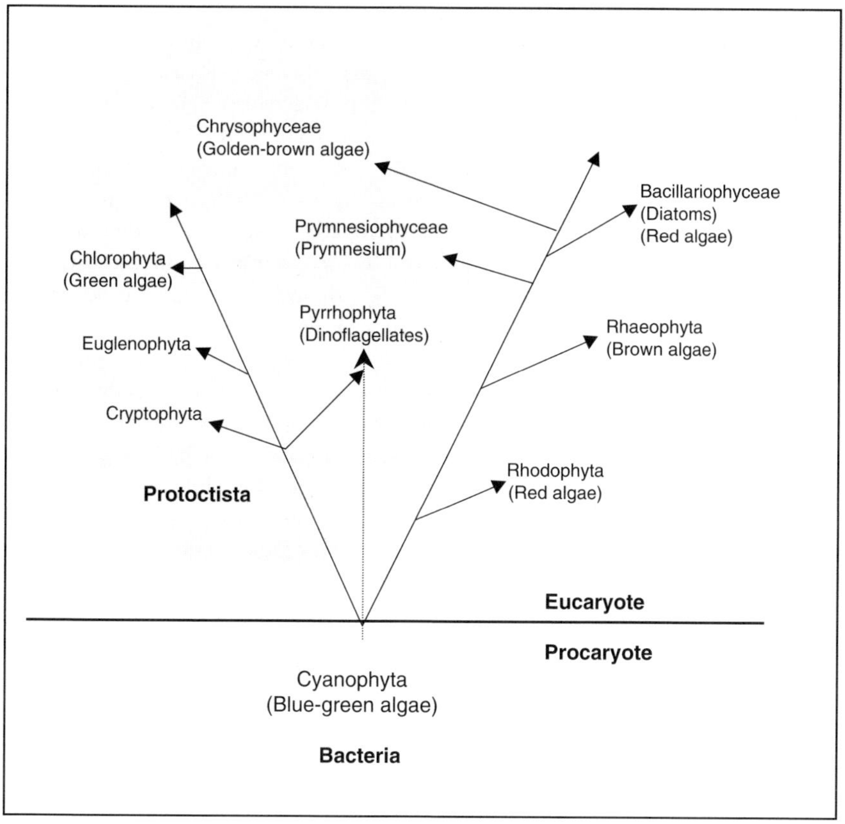

Figure 4.1 Approximate phylogenic relationship of algae.

because of their procaryotic nature. Nevertheless, the organisms are photosynthetic and share many algal characteristics with the eucaryotic counterparts. Moreover, it is generally believed that most photosynthetic algae have their phylogenic origin in Cyanophyta. Therefore, they are included in this review.

With tens of thousands unexplored species and an infinite number of possible chemovars, marine microalgae seem to be a very promising source of useful compounds. Also, there is strong evidence that many interesting compounds found in marine environments have their origins in microalgae. There is widespread speculation that many of the cyclic peptides found in tunicates and other marine invertebrates have their origin in symbiotic blue-greens or closely related organisms, prochlorons (Lewin and Cheng, 1989). For example, it is speculated that the symbiotic prochloron in the tunicate, *Didemnum* sp. is totally or partially responsible for the production of didemnins. With a few exceptions, it is not feasible to do chemical work with material from natural population of marine microalgae. At present, many important organisms remain unculturable despite enormous efforts.

From 1960 to 1982 some 16,000 marine organism-derived samples were screened for antitumor activity, mainly by the NCI. In the early 1980s, the NCI program was discontinued because it was perceived that few novel active leads were being isolated from natural sources. Of particular concern was the failure to yield agents possessing activity against the solid tumor disease types. This apparent failure might, however, be attributed more to the nature of the primary screens being used at the time, rather than to a deficiency of nature.

During 1985–90 the NCI developed a new *in vitro* screen based upon a diverse panel of human tumor cell lines (Boyd, 1997). The screen strategy comprised 60 human cancer cell lines derived from nine cancer types, organized into sub-panels representing leukemia, lung, colon, CNS, melanoma, ovarian, renal, prostate and breast. In early 1999, a pre-screen comprising three cell lines (MCF-7 (breast), NCI H460 (lung), SK268 (CNS)), which detected >95% of the materials found to exhibit activity in the 60 cell line screen was introduced. With the development of this new *in vitro* cell line screening strategy, the NCI once again turned to nature as a potential source of novel anticancer agents and a new natural products acquisition program was implemented in 1986.

In this chapter are reviewed algal extracts and isolated metabolites with cytotoxic and antineoplastic activity and potential for pharmaceutical exploitation. The data concerning the activities exhibited from crude extracts or mixtures are summarized in Tables 4.1–4.4 organized on the phylogenetic basis of the source organism and brief description of the activities is included in the tables. Reports on the cytotoxicity or antineoplastic activity of isolated algal metabolites are organized in Tables 4.5–4.8 with emphasis on the chemical nature. The chemical structures, the exhibited activity and mode of action are briefly discussed in the text with reference to the original articles. This review covers the literature till February 2002.

4.2 Cytotoxic metabolites from chlorophyta

Dimethylmethane derivatives **C-1**, **C-2** and **C-3** isolated from the extract of *Avrainvillea rawsonii* exhibited moderate inhibition of the IMPDH enzyme, which is involved in cell proliferation. The IC_{50} in μM, was 18.0, 10.0 and 7.4, respectively.

From the green alga *Bryopsis* sp. a bioactive depsipeptide, Kahalalide F (C-4), was isolated from the ethanolic extract. This compound shows selectivity against solid tumor cell lines. The IC_{50} values against A-549, HT-29, LOVO, P-388, KB and CV-1 cell lines are 2.5, 0.25, <1.0, 10, >10 and 0.25 $\mu g\,mL^{-1}$, respectively.

Recent data presented at the International Conference on Molecular Targets and Cancer Therapeutics, suggest that Kahalalide F (KF) is a novel anticancer compound with potential in the treatment of refractory ovarian and prostate cancers and leukemia. KF is one of a family of novel dehydroaminobutyric acid-containing peptides, which have shown activity in a number of solid tumor models.

At the moment an ongoing Phase I clinical and pharmacokinetic study in patients with advanced, metastatic, androgen-refractory prostate cancer is held. In this study, KF has been administered as an intravenous 1-h infusion on 5 consecutive days every 3 weeks. So far, the study has included 12 patients (across 6 dose levels), using an equivalent total dose of 100–2830 $\mu g\,m^{-2}$. To date, the schedule has been well tolerated, though adverse events include rapidly reversible mild headache, fatigue, and reversible transaminitis. The only drug toxicity observed so far was rapidly reversible Grade 3 transaminitis at 320 $\mu g\,m^{-2}$ day. Clinical benefit associated with pain relief was expressed by a decrease in *prostate specific antigen (PSA)* of over 50%. Pharmacokinetic analysis has shown KF to be rapidly eliminated, with potentially active concentrations being reached using a dosage of 425 $\mu g\,m^{-2}$ day during five consecutive days. No metabolites have been found, and the maximum tolerated dose has not yet been reached.

Table 4.1 Chlorophyta extracts

Source	*Chemistry*	*Activity*	*Literature*
Anadyomene menziesii	Aqueous extract	Against KB cell line system	Hodgson, 1984
Anadyomene stellata	Aqueous extract	Against KB cell line system	
	Chloroform extract	Against PS cell culture	
Caulerpa prolifera	Extract	Against PS cell culture	
Caulerpa racemosa var. *peltata*	Extract	Against P-388 lymphocytic leukemia Against Ehrlich ascites tumor systems	Kashiwagi *et al.*, 1980
Caulerpa racemosa var. *laete-virense*	Methanolic extract	Against L-1210 mouse leukemia cell lines	Harada *et al.*, 1997
Caulerpa sertularioides	Methanolic extract	Strong *in vitro* telomerase inhibiting activity when added to MOLT-4 cell culture at a level of 1.25% (v/v)	Kanegawa *et al.*, 2000
	Extract	Against PS cell culture	Hodgson, 1984
Caulerpa taxifolia	Aqueous extract	Lethal in mice at $1\,g\,kg^{-1}$ (winter and spring extract) Cytotoxic activity against the fibroplastic cell line BHK21/C13 with an IC_{50} $800 \pm 28\,\mu g\,mL^{-1}$ (winter extract)	Fischel *et al.*, 1995; Lemée *et al.*, 1993a
	Methanolic extract	Lethal in mice in 12 h at $1\,g\,kg^{-1}$ (summer extract) Cytotoxic activity against the fibroplastic cell line BHK21/C13 with an $IC_{50} = 250 \pm 20\,\mu g\,mL^{-1}$ (winter extract); $150 \pm 14\,\mu g\,mL^{-1}$ (summer extract) Toxicity against sea urchin eggs with an $IC_{50} = 65 \pm 9\,\mu g\,mL^{-1}$ (autumn extract); $330 \pm 15\,\mu g\,mL^{-1}$ (winter extract)	
	Dichloromethane phase	Lethal in mice in 12 and 24 h at 150 and $75\,mg\,kg^{-1}$, respectively (autumn extract) Toxicity against sea urchin eggs with an $IC_{50} = 26 \pm 8\,\mu g\,mL^{-1}$ (autumn extract)	
	Ether phase	Lethal in mice in 12 and 24 h at 200 and $150\,mg\,kg^{-1}$, respectively (autumn extract) Toxicity against sea urchin eggs with an $IC_{50} = 16 \pm 3\,\mu g\,mL^{-1}$ (autumn extract)	

(continued)

Table 4.1 (Continued)

Source	*Chemistry*	*Activity*	*Literature*
Caulerpa verticillata	Extract	Against PS cell culture	Hodgson 1984
Cladophoropsis vaucheriaeformis	Methanolic extract	Cytostatic activity against L-1210 and P-388 mouse leukemia cell lines, 95% inhibition of growth rate at 50 $\mu g\,mL^{-1}$	Harada *et al.*, 1997; Harada and Kamei, 1998
Cladophoropsis zollingeri	Methanolic extract	Medium *in vitro* telomerase inhibiting activity when added to MOLT-4 cell culture	Kanegawa *et al.*, 2000
Codium pugniformis	Purified aqueous extract	Against Ehrlich ascites tumor systems Against solid tumors produced by Elrlich carcinoma Against Sarcoma-180	Nakazawa *et al.*, 1976a
Enteromorpha prolifera	Methanolic extract	63.7% inhibition of Trp-P-1-induced umu C gene expression of *Salmonella Typhimurium* (TA 1535/pSK 1002) and 90.6% inhibition of TPA-dependent ornithine decarboxylase induction in BALB/c 3T3 fibroblast cells	Okai *et al.*, 1994
Halicoryne wrightii	Methanolic extract	Medium *in vitro* telomerase inhibiting activity when added to MOLT-4 cell culture	Kanegawa *et al.*, 2000
Halimeda discoidea	Methanolic extract	Against L-1210 and P-388 mouse leukemia cell lines 90% inhibition of growth rate at 12.5 $\mu g\,mL^{-1}$	Harada *et al.*, 1997; Harada and Kamei, 1998
Halimeda macroloba	Methanolic extract	Against L-1210 mouse leukemia cell lines	Harada *et al.*, 1997
Halimeda sp.	Extract	Against P-388 lymphocytic leukemia Against Ehrlich ascites tumor systems	Kashiwagi *et al.*, 1980
Hizikia fusiformis	Aqueous extract	Strong immunomodulating activity on human lymphocytes	Shan *et al.*, 1999
Meristotheca papulosa	Aqueous extract	Strong immunomodulating activity on human lymphocytes	Shan *et al.*, 1999
Monostroma nitidium	Non-dialyzable fraction sulfated polysaccharides	Against L-1210 Leukemia *In vivo*, 56% inhibition on tumor growth with 400 $mg\,kg^{-1}$28 days	Yamamoto *et al.*, 1982 Noda *et al.*, 1982
Tydemania expeditionis	Extract	Against P-388 lymphocytic leukemia Ehrlich ascites tumor systems	Kashiwagi *et al.*, 1980
Udotea geppii	Extract	Against P-388 lymphocytic leukemia Against Ehrlich ascites tumor systems	Kashiwagi *et al.*, 1980
Ulva lactuca	Ulvan oligosaccharides	Modification of the adhesion phase and the proliferation of normal colonic and undifferentiated HT-29 cells	Kaeffer *et al.*, 1999

Table 4.2 Rhodophyta extracts

Source	*Chemistry*	*Activity*	*Literature*
Ahnfeltia paradox	Methanolic extract	Medium *in vitro* telomerase inhibiting activity when added to MOLT-4 cell culture	Kanegawa *et al.*, 2000
Amphiroa zonata	Methanolic extract	Selective cytotoxicity to all leukemic cell lines at concentrations 15–375 $\mu g\, mL^{-1}$ Against murine leukemic cells L-1210 Against human leukemic cells K-562, HL60, MOLT-4, Raji, WIL2-NS	Harada and Kamei, 1997
Acrosorium flabellatum	PBS extract	Medium *in vitro* telomerase inhibiting activity when added to MOLT-4 cell culture	Kanegawa *et al.*, 2000
Bangia sp.	Extract	Against P-388 lymphocytic leukemia Against Ehrlich ascites tumor systems	Kashiwagi *et al.*, 1980
Chondria crassicaulis	Methanolic extract	Against L-1210 mouse leukemia cell lines	Harada *et al.*, 1997
Chondrus occellatus	PBS extract	Against L-1210 mouse leukemia cell lines	Harada *et al.*, 1997
Cryptomenia crenulata	Extract	Against P-388 lymphocytic leukemia Against Ehrlich ascites tumor systems	Kashiwagi *et al.*, 1980
Eucheuma muricatum	Aqueous extract	Weak immunomodulating activity on human lymphocytes	Shan *et al.*, 1999
Galaxaura robusta	Methanolic extract	Medium *in vitro* telomerase inhibiting activity when added to MOLT-4 cell culture	Kanegawa *et al.*, 2000
Galaxaura falcata	Methanolic extract	Medium *in vitro* telomerase inhibiting activity when added to MOLT-4 cell culture	Kanegawa *et al.*, 2000
Gloiopeltis tenax	Water extract funoran, sulfated polysaccharide	Significantly inhibited the growth of Ehrlich ascites carcinoma and solid Ehrlich, Meth-A fibrosarcoma, and Sarcoma-180 tumors	Ren *et al.*, 1995
Gracilaria salicornia	Extract	Against P-388 lymphocytic leukemia Against Ehrlich ascites tumor systems	Kashiwagi *et al.*, 1980
Herposiphonia arcuata	Extract	Against P-388 lymphocytic leukemia Against Ehrlich ascites tumor systems	
Laurencia papillosa	Methanolic extract	Medium *in vitro* telomerase inhibiting activity when added to MOLT-4 cell culture	Kanegawa *et al.*, 2000
Laurencia yamadae	Methanolic extract	Medium *in vitro* telomerase inhibiting activity when added to MOLT-4 cell culture	Kanegawa *et al.*, 2000

(*continued*)

Table 4.2 (Continued)

Source	*Chemistry*	*Activity*	*Literature*
Meristotheca coacta	Methanolic extract	Medium *in vitro* telomerase inhibiting activity when added to MOLT-4 cell culture	Kanegawa *et al.*, 2000
Meristotheca papulosa	Water extract	Weak immunomodulating activity on human lymphocytes	Shan *et al.*, 1999
Plocamium telfairiae	Methanolic extract	Against L-1210 mouse leukemia cell lines	Harada *et al.*, 1997
Porphyra tenera	Methanolic extract	54.4% inhibition of Trp-P-1-induced umu C gene expression of *Salmonella Typhimurium* (TA 1535/pSK 1002) and 92.4% inhibition of TPA-dependent ornithine decarboxylase induction in BALB/c 3T3 fibroblast cells	Okai *et al.*, 1994
	Extracts	Inhibition to mutagenicity produced by 1,2-dimethylhydrazine and other carcinogens Inhibition of mammary tumors induced by 1,2-dimethylhydrazine	Reddy *et al.*, 1984; Teas, 1983; Teas *et al.*, 1984; Yamamoto and Maruyama, 1985 Yamamoto *et al.*, 1987
	Methanolic extract mainly β-carotene, chlorophyll-α and lutein	Suppressive effect on mutagen-induced umu C gene expression in *Salmonella Typhimurium* (TA 1535/pSK 1002) Additive effect of these pigments (inhibition 19.6–30.8% at 20 $\mu g\,mL^{-1}$ of each compound, inhibition 42.8% at the same final concentration of the combined pigments)	Okai *et al.*, 1996
Porphyra yezoensis	Porphyran, phospholipid	*In vivo* inhibition on tumor growth rate 45.3–58.4% with 6.7 $mg\,kg^{-1}$7 days	Noda *et al.*, 1982
Solieria robusta	Glycoproteins	*In vitro* against mouse leukemia cells L-1210 and mouse FM 3A tumor cells	Hori *et al.*, 1988
Spyridia filamentosa	Extract	Against P-388 lymphocytic leukemia Against Ehrlich ascites tumor systems	Kashiwagi *et al.*, 1980

Table 4.3 Phaeophyta extracts

Source	*Chemistry*	*Activity*	*Literature*
Agarum crathrum	Methanolic extract	*In vitro* promoting activity of human interferon β production	Nakano *et al.*, 1997
Ascophyllum nodosum	Fucoidan extract	Inhibition of cell proliferation in both *in vitro* and *in vivo* bronchopulmonary carcinoma models	Riou *et al.*, 1996
Chordaria flagelliformis	PBS extract	Against L-1210 leukemia in mice	Harada *et al.*, 1997
Colpomenia peregrina	Ethereal extract, containing fatty acids and fucoxanthin	Against He-La cell culture	Biard and Verbist 1981
Dilophus okamurae	Methanolic extract	Strong cytotoxicity to leukemic cell lines at concentrations 50 $\mu g\,mL^{-1}$ Against murine leukemic cells L-1210 and human leukemic cells HL60 and MOLT-4	Harada and Kamei, 1997; Harada *et al.*, 1997
Ecklonia cava	PBS extract	Against L-1210 leukemia in mice	Harada *et al.*, 1997
	Non-dialyzable fraction Crude fucoidin	Against L-1210 leukemia in mice	Yamamoto *et al.*, 1987; Yamamoto *et al.*, 1982
Eisenia bicyclis	Non-dialyzable fraction of aqueous extract	Against L-1210 leukemia Against Sarcoma –180	Takahashi, 1983; Usui *et al.*, 1980;
	Crude fucoidin	Enhanced host defense mechanism to neoplasia	Yamamoto *et al.*, 1984a, 1987
	PBS extract	Against L-1210 leukaemia in mice	Harada *et al.*, 1997
	Extracts	Inhibition to mutagenicity produced by 1,2-dimethylhydrazine and other carcinogens	Reddy *et al.*, 1984; Teas, 1983; Teas *et al.*, 1984; Yamamoto and Maruyama, 1985
Isige sinicola	Methanolic extract	Medium *in vitro* telomerase inhibiting activity when added to MOLT-4 cell culture	Kanegawa *et al.*, 2000
Laminaria angustata	Extracts	Inhibition to mutagenicity produced by 1,2-dimethyl hydrazine and other carcinogens	Reddy *et al.*, 1984; Teas 1983; Teas *et al.*, 1984; Yamamoto and Maruyama, 1985
	Non-dialyzed part of aqueous extract	94.5% inhibition of Sarcoma –180 Against P-388 lymphocytic leukemia	Yamamoto *et al.*, 1974
	Methanolic extract	31.8% inhibition of Trp-P-1-induced umu C gene expression of *Salmonella typhimurium* (TA 1535/pSK 1002) and 86.6% inhibition of TPA-dependent ornithine decarboxylase induction in BALB/c 3T3 fibroblast cells	Okai *et al.*, 1994

(*continued*)

Table 4.3 (Continued)

Source	*Chemistry*	*Activity*	*Literature*
Laminaria angustata var. *longissima*	Non-dialyzed part of aqueous extract	92.3% inhibition of Sarcoma-180	Yamamoto *et al.*, 1974, 1982, 1986
	Sulfated polysaccharide	Against P-388 lymphocytic leukemia	
	Fractions of aqueous extract containing polysaccharides and nucleic acids	Against Meth-A, B-16 Melanoma and Sarcoma-180 Against L-1210 leukemia *In vitro* against L-1210 and He-La cell lines	Suzuki *et al.*, 1980
	Crude fucoidin	Against L-1210 Leukemia in mice	Maruyama *et al.*, 1987;
	Fucoidin containing fractions of aqueous extracts	Against Sarcoma-180	Yamamoto *et al.*, 1984a
	Extracts	Inhibition to mutagenicity produced by 1,2-dimethylhydrazine and other carcinogens	Reddy *et al.*, 1984; Teas, 1983; Teas *et al.*, 1984; Yamamoto and Maruyama, 1985
Laminaria cloustoni	Sulfated and degraded laminarin	Against tumors of Sarcoma-180	Fomina *et al.*, 1966; Jolles *et al.*, 1963
Laminaria japonica	Non-dialyzed part of aqueous extract	Against L-1210 leukemia in mice Against Sarcoma-180	Yamamoto *et al.*, 1982, 1986
	Sulfated polysaccharide		
Laminaria japonica var. *ochotensis*	Crude fucoidin	Against L-1210 Leukemia in mice	Maruyama *et al.*, 1987;
	Fucoidin containing fractions of aqueous extracts	Against Sarcoma-180	Yamamoto *et al.*, 1984a
	Extracts	Against mammary tumorigenesis	Yamamoto *et al.*, 1987
Laminaria religiosa	Extracts	Against mammary tumorigenesis	Yamamoto *et al.*, 1987
Macrocystis pyrifera	Extract	Against P-388 lymphocytic leukemia Against Ehrlich ascites tumor systems	Kashiwagi *et al.*, 1980

Sargassum fulvellum	Non-dialyzable fraction of the water extract	89.4% inhibition on Sarcoma-180 tumors	Yamamoto *et al.*, 1974, 1977
	Polysaccharide components		
	Acrine metabolites either sulfated epdidoglycuronoglycan or sulfated glycuronoglycan		
	D-manno-L-gulonoglycans	Neoplasm inhibitor activity	Meiun, 1981
	Sodium alginate	Against Sarcoma-180 in mice Against Ehrlich ascites, against IMC carcinomas Interferon-inducing activity	Fugihara *et al.*, 1984a,b
Sargassum hemiphyllum	Fractions of dialyzed water extracts containing polysaccharides and a sugar-containing protein	Against Ehrlich ascites and Sarcoma-180 tumors Host-mediated effects	Nakazawa *et al.*, 1974, 1976b; Nakazawa and Ikeda, 1972
	Methanolic extract	*In vitro* promoting activity of human interferon β production	Nakano *et al.*, 1997
Sargassum horneri	Fractions of dialyzed aqueous extracts containing polysaccharides and a sugar-containing protein	Against Ehrlich ascites and Sarcoma-180 tumors Host-mediated effects	Nakazawa *et al.*, 1974, 1976b; Nakazawa and Ikeda, 1972
Sargassum kjellmanianum	Non-dialyzable fraction of water extracts	Against Sarcoma-180 ascites Host-mediated mechanism	Jiang *et al.*, 1986; Yamamoto *et al.*, 1981
	Polysaccharide fraction		Nagumo, 1983
	Polysulfated polysaccharide fractions containing L-fucose	Against L-1210 tumor growth in mice	Yamamoto *et al.*, 1984b

(*continued*)

Table 4.3 (Continued)

Source	*Chemistry*	*Activity*	*Literature*
Sargassum ringgoldianum	Fucoidan, neutral lipid, glycolipid, phospholipid, polysaccharide	Inhibition 36.1–78.1% *in vivo* in mice with 40 mg kg^{-1} daily 7 days	Yamamoto *et al.*, 1984b
Sargassum thunbergii	Non-dialyzable fraction of water extracts Polysaccharide fraction	Antitumor effect on Ehrlich ascites carcinoma Enhancement of the immune response Against Sarcoma-180 ascites Host-mediated mechanism	Fujii *et al.*, 1975; Ito and Suriura, 1976; Ito and Suriura, 1976; Jiang *et al.*, 1986, Nagumo, 1983; Yamamoto *et al.*, 1981
	Polysaccharides especially fucoidan (sulfated polysaccharide, a hexouronic acid containing L-fucan sulfate)	Antitumor effect on Ehrlich ascites carcinoma in mice Enhancement of phagocytosis	Itoh *et al.*, 1993
	Fucoidan (a hexouronic acid containing L-fucan sulfate)	Inhibition of lung metastases Combination treatment with fucoidan and 5-fluorouracil inhibits significantly the lung metastases	Itoh *et al.*, 1995
Sargassum tortile	Fractions of dialyzed aqueous extracts containing polysaccharides and a sugar-containing protein	Against Ehrlich ascites and Sarcoma-180 tumors Host-mediated effects	Nakazawa *et al.*, 1974,1976b; Nakazawa and Ikeda, 1972
Sargassum yendoi	Methanolic extract	Against L-1210 leukemia in mice	Harada *et al.*, 1997
Scytosiphon lomentaria	PBS extract	Against L-1210 leukemia in mice	Harada *et al.*, 1997
Spatoglossum schmittii	Spatol	Antitumor activity in the urchin egg assay Against T242 Melanoma and 224C Astrocytoma neoplastic cell lines	Gerwick *et al.*, 1980

Stypopodium zonale	Chloroform and methanol extracts	Against PS cell cultures	Hodgson, 1984
Undaria pinnantifida	Ethanol precipitate of the aqueous extract	Against intraperitoneally implanted Lewis lung carcinoma (LCC) in syngeneic mice	Furusawa and Furusawa, 1990
	Partially purified polysaccharide	95% increase in life span (ILS)	
	composed of uronic acid, fucose and galactose at a ratio of 3 : 1 : 1	Greater ILS when combined with low doses of chemotherapeuticals (Adriamycin, cisplatin, 5-fluoro-uracil and vincristine)	
	Water insoluble fraction	Against LCC	Furusawa and Furusawa, 1985
	Mainly polysaccharide	Moderate prophylactic activity against LCC in allogeneic mice	
		Enhancement of natural cytolic activity of peritoneal macrophages against KB cells	
		Synergistic activity with standard chemotherapeuticals	
	Cold water extract	Against spontaneous AKR T cell leukemia in mice	Furusawa and Furusawa, 1988
	80% polysaccharides	Anti-LCC activity superior to that of the synthetic immunomodulator isoprinosine	Furusawa and Furusawa, 1989
	Polysaccharides, Fucoidan	Against LCC	Furusawa and Furusawa, 1985; Noda *et al.*, 1990
	Methanolic extract	33.0% inhibition of Trp-P-1-induced umu C gene expression of *Salmonella typhimurium* (TA 1535/pSK 1002) and 93.9% inhibition of TPA-dependent ornithine decarboxylase induction in BALB/c 3T3 fibroblast cells	Okai *et al.*, 1994

Table 4.4 Microalgae extracts

Source	*Chemistry*	*Activity*	*Literature*
Anacystis dimidata	Mixture of extracts	Against P-388 lymphocytic leukemia Against Ehrlich ascites tumor systems	Kashiwagi *et al.*, 1980
Aphanococcus biformis	Mixture of extracts	Against P-388 lymphocytic leukemia Against Ehrlich ascites tumor systems	Kashiwagi *et al.*, 1980
Chlorella vulgaris	Extract	Against Syngeneic ascites tumor cells Oral administration	Konishi *et al.*, 1985; Nomoto *et al.*, 1983; Soeder, 1976; Tanaka *et al.*, 1984, 1990a,b, 1997
Chlorella vulgaris strain CK22	Glycoprotein extract	Antitumor effect against both spontaneous and experimentally induced metastasis in mice	Konishi *et al.*, 1996; Noda *et al.*, 1996; Tanaka *et al.*, 1998
	Consists of 6-linked β1-6galactopyranose-rich carbohydrate (70%) and protein (30%)	Antimetastatic activity through T cell activation in lymphoid organs and enhancement of recruitment of these cells to the tumor sites. Protective effect on 5- fluorouracil-induced myelosuppression and indigenous infection in mice	
Chlorella sp.	Carbohydrate fraction	Inhibitory effect toward tumor promotion	Nomoto *et al.*, 1983;
	A-D-glucan and α-L-arabino-α-L-rhamno-α-D-galactan		Miyazawa *et al.*, 1988 Mizuno *et al.*, 1980
	Glycoproteins	*In vitro* against mouse lymphocytic leukemia cells *In vivo* against Sarcoma-180	Shinho, 1986, 1987
Chroococcus minor	Mixture of extracts	Against P-388 lymphocytic leukemia Against Ehrlich ascites tumor systems	Kashiwagi *et al.*, 1980
Entophysalis deusta	Mixture of extracts	Against P-388 lymphocytic leukemia Against Ehrlich ascites tumor systems	Kashiwagi *et al.*, 1980
Haslea ostrearia	Pigment containing aqueous extract	Against cell proliferation of solid tumors, lung carcinoma (NSCLC-N6) $IC_{50} = 30.2\,\mu g\,mL^{-1}$, kidney carcinoma (E39) $IC_{50} = 34.2\,\mu g\,mL^{-1}$ and melanoma (M96) $IC_{50} = 57.8\,\mu g\,mL^{-1}$ *In vivo* antitumor activity on mice	Carbonnelle *et al.*, 1999
Hormothamnion enteromorphoides	Peptide Hormonothamnion A	Against human lung carcinoma SW1271 ($IC_{50} = 0.2\,\mu g\,mL^{-1}$), carcinoma A529 ($IC_{50} =$) $0.16\,\mu g\,mL^{-1}$ Murine Melanoma B16-F10 ($IC_{50} = 0.13\,\mu g\,mL^{-1}$), Human colon HCT-116 ($IC_{50} = 0.72$ $0.13\,\mu g\,mL^{-1}$)	Gerwick, 1989; Gerwick *et al.*, 1989

Lyngbya confervoides	Extract	Against P-388 lymphocytic leukemia Against Ehrlich ascites tumor systems	Kashiwagi *et al.*, 1980
Lyngbya gracilis	Chloroform extract Debromoaplysiatoxin	Against P-388 lymphocytic leukemia	Mynderse *et al.*, 1977
Lyngbya majuscula	Extract	Against P-388 lymphocytic leukemia Against Ehrlich ascites tumor systems	Kashiwagi *et al.*, 1980
	Aplysiatoxin, Lyngbyatoxin A	Tumor promoters	Moore, 1982
Lyngbya sp.	Extract	Against P-388 lymphocytic leukemia Against Ehrlich ascites tumor systems	Kashiwagi *et al.*, 1980
Oscillatoria annae	Extract	Against P-388 lymphocytic leukemia Against Ehrlich ascites tumor systems	
Oscillatoria foreaui	Mixture of extracts	Against P-388 lymphocytic leukemia Against Ehrlich ascites tumor systems	
Oscillatoria nigroviridis	Chloroform extract Debromoaplysiatoxin 31-nor-debromoaplysiatoxin	*In vivo* against P-388 lymphocytic leukemia	Mynderse and Moore, 1978
Oscillatoria sp.	Extract	Against P-388 lymphocytic leukemia Against Ehrlich ascites tumor systems	Kashiwagi *et al.*, 1980
Phormidium crosbyanum	Extract	Against P-388 lymphocytic leukemia Against Ehrlich ascites tumor systems	
Phormidium sp.	Extract	Against P-388 lymphocytic leukemia Against Ehrlich ascites tumor systems	
Rivularia atra	Mixture of extracts	Against P-388 lymphocytic leukemia Against Ehrlich ascites tumor systems	
Schizothrix calcicola	Chloroform extract Debromoaplysiatoxin 31-nor-debromoaplysiatoxin	*In vivo* against P-388 lymphocytic leukemia	Mynderse and Moore, 1978
	Extract	Against P-388 lymphocytic leukemia Against Ehrlich ascites tumor systems	Kashiwagi *et al.*, 1980
Schizothrix sp.	Extract	Against P-388 lymphocytic leukemia Against Ehrlich ascites tumor systems	
Skeletonema costatum	Organic extract	*In vitro* inhibition of lung carcinoma (NSCLC-N6) cell line proliferation by inducing terminal differentiation	Bergé *et al.*, 1997
Symploca muscorum	Chloroform extract	*In vivo* against P-388 lymphocytic leukemia	Mynderse *et al.*, 1977
Tolypothrix crosbyanum var. *chlorata*	Extract	Against P-388 lymphocytic leukemia Against Ehrlich ascites tumor systems	Kashiwagi *et al.*, 1980

Caulerpenyne (**C-5**) isolated from *Caulerpa taxifolia* has been shown to be cytotoxic against KB cells and fibroblasts from hamsters. Caulerpenyne along with other drugs representative of the major classes of anticancer products was tested against eight cancer cell lines of human origin. Caulerpenyne demonstrated growth inhibitory effects in all cases with some variability between cell lines; this inter-cell variability was, however, less marked than that observed with the anticancer drug tested. Cells of colorectal cancer origin were the most sensitive to the presence of Caulerpenyne. The activity was of the same order or greater than that obtained from, cisplatinum and fotemustine. In particular, Caulerpenyne does not affect the microfilament-dependent processes of fertilization and cytokinesis and allows the beginning of mitosis, but prevents normal DNA replication and results in metaphase-like arrest of sea urchin embryos. Caulerpenyne (**C-5**) is not lethal in mice, although it displays cytotoxic activity against the fibroblastic cell line BHK21/C13 with an $IC_{50} = 15 \pm 2\,\mu g\,mL^{-1}$, as well as toxicity against sea urchin eggs with an $IC_{50} = 16 \pm 2\,\mu g\,mL^{-1}$.

Taxifolial A (**C-6**) although is structurally closely related to Caulerpenyne (**C-5**), it is less toxic in the sea-urchin test with an $IC_{50} = 28 \pm 1\,\mu g\,mL^{-1}$.

10,11-Epoxycaulerpenyne (**C-8**) is weakly active on the sea urchin eggs assay but lethal on mice at $75\,\mu g\,kg^{-1}$. According to the classification of Hodgson (1987) this compound is very toxic.

Taxifolial D (**C-7**), the only example of monoterpene isolated from *C. taxifolia*, is not active on fibroblasts and has not been tested on mice.

Clerosterol (**C-9**) and five oxygenated derivatives (**C-10** to **C-14**) were isolated from the green alga *Codium arabieum*. The cytotoxicity of these compounds was tested against the cancer cell lines, P-388, KB, A-549 and HT-29. Clerosterol exhibited significant activity against P-388 cells (ED_{50} $1.7\,\mu g\,mL^{-1}$) and was the most active against A-549 cells (ED_{50} $0.3\,\mu g\,mL^{-1}$) among the compounds tested. However, Clerosterol was inactive against the growth of KB and HT-29 cells. All oxidized products (**C-10** to **C-14**) showed significant activity against the growth of the four mentioned cancer cell lines, indicating that oxidation increases the activity of Clerosterol.

Cymobarbatol (**C-15**) and 4-isocymobarbatol (**C-16**) were isolated from the marine green alga *Cymopolia barbata*. Both compounds exhibited strong inhibition of the mutagenicity of 2-aminoanthracene and ethyl methanesulfonate toward, the T-98 strain with a metabolic activator and T-100.

Species of the genus *Halimeda* were found to contain significant amounts (~15% of the dichloromethane extracts) of Halimedatrial (**C-17**), which exhibits cytotoxic activity in laboratory bioassays. At $1\,\mu g\,mL^{-1}$, Halimedatrial completely inhibited cell division for the first cleavage of fertilized sea urchin eggs and the motility of sea urchin sperm.

Three halogenated sesquiterpene (**C-18** to **C-20**) were isolated from the green alga *Neomeris annulata*. Their cytotoxic activity was indicated by their toxicity to brine shrimp. LD_{50} values were determined for **C-18**, **C-19** and **C-20** to be 9, 8 and $16\,\mu g\,mL^{-1}$, respectively.

Sulfated cycloartanol derivatives (**C-21** to **C-23**) from the green alga *Tydemania expeditionis* were identified as inhibitors of $pp60^{v\text{-}src}$, the oncogenic protein tyrosine kinase encoded by Rous sarcoma virus. Protein tyrosine kinases comprise a large family of enzymes that regulate cell growth and intracellular signaling pathways. Inhibitors of these enzymes may have utility in cancer and other hyperproliferative conditions. Cycloartanol sulfates **C-21**, **C-22** and **C-23** showed IC_{50}s of 32, 100 and $39\,\mu M$ in the $pp60^{v\text{-}src}$ assay.

Ulvans, from *Ulva lactuca*, constitute a dietary fiber structurally similar to the mammalian glycosaminoglycans. Desulfated, reduced and desulfated-reduced polysaccharides

Table 4.5 Cytotoxic metabolites from chlorophyta

Source	*Metabolite*	*Code*	*Literature*
Avrainvillea rawsonii	Avrainvilleol	C-1	Chen and Gerwick, 1994
	Rawsonol	C-2	
	Isorawsonol	C-3	
Bryopsis sp.	Kahalide F	C-4	Hamann and Scheuer, 1993; Hamann *et al.*, 1996; Garcia-Rocha *et al.*, 1996; Goetz *et al.*, 1999
Caulerrpa taxifolia	Caulerpenyne	C-5	Fischel *et al.*, 1994, 1995; Pesando *et al.*, 1996, 1998
	Caulerpenyne	C-5	Lemée *et al.*, 1993b
	Taxifolial A	C-6	
	Taxifolial D	C-7	
	10,11-Epoxy-caulerpenyne	C-8	
Codium arabieum	Clerosterol	C-9	Sheu *et al.*, 1995
	Oxygenated Clerosterols	C-10 to C-14	
Cymopolia barbata	Cymobarbatol	C-15	Wall *et al.*, 1989
	4-Isocymobarbatol	C-16	
Halimeda sp.	Halimedatrial	C-17	Paul and Fenical, 1983
H. tuna			
H. opuntia			
H. incrassata			
H. simulans			
H. scabra			
H. copiosa			
Neomeris annulata	Halogenated sesquiterpenes	C-18 to C-20	Barnekow *et al.*, 1989
Tydemania expeditionis	Sulfated cycloartanols	C-21 to C-23	Govindan *et al.*, 1994
Ulva lactuca	Ulvan oligosaccharides		Kaeffer *et al.*, 1999

were examined on the adhesion, proliferation and differentiation of normal or tumoral colonic epithelial cells cultured in conventional or rotating bioreactor culture conditions. In conventional culture conditions, Ulvan modified the adhesion phase and the proliferation of normal colonic sells and undifferentiated HT-29 cells.

4.3 Cytotoxic metabolites from rhodophyta

The brine shrimp toxicity bioassay was used to direct the fractionation of the red alga *Ceratodictyon spongiosum* extract. This process afforded two stable conformers of a cyclic heptapeptide, *cis,cis*- and *trans,trans*- Ceratospongamide (**R**-1 and **R**-2).

Five oxygenated Desmosterols (**R**-3 to **R**-7) were isolated from the red alga *Galaxaura marginata*, which exhibited significant cytotoxicity to P-388, KB, A-549 and HT-29 cancer cell lines. Even though Desmosterol was not cytotoxic, the oxidized products were quite

Cytotoxic metabolites from chlorophyta

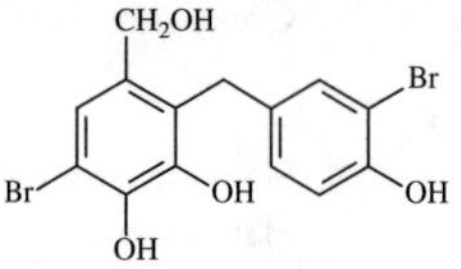
C-1

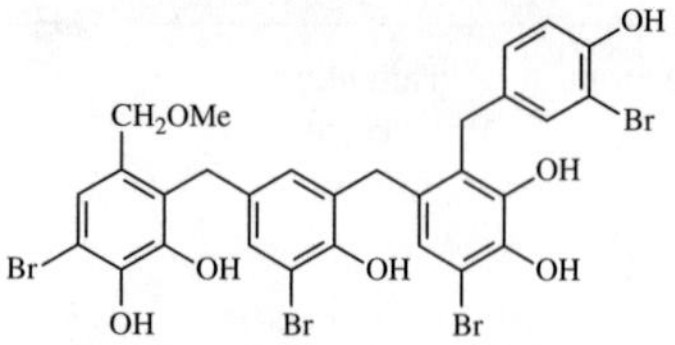
C-2

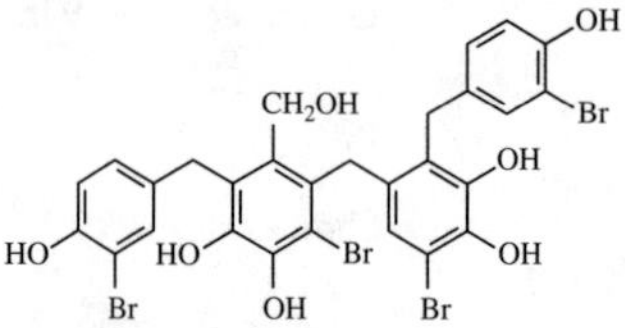
C-3

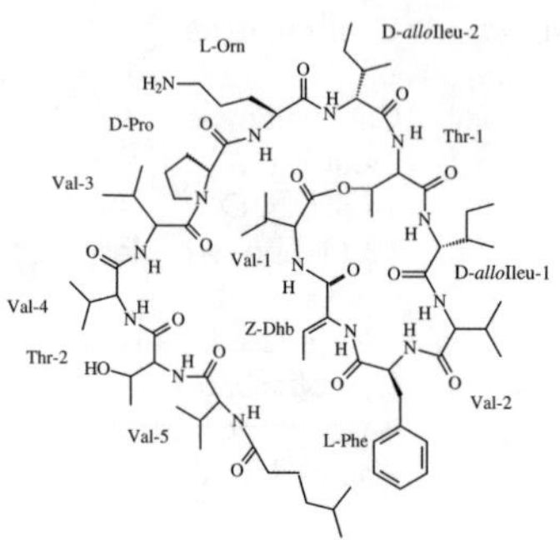
C-4

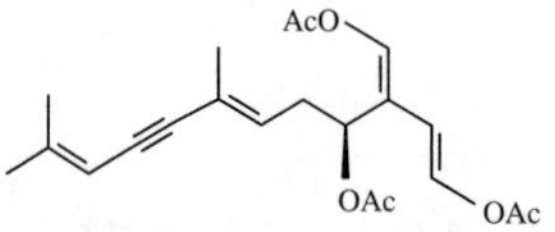
C-5

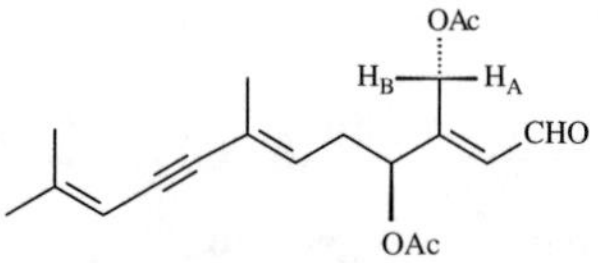
C-6

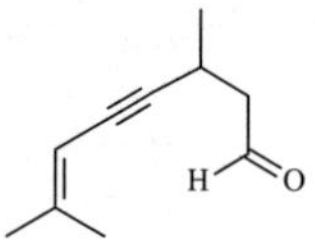
C-7

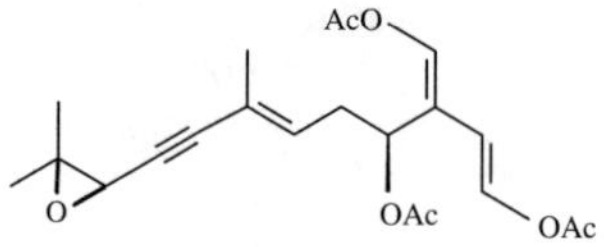
C-8

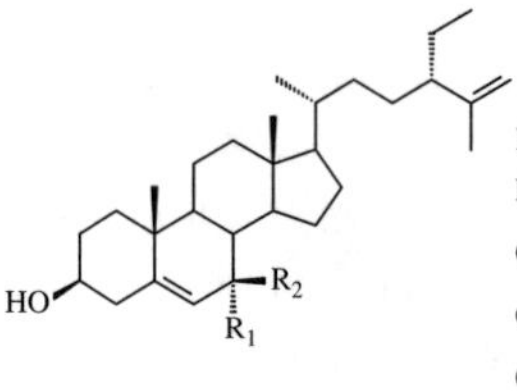

R_1	R_2	
H	H	**C-9**
O	O	**C-10**
OOH	H	**C-11**
OH	H	**C-12**

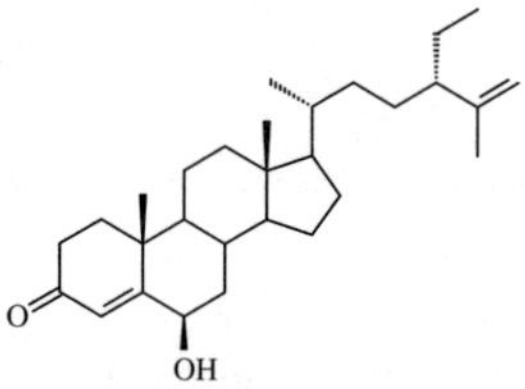
C-13

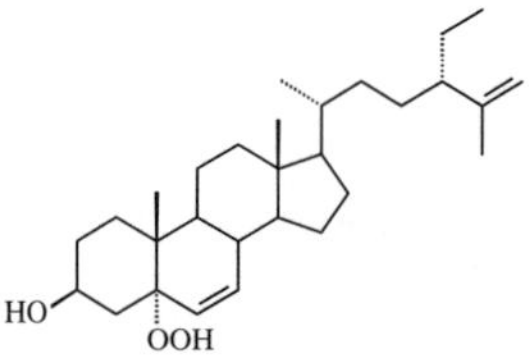
C-14

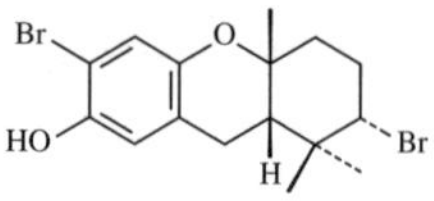
C-15

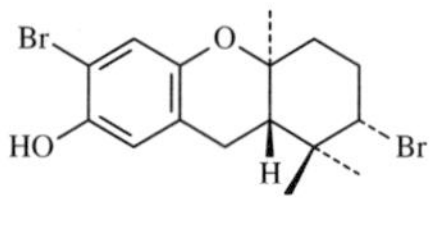

C-16

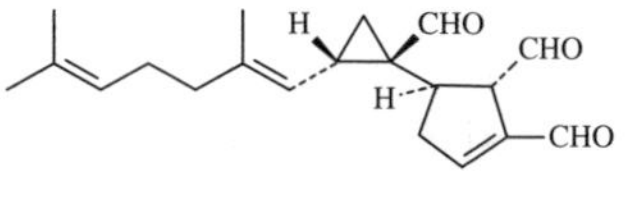

C-17

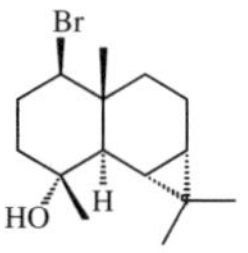

C-18

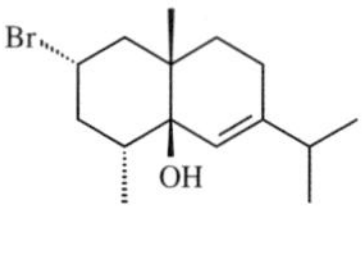

C-19

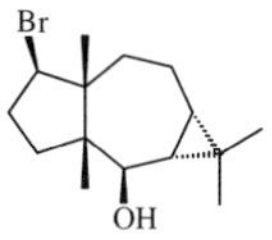

C-20

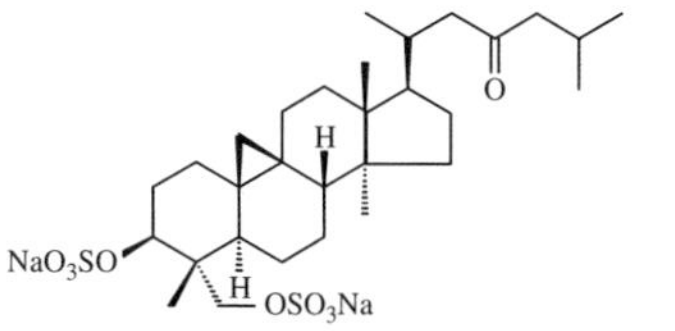

C-21

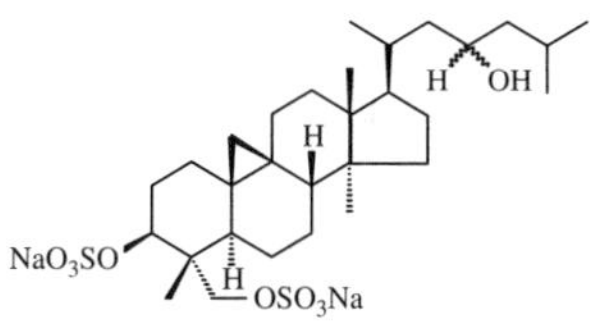

C-22

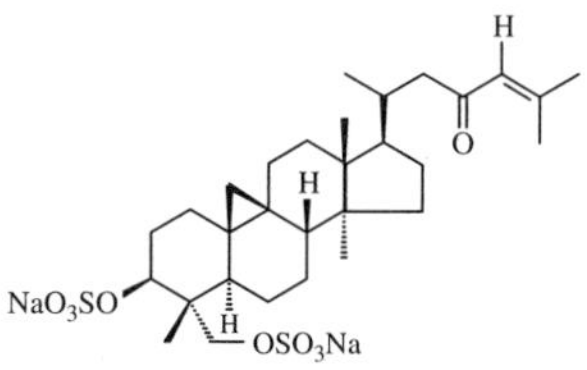

C-23

cytotoxic, indicating that oxidation increases the activity. Four additional oxygenated desmosterols (**R-8** to **R-11**) were isolated from the same organism and exhibited significant cytotoxicity against P-388, KB, A-549 and HT-29 cancer cell lines, with ED_{50} values within the range of 0.11–2.37 $\mu g\, mL^{-1}$.

From the red algae *Gigartina tenella* a sulfolipid (**R-12**) that belongs to the class of sulfoquinovosyldiacyl glycerol was isolated. The compound potently inhibited the activities of mammalian DNA polymerase α and β and terminal deoxynucleotidyl transferase (TdT), and enhanced the cytotoxicity of bleomycin. Complete inhibition doses of each were achieved at 1.0–2.0 μM for polymerase α and TdT and 7.5 μM for polymerase β.

Three new Malyngamides: Malyngamide M (**R-13**), Malyngamide N (**R-14**) and Malyngamide I acetate (**R-15**) were isolated from the Hawaiian red alga *Gracilaria coronopifolia*. Malyngamide N and Malyngamide I acetate showed moderate cytotoxicity to mouse neuroblastoma (NB) cells in the tissue culture. The IC_{50} values of **R-14** and **R-15** were 12 μM (4.9 $\mu g\, mL^{-1}$) and 12 μM (7.1 $\mu g\, mL^{-1}$), respectively. In contrast Malyngamide M showed rather weak cytotoxicity to NB cells ($IC_{50} > 20 \mu M$). Malyngamides are known as metabolites of blue green algae, in particular *Lyngbya majuscula*. Furthermore it has been reported that epiphytes such as blue green algae grow on *Gracilaria*. Therefore the true origin of **R-13** to **R-15** is likely a blue green alga that grows on *Gracilaria coronopifolia*.

A cytotoxic oxysterol, 16β-hydroxy-5a-cholestane-3,6-dione (**R-16**) was isolated from the red alga *Jania rubens* and was found to be significantly cytotoxic towards the KB tumor cell line with an ID_{50} value 0.5 $\mu g\, mL^{-1}$.

Callicladol (**R-17**), a brominated metabolite has been isolated from the red alga *Laurencia calliclada*. This compound displayed cytotoxic activity *in vitro* against P-388 murine leukemia cell with IC_{50} value 1.75 $\mu g\,mL^{-1}$.

Six chamigrane derivatives (**R-18** to **R-23**) isolated from *Laurencia cartilaginea*, were screened for toxicity. All metabolites have shown remarkable results against various cancer cell lines at low concentrations, especially to HT-29. The IC_{50} values for the compounds **R-18** to **R-23** were 1.0, 1.0, 1.0, 1.0, 5.0 and 5.0 $\mu g\,mL^{-1}$ for the P-388 cell line, 0.1, 1.0, 1.0, 1.0, 5.0 and 1.0 $\mu g\,mL^{-1}$ for the A-549 cell line, 0.1, 0.025, 0.025, 0.25, 0.5 and 0.25 $\mu g\,mL^{-1}$ for the HT-29 cell line and 0.1, 1.0, 1.0, 1.0, 10.0 and 1.0 $\mu g\,mL^{-1}$ for the MEL-28 cell line, respectively.

Majapolene A (**R-24**), a dioxabicyclo[2.2.2]-alkene, was isolated from the red alga *Laurencia majuscula*. It displayed modest mean response parameter values for all NCI 60-cell lines of 0.4 μM for GI_{50} (50% net growth inhibition, relative to controls), 0.9 μM for TGI (net total growth inhibition) and 2.8 μM for LC_{50} (50% net cell death).

Thyrsiferyl 23-acetate (**R-25**) has been isolated from the red alga *Laurencia obtusa*, which showed strong cytotoxicity against mammalian cells. Actually, TF23A is a specific inhibitor of protein phosphatase 2A (PP2A) activity.

Red seaweeds of genus *Laurencia* is known to produce interesting active polyether squalene-derived metabolites, which possess strong cytotoxic properties. Mechanisms of growth inhibition by the novel marine compound Dehydrothyrsiferol (DHT) (**R-26**), isolated from the red alga *Laurencia viridis* and *Laurencia pinnatifida*, were investigated in a sensitive and an MDR^{+} human epidermoid cancer cell line. DHT was found to circumvent multidrug resistance mediated by P-glycoprotein. Cell cycle analysis revealed an accumulation in S-phase. Growth inhibition in KB cancer cells is not mediated by apoptosis but by growth retardation. The IC_{50} values of DHT in all investigated cell lines were, although in the μM range, found to be higher than the ones determined for the clinically established chemotherapeutic compound Doxorubicin and the cytotoxic compound Colchicine. The IC_{50} values determined in tumor cell lines derived from different primary tissues support the notion that the cytotoxicity mediated by DHT may be more tissue related than correlated to a single mechanism of growth inhibition throughout the various cancer systems.

Screening for cytotoxicity was performed on compounds **R-26** to **R-36** with a battery of cultured tumor cell lines: P-388, suspension culture of a lymphoid neoplasm from a DBA/2 mouse; A-549, monolayer culture of a human lung carcinoma; HT-29, monolayer culture of a human colon carcinoma; MEL-28, monolayer culture of a human melanoma. This assay proved them to possess a potent and selective activity against P-388 cells.

Compounds Thyrsiferol (**R-27**), Dehydrothyrsiferol (**R-26**), Dehydrovenustatriol (**R-28**), Isodehydrothyrsiferol (**R-31**) and Thyrsenol B (**R-36**) had $IC_{50} = 0.01\,\mu g\,mL^{-1}$. This activity was significantly higher than that of 15–16-dehydrovenustatriol (**R-29**), Thyrsenol A (**R-35**), ($IC_{50} = 0.25\,\mu g\,mL^{-1}$), 16-hydroxydehydrothyrsiferol (**R-32**), 10-epi-15–16-dehydrothyrsiferol (**R-33**), ($IC_{50} = 0.50\,\mu g\,mL^{-1}$), 10-epidehydrothyrsiferol (**R-34**) ($IC_{50} = 1.00\,\mu g\,mL^{-1}$) and predehydrovenustatriol acetate (**R-30**) ($IC_{50} = 1.20\,\mu g\,mL^{-1}$), establishing that small chemical changes in the molecule greatly affect the cytotoxicity. Moreover compound **R-31** showed selective activity against P-388 mouse lymphoid neoplasm.

Martiriol (**R-37**) along with three other derivatives of dehydrothyrsiferols (**R-38** to **R-40**) were isolated from *Laurencia viridis* and tested for their cytotoxicity against different cancer cell lines. The results showed that Martiriol (**R-37**) was inactive at concentrations lower than 10 $\mu g\,mL^{-1}$ and compounds **R-38** to **R-40** were inactive at concentrations lower than 1 $\mu g\,mL^{-1}$.

From the tropical marine red alga *Plocamium hamatum* two polyhalogenated monoterpenes (**R-41**, **R-42**) were isolated. Compound **R-41** was moderately cytotoxic (IC_{50}: Lu1 12.9 μg mL^{-1}, KB 13.3 μg mL^{-1}, ZR-75–1 7.8 μg mL^{-1}) as was compound **R-42** (IC_{50}: KB-V (-VBL) 5.3 μg mL^{-1}, KB 12.4 μg mL^{-1}, LNCaP 14.8 μg mL^{-1}). An array of similar halogenated monoterpenes has been isolated by other researchers from *Plocamium* sp. According to Mynderse and Faulker (1978) the observed chemical variability is not caused from extraction decompositions but is depended on the algae geographic location.

The polyhalogenated acyclic monoterpene Halomon, (**R-43**) was obtained as a major component of the organic extract of the red algae *Portieria hornemannii*. It exhibited highly differential cytotoxicity against the NCI's new *in vitro* human tumor cell line screening panel; brain tumor, renal, and colon tumor cell lines were most sensitive, while leukemia and melanoma cell lines were relatively less sensitive. On the basis of its unprecedented cytotoxicity profile on the NCI primary screen this compound has been selected by the NCI Decision Network Committee for preclinical drug development. Pharmacological studies of Halomon have been conducted concerning the *in vitro* metabolism, pharmacokinetics, bioavailability and tissue distribution in mice.

A second collection of *Portieria hornemannii* yielded a monocyclic 3-halogenated monoterpene (1-[3-(1-chloro-2(E)-propenyl)]-2,4-dichloro-3,3-dimethylcyclohex-5-ene, **R-44**), which proved to be one order of magnitude less potent than **R-43** and devoid of differential activity.

Isohalomon **R-45**, an isomer of Halomon, **R-43**, with a diatropic rearrangement of the halogens at C-6 and C-7, dehydrobromo derivative of isohalomon **R-46**, dehydrochloro derivative of Halomon **R-47** and the monocyclic halogenated monoterpene **R-48** uniformly exhibited the unique differential cytotoxicity profile reported earlier for Halomon against the NCI panel of 60 human tumor cell lines, with comparable panel-averaged potency.

The monocyclic halogenated monoterpene **R-48** was more comparable in overall (panel-averaged) potency to Halomon, however, there was little differential response of the cell lines, and consequently no significant correlation to the profile of the Halomon **R-43**. Mean panel response (Values $\times 10^{-6}$M): **R-43** $GI_{50} = 0.676$, $LC_{50} = 11.5$; **R-44** $GI_{50} = 20.0$, $LC_{50} > 100$; **R-45** $GI_{50} = 1.32$, $LC_{50} = 16.2$; **R-46** $GI_{50} = 0.741$, $LC_{50} = 17.0$; **R-47** $GI_{50} = 0.691$, $LC_{50} = 13.5$; **R-48** $GI_{50} = 1.15$, $LC_{50} = 20.0$.

A structure/activity relationship study with compounds **R-43**, and **R-45** to **R-48** exhibited a similar cytotoxicity profile, displaying higher activity than **R-49** to **R-53**. These results suggest that halogen on C-6 is essential for this characteristic activity profile.

Three agglutinins have been isolated from the aqueous ethanolic extract of the marine red alga *Solieria robusta*. These proteins, designated solnins A, B and C, were monomeric glycoproteins with a similar MW and they share predominant amino acids as Gly, Asx and Glx. Solnins showed mitogenic activity for mouse splenic lymphocytes, while they inhibited the growth *in vitro* of mouse leukemia cells L-1210 and mouse FM3A tumor cells.

Four sulfated triterpenoids **R-54** to **R-57** were isolated from brine shrimp-toxic fractions of the methanolic extract of the red alga *Tricleocarpa fragilis*. Compounds **R-54** and **R-55** were the most active, showing $55.7 \pm 8.7\%$ and $47.1 \pm 15.1\%$ immobilization of brine shrimp respectively, at 17 μg mL^{-1}. Compounds **R-56** and **R-57** showed $39.1 \pm 11.0\%$ and $35.5 \pm 12.8\%$ immobilization respectively, at 50 μg mL^{-1}. Toxicity toward P-388, A-549, MEL-28 and HT-29 cell lines was also evaluated. IC_{50} values for **R-54** and **R-55** were > 10 μg mL^{-1} and for **R-56** and **R-57** > 1 μg mL^{-1} against all cell lines tested.

Table 4.6 Cytotoxic metabolites from rhodophyta

Source	*Metabolite*	*Code*	*Literature*
Ceratodictyon spongiosum	*cis,cis*-Ceratospongamide	R-1	Tan *et al.*, 2000a
	trans,trans-Ceratospongamide	R-2	
Galaxaura marginata	Oxygenated desmosterols	R-3 to R-7	Sheu *et al.*, 1996
		R-8 to R-11	Sheu *et al.*, 1997a
Gigartina tenella	Sulfoquinovosyldiacyl glycerol	R-12	Ohta *et al.*, 1999
Gracilaria coronopifolia	Malyngamide M	R-13	Kan *et al.*, 1998
	Malyngamide N	R-14	
	Malyngamide I acetate	R-15	
Jania rubens	16β-Hydroxy-5a-cholestane-3,6-dione	R-16	Ktari *et al.*, 2000
Laurencia calliclada	Callicladol	R-17	Suzuki *et al.*, 1995
Laurencia cartilaginea	Chamigrane deriv.	R-18 to R-21	Juagdan *et al.*, 1997
	Ma'ilione	R-22	
	Allo-isoobtusol	R-23	
Laurencia majuscula	Majapolene A	R-24	Erickson *et al.*, 1995
Laurencia obtusa	Thyrsiferyl 23-acetate	R-25	Matsuzawa *et al.*, 1994
Laurencia viridis	Dehydrothyrsiferol (DHT)	R-26	Pec *et al.*, 1998, 1999
	Dehydrothyrsiferol (DHT)	R-26	Fernández *et al.*, 1998
	Thyrsiferol	R-27	
	Dehydrovenustatriol	R-28	
	15–16 Dehydrovenustatriol	R-29	
	Predehydrovenustatriol acetate	R-30	
	Isodehydrothyrsiferol	R-31	
	16-Hydroxydehydrothyrsiferol	R-32	
	10-epi-15,16 Dehydrothyrsiferol	R-33	
	10-epi-Dehydrothyrsiferol	R-34	
	Thyrsenol A	R-35	Norte *et al.*, 1996, 1997
	Thyrsenol B	R-36	
	Martiriol	R-37	Manriquez *et al.*, 2001
	Dehydrothyrsiferol derivatives	R-38 to R-40	

Plocamium hamatum	Polyhalogenated monoterpenes	R-41, R-42	Coll *et al.*, 1988 Koenig *et al.*, 1999
Portieria hornemannii	6(R)-Bromo-3(S)-(bromomethyl)-7-methyl-2,3,7-trichloro-1-octene (Halomon)	R-43	Fuller *et al.*, 1992
	1-[3-(1-chloro-2(E)-propenyl)]-2,4-dichloro-3,3-dimethylcyclohex-5-ene	R-44	
Portieria hornemannii	Isohalomon	R-45	Fuller *et al.*, 1994
	Dehydrobromo derivative of Isohalomon	R-46	Egorin *et al.*, 1996, 1997
	Dehydrochloro derivative of Halomon	R-47	
	Monocyclic halogenated monoterpene	R-48	
	Acyclic halogenated monoterpene	R-49	
	Acyclic halogenated monoterpene	R-50	
	Acyclic halogenated monoterpene	R-51	
	Acyclic halogenated monoterpene	R-52	
	Monocyclic halogenated monoterpene	R-53	
Solieria robusta	Isoagglutinins		Hori *et al.*, 1988
	Solnins A-C		
Tricleocarpa fragilis	Triterpenoid sulfates	R-54 to R-57	Horgen *et al.*, 2000

Cytotoxic metabolites from rhodophyta

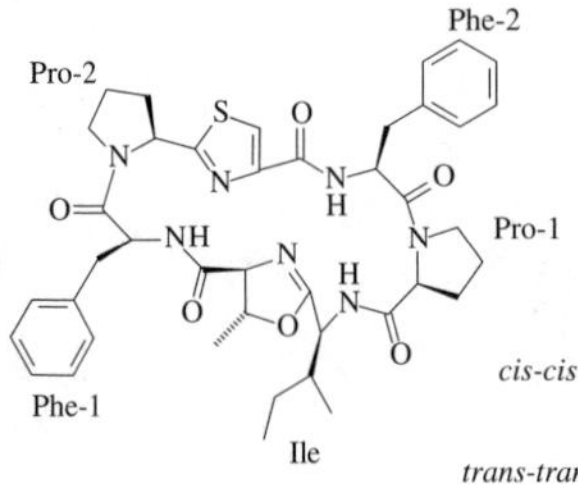
cis-cis **R-1**
trans-trans **R-2**

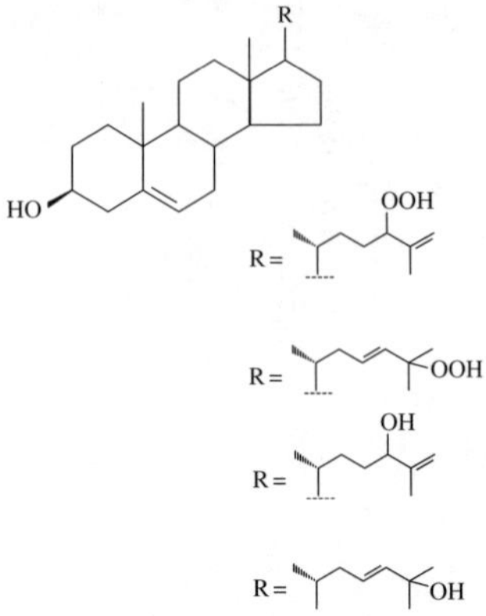
R-3
R-4
R-5
R-6

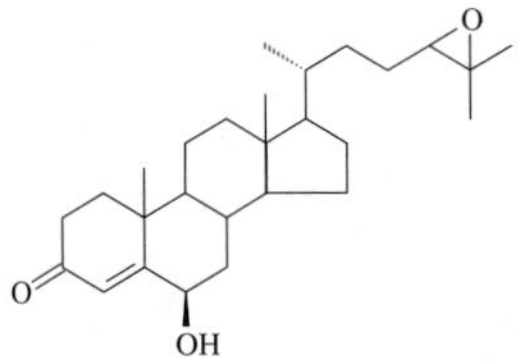
R-7

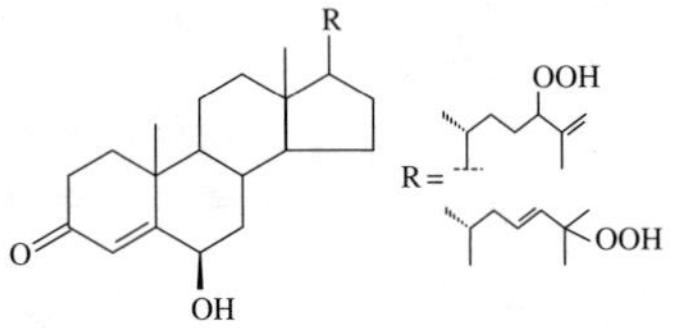
R-8
R-9

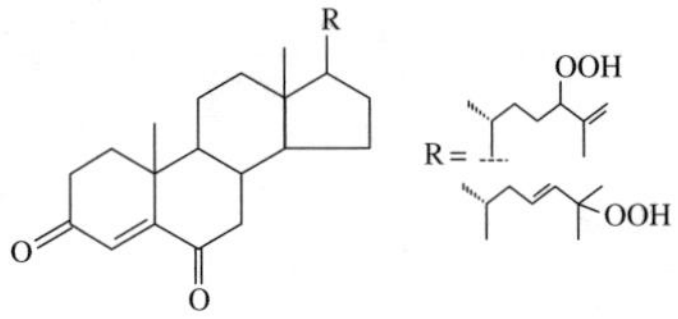
R-10
R-11

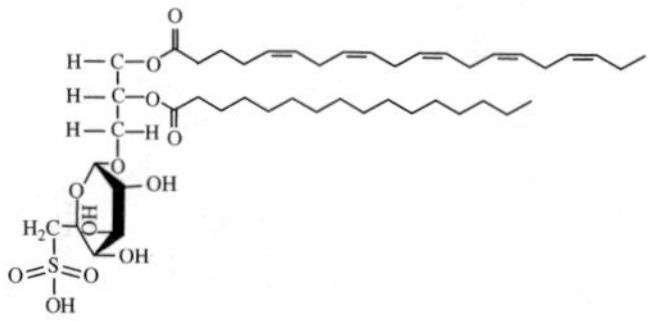
R-12

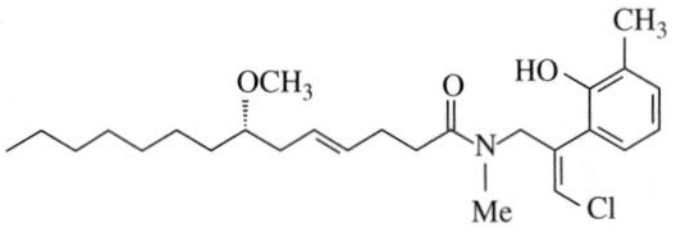
R-13

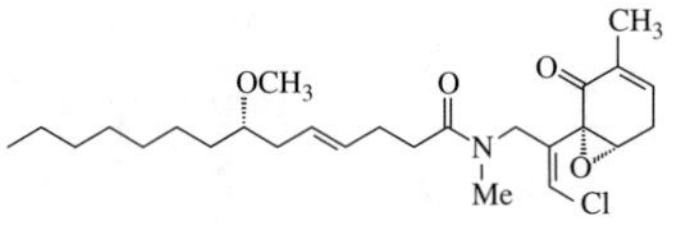
R-14

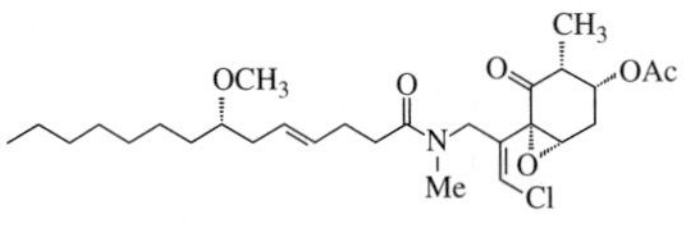
R-15

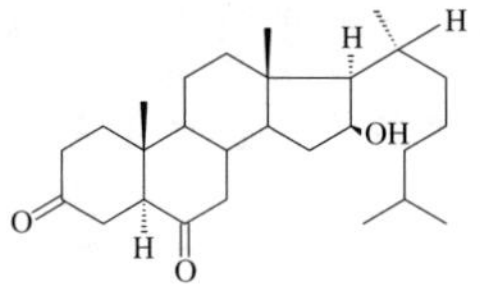
R-16

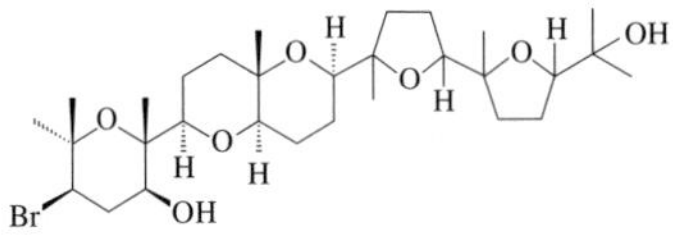
R-17

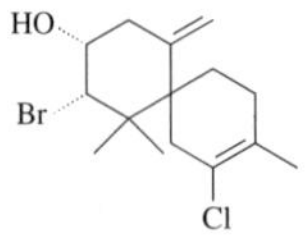
R-18

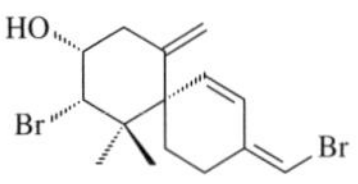
R-19

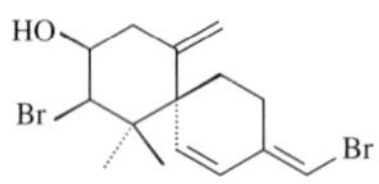
R-20

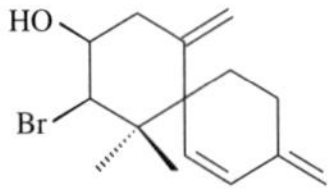
R-21

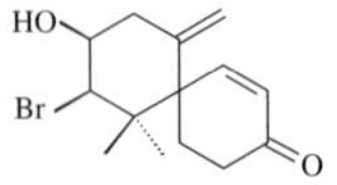
R-22

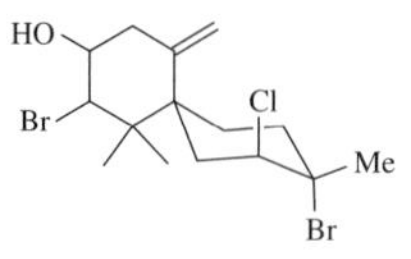
R-23

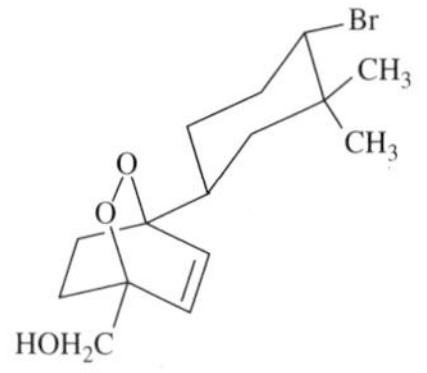
R-24

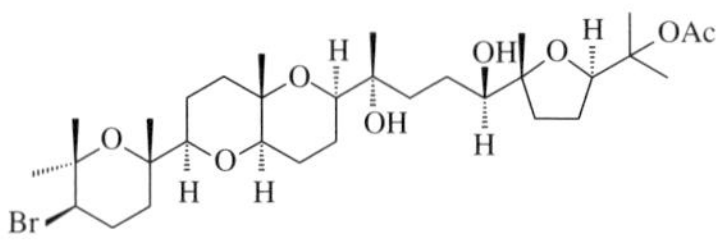
R-25

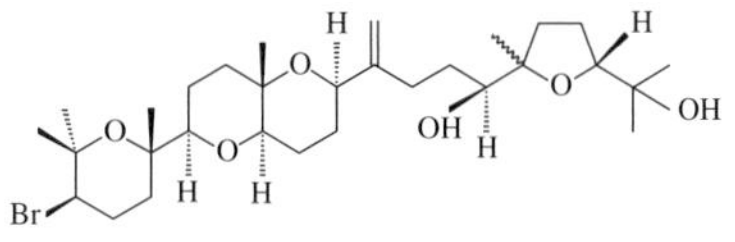
R-26

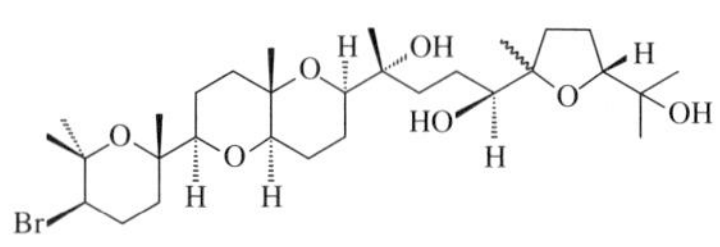
R-27

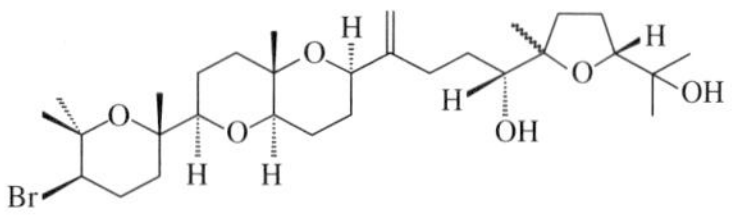
R-28

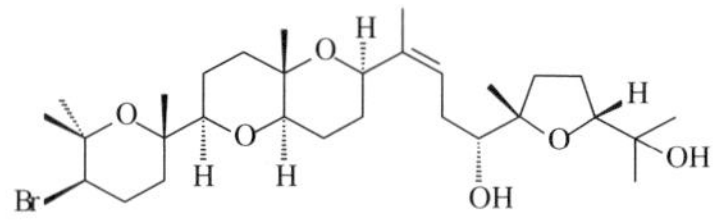
R-29

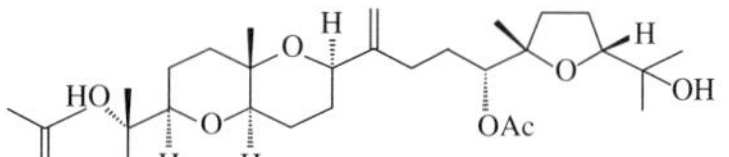
R-30

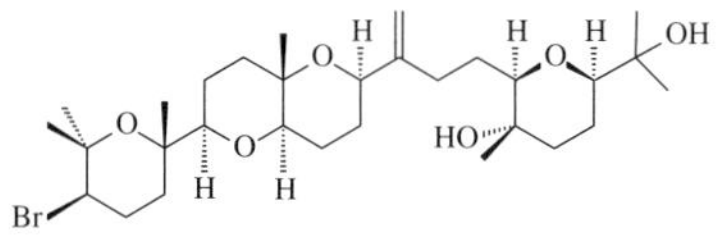
R-31

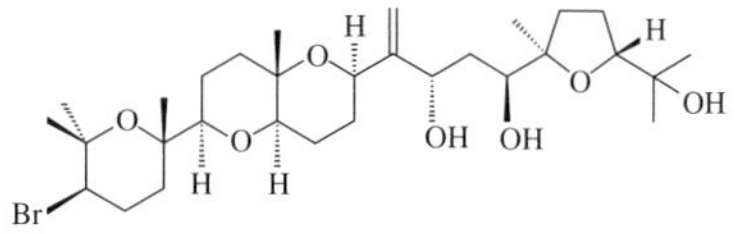
R-32

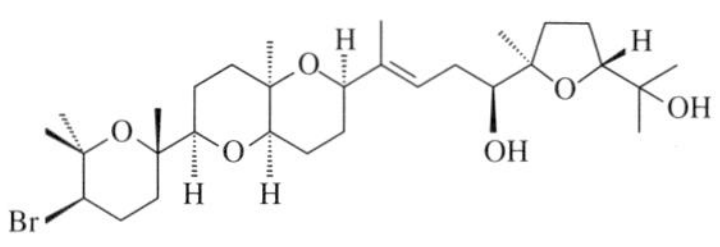
R-33

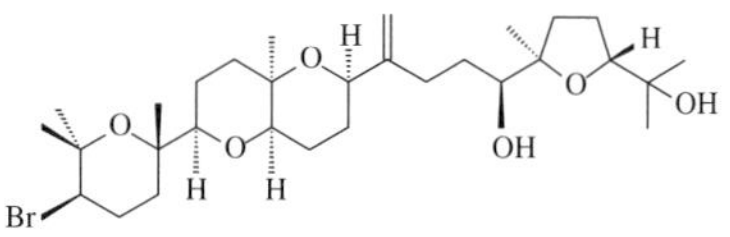
R-34

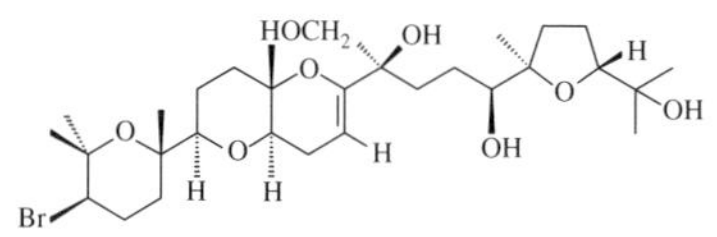
R-35

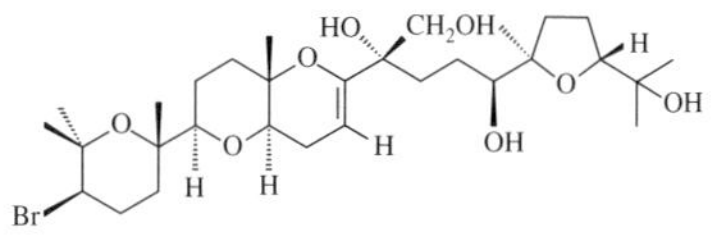
R-36

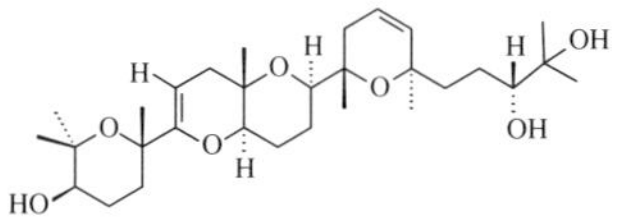
R-37

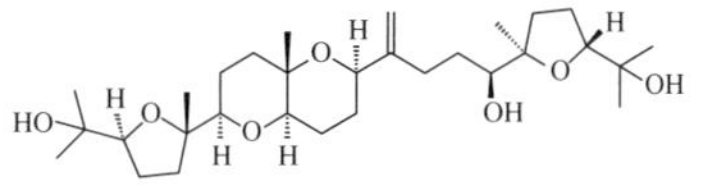
R-38

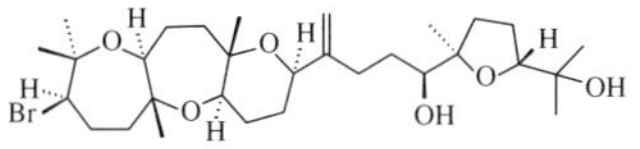
R-39

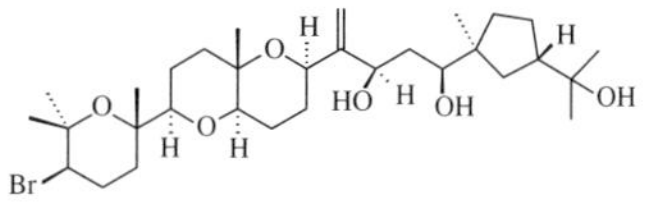
R-40

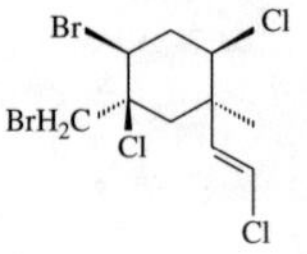

R-41

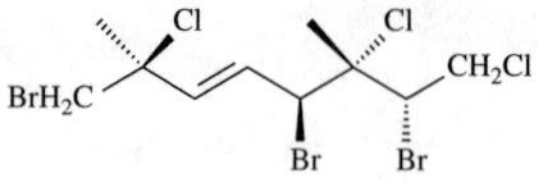

R-42

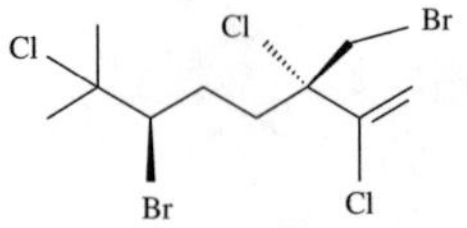

R-43

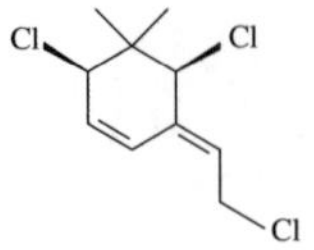

R-44

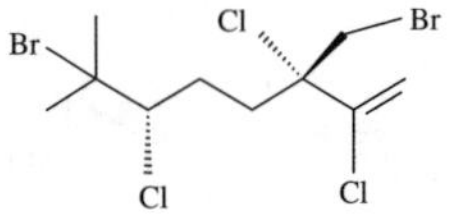

R-45

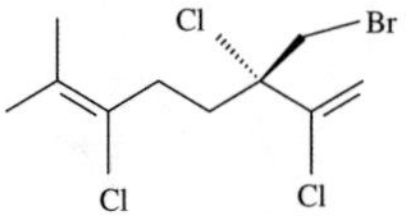

R-46

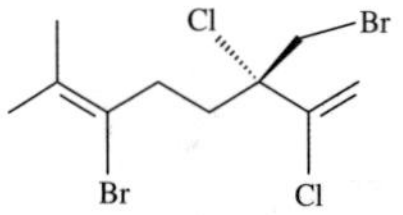

R-47

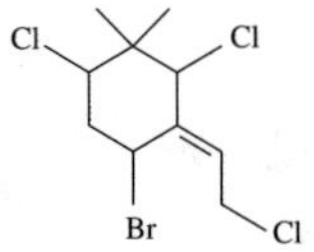

R-48

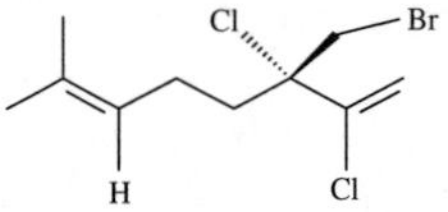

R-49

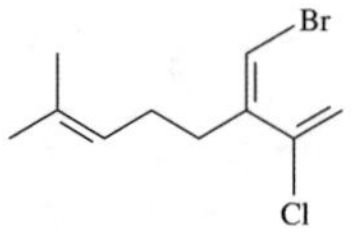

R-50

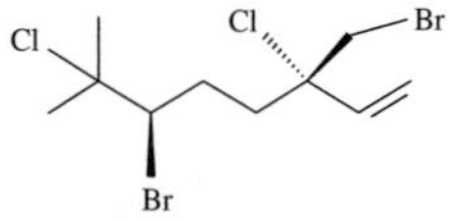

R-51

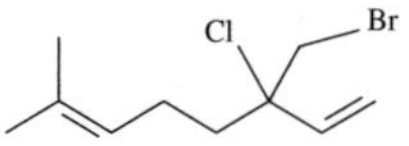

R-52

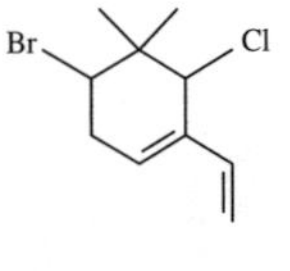

R-53

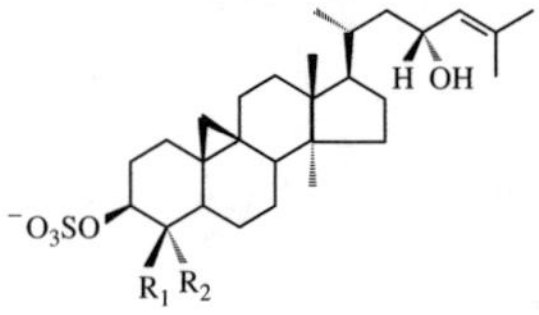

R_1	R_2	
CH_3	$COOCH_3$	**R-54**
CH_3	CH_2OH	**R-55**
H	$COOCH_3$	**R-56**

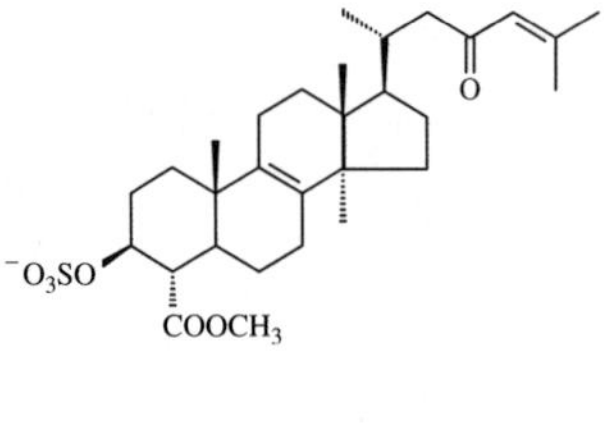

R-57

4.4 Cytotoxic metabolites from phaeophyta

From the brown algae *Bifurcaria bifurcata* five linear diterpenes (**B-1** to **B-5**) and two terminally cyclized derivatives (**B-6**, **B-7**) were isolated and revealed potent cytotoxicity to fertilized sea urchin eggs. Bifurcanol (**B-4**) and bifurcane (**B-6**) were the most active from the compounds tested with an ED_{50} 4 and 12 $\mu g\,mL^{-1}$, respectively. Eleganediol (**B-1**), 12-(S)-hydroxygeranylgeraniol (**B-2**) and 12-(S)-hydroxy-geranylgeranic acid (**B-3**) exhibited an ED_{50} 36, 18 and 60 $\mu g\,mL^{-1}$, respectively, while the two compounds (**B-5** and **B-7**) did not exhibit significant cytotoxic activity.

The Et_2O extract of *Cystoseira mediterranea*, containing meroterpenoids, possess antineoplastic activity attributable to Mediterraneol A, one of its major components. Mediterraneol A (**B-8**), Mediterraneone (**B-9**) and Cystoseirol (**B-10**) were tested by the crown-gall potato disc bioassay, as a high correlation between this test and the mouse P-388 leukemia protocol has been demonstrated. While Didemnin B, a potent antitumor cyclic depsipeptide, inhibited 100% the tumor growth (number of tumors per leaf disc), Mediterraneol A, Mediterraneone and Cystoseirol inhibited tumor growth by 88%, 76% and 73%, respectively.

Four meroterpenes have been isolated from the brown alga *Cystoseira usneoides*, Usneoidone E (**B-11**), Usneoidone Z (**B-12**), Usneoidol E (**B-13**) and Usneoidol Z (**B-14**). The antitumoral activity of compound **B-11** and **B-12** was tested against P-388, A-549, HeLa and B-16 cell lines with an IC_{50} 0.8, 1.25, 1.0 and 1.0 $\mu g\,mL^{-1}$ and 1.5, 1.4, 1.3 and 1.5 $\mu g\,mL^{-1}$, respectively. The other two compounds were tested against P-388, L-1210 and A-549 cell lines and were also found to be cytotoxic.

Bicyclic diterpenes, which possess a decalin skeleton, have been isolated from the brown algae *Dictyota dichotoma* and *Pachydictyon coriaceum* and their cytotoxicity was tested against murine B16 melanoma cells. It was found that Dictyotin A (**B-15**), Dictyotin B (**B-16**), Dictyotin C (**B-17**), Dictyotin B methyl ether (**B-32**) and Dictyotin D methyl ether (**B-33**) had IC_{50} values 8, 3, 15, 10 and 19 $\mu g\,mL^{-1}$, respectively.

Xenicane and norxenicane diterpenes (**B-18** to **B-21**) have been isolated from the brown alga *Dictyota dichotoma* and their cytotoxicity was tested against murine B16 melanoma cells. It was found that 4-acetoxydictyolactone (**B-18**), Dictyotalide A (**B-19**), Dictyotalide B (**B-20**) and nordictyotalide (**B-21**) had IC_{50} values 1.57, 2.57, 0.58 and 1.58 $\mu g\,mL^{-1}$, respectively.

Four Dolabellane (**B-22** – **B-25**) and one hydroazulenoid (**B-26**) diterpenes, isolated from *Dictyota dichotoma*, were tested against the following cancer cell lines: P-388 mouse lymphoma, A-549 Human Lung Carcinoma, HT-29 Human Colon Carcinoma and MEL-28 Human Melanoma. Compounds **B-23** to **B-26** were mildly active with ED_{50} 5 $\mu g\,mL^{-1}$ in all cases, whereas **B-22** exhibited the greatest activity with ED_{50} equal to 1.2 $\mu g\,mL^{-1}$ against P-388 and A-549 tumor cell lines and 2.5 $\mu g\,mL^{-1}$ against HT-29 and MEL-28 tumor cell lines. Dolabellane **B-27** was found to possess interesting bioactivities among them cytotoxicity against KB cancer cells.

Metabolites Dilopholide (**B-28**), hydroxyacetyldictyolal (**B-29**), acetylcoriacenone (**B-31**), and isoacetylcoriacenone (**B-30**) were isolated from the brown alga *Dilophus ligulatus*. These metabolites displayed cytotoxic activity to several types of mammalian cells in culture (KB, P-388, P-388/DOX, and NSCLC-N6). Especially, Dilopholide (**B-28**) showed significant cytotoxic activity (ED_{50} $<4\,\mu g\,mL^{-1}$) against KB (human nasopharynx carcinoma), NSCLC-N6 (human lung carcinoma) cells, and P-388 (murine leukemia) cells.

24-Hydroperoxy-24 –vinyl cholesterol (**B-34**) was isolated from the dichloromethane extract of the brown alga *Padina pavonica* and was found to be cytotoxic toward the KB tumor cell line. The ID_{50} was approximately 6.5 $\mu g\,mL^{-1}$ (14. $10^{-3}\,\mu M$).

Fucoidan (GIV-A) **B-35**, a hexouronic acid containing L-fucan sulfate was isolated from *Sargassum thunbergii* and showed antimetastatic effect when examined on an experimental model of lung metastases induced by LLC in mice.

It is speculated that the antitumor action of GIV-A may be correlated with the activation of complement C3 macrophages and reticuloendothelial system, and the enhancement of antiboby-producing capacity and cell-mediated immunity. This seems to be favorable for cancer immunotherapy.

Bioassay-directed fractionation of the methanolic extract of the marine brown alga *Sargassum tortile* has led to the isolation and characterization of eight compounds which include the chromenes Sargaol (**B-36**), Sargadiol-I (**B-37**), Sargadiol-II (**B-38**), Sargasal-I (**B-39**), Sargasal-II (**B-40**), hydroxysargaquinone (**B-41**), Kjellmanianone (**B-42**) and Fucosterol (**B-43**). Among them, hydroxysargaquinone (**B-41**) and Sargasals-I and −II (**B-37**, **B-38**) demonstrated significant ($ED_{50} = 0.7\,\mu g\,mL^{-1}$) and marginal ($ED_{50} = 5.8$ and $5.7\,\mu g\,mL^{-1}$) cytotoxicity against cultured P-388 lymphotic leukemia cells, respectively, while the other compounds showed moderate activity.

Spatol (**B-44**) was isolated from the brown seaweed *Spatoglossum schmittii* and showed an $ED_{50} = 1.2\,\mu g\,mL^{-1}$ in the urchin egg assay. Further, at the preliminary cell culture testing concentration of $16\,\mu g\,mL^{-1}$ of Spatol completely inhibits cell division in human T242 melanoma and 224C astrocytoma neoplastic cell lines.

14-Keto-stypodiol diacetate (SDA) (**B-45**) was isolated from the brown alga *Stypopodium flabelliforme* and its effect on the cell growth and tumor invasive behavior of DU-145 human prostate cells was studied. SDA at concentrations of 45 μM decreased cell growth by 61%. This compound induces mitotic arrest of tumor cells, an effect that could be associated to alterations in the normal microtubule assembly process. SDA disrupts the normal organization of the microtubule cytoskeleton in the DU145 cell line as revealed by immunofluorescence studies. It affects protease secretion and the *in vitro* invasive capacity, both properties of cells from metastases.

The different effects of SDA, the microtubule assembly inhibition together with its cellular effects in arresting mitosis and blocking protease secretion mechanisms and cell invasion, suggest that SDA interferes with the tumoral activity of these prostatic cancer cells.

(-)-Stypoldione (**B-46**) was isolated from the brown algae *Stypodium zonale* and proven to be an interesting cytotoxic metabolite. Stypoldione inhibits microtubule polymerization, and sperm motility, in contrast to the properties of other microtubule assembly inhibitors. This metabolite seems to prolong the survival time of mice injected with tumor cells, showing relatively little cytotoxicity itself. Actually, using tumor cells derived from P-388 lympholytic leukemia cells injected into BDF1 or CDF1 mice, and drug treatment up to 30 days, a 42% increase in survival time in mice treated with stypoldione was observed.

Four oxygenated Fucosterols were isolated from the brown alga *Turbinaria conoides* and were tested for cytotoxicity against P-388, KB, A-549 and HT-29 cancer cell lines. Steroid **B-50** exhibited significant cytotoxicity against the above four cancer cell lines ($ED_{50} < 2\,\mu g\,mL^{-1}$). Compounds **B-47** to **B-49** exhibited significant activity against the growth of P-388, A-549 and HT-29 cancer cells, and moderate cytotoxicity toward KB cells.

Turbinaric acid, a secosqualene carboxylic acid, (**B-51**) isolated from the brown alga *Turbinaria ornate* exhibited cytotoxicity against murine melanoma and human colon carcinoma cells at $26.6\,\mu g\,mL^{-1}$ and $12.5\,\mu g\,mL^{-1}$, respectively.

Two hydroperoxysterols (**B-34** and **B-52**) and Fucosterol (**B-43**) were isolated from the extracts of *Turbinaria ornata*. The cytotoxic activities of these metabolites against KB, P-388, A-549 and HT-29 cell lines were assayed by a modification of the MTT colorimetric method. The results showed that steroids **B-34**, **B-52** and **B-43** were active against the growth of P-388 cells. Fucosterol **B-43** was not cytotoxic against KB, A-549 and HT-29 cells, however oxygenated sterols **B-34** and **B-52** were moderately cytotoxic.

Table 4.7 Cytotoxic metabolites from phaeophyta

Source	*Metabolite*	*Code*	*Literature*
Bifurcaria bifurcata	Eleganediol	B-1	Valls *et al.*, 1993
	12-(S)-Hydroxygeranylgeraniol	B-2	
	12-(S)-Hydroxygeranylgeranic acid	B-3	
	Bifurcanol	B-4	
	Eleganolone	B-5	
	Bifurcane	B-6	Valls *et al.*, 1995
	Epoxyeleganolactone	B-7	
Cystoseira mediterranea	Mediterraneol	B-8	Fadli *et al.*, 1991
	Cystoseirol	B-9	
	Mediterraneone	B-10	
Cystoseira usneoides	Usneoidone E	B-11	Urones *et al.*, 1992a
	Usneoidone Z	B-12	
	Usneoidol E	B-13	Urones *et al.*, 1992b
	Usneoidol Z	B-14	
Dictyota dichotoma	Dictyotin A	B-15	Ishitsuka *et al.*, 1990a
	Dictyotin B	B-16	
	Dictyotin C	B-17	
	4-Acetoxydictyolactone	B-18	Ishitsuka *et al.*, 1988, 1990b
	Dictyotalide A	B-19	
	Dictyotalide B	B-20	
	Nordictyotalide	B-21	
	Dolabellane and	B-22 to B-25	Durán *et al.*, 1997
	Hydroazulenoid diterpenes	B-26	
	Dolabellane	B-27	Piattelli *et al.*, 1995
Dilophus ligulatus	Dilopholide	B-28	Bouaicha *et al.*, 1993a,b
	Hydroxyacetyldictyolal	B-29	
	Isoacetylcoriacenone	B-30	
	Acetylcoriacenone	B-31	
Pachydictyon coriaceum	Dictyotin B methyl ether	B-32	Ishitsuka *et al.*, 1990a
	Dictyotin D methyl ether	B-33	
Padina pavonica	24-Hydroperoxy-24-vinyl-cholesterol	B-34	Ktari and Guyot, 1999

(*continued*)

Table 4.7 (Continued)

Source	*Metabolite*	*Code*	*Literature*
Sargassum thunbergii	Fucoidan	B-35	Itoh *et al.*, 1993; 1995; Zhuang *et al.*, 1995
Sargassum tortile	Sargaol	B-36	Numata *et al.*, 1992
	Sargadiol-I	B-37	
	Sargadiol-II	B-38	
	Sargasal-I	B-39	
	Sargasal-II	B-40	
	Hydroxysargaquinone	B-41	
	Kjellmanianone	B-42	
	Fucosterol	B-43	
Spatoglossum schmittii	Spatol	B-44	Gerwick *et al.*, 1980
Stypopodium flabelliforme	14-Keto-stypodiol diacetate	B-45	Depix *et al.*, 1998
Stypopodium zonale	Stypoldione	B-46	Mori and Koga, 1992; Gerwick and Fenical, 1981
	Stypoldione	B-46	O'Brien *et al.*, 1984
Turbinaria conoides	Oxygenated fucosterols	B-47 to B-50	Sheu *et al.*, 1999
Turbinaria ornata	Turbinaric acid	B-51	Asari *et al.*, 1989
	24-Hydroperoxy-24-vinyl-cholesterol	B-34	Sheu *et al.*, 1997b
	29-Hydroperoxystigmasta-5,24(28)-dien-3b-ol	B-52	
	Fucosterol	B-43	

Cytotoxic metabolites from phaeophyta

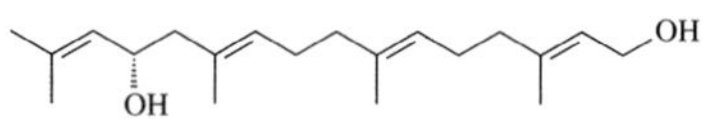

B-1

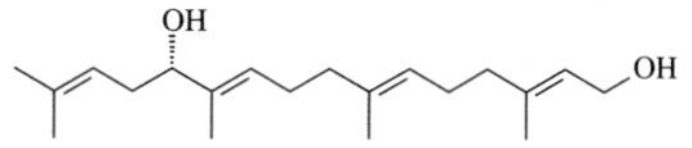

B-2

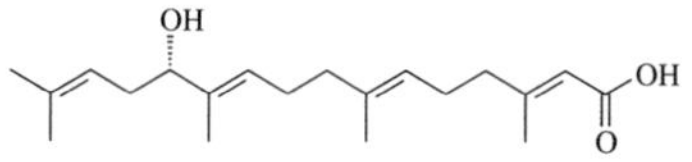

B-3

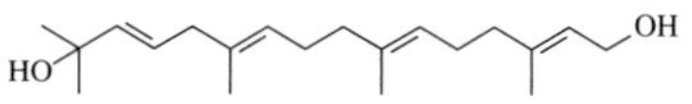

B-4

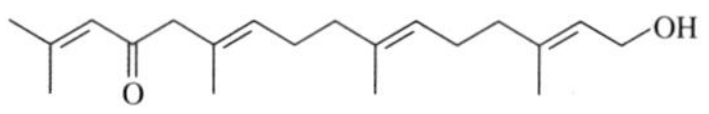

B-5

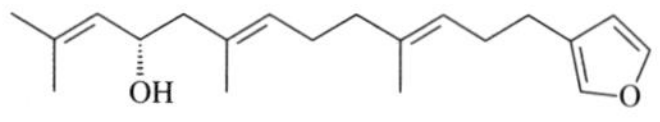

B-6

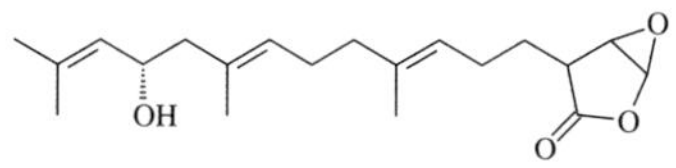

B-7

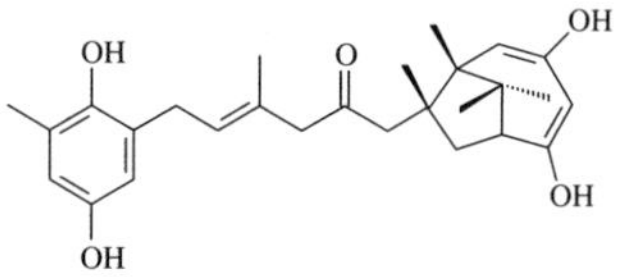

B-8

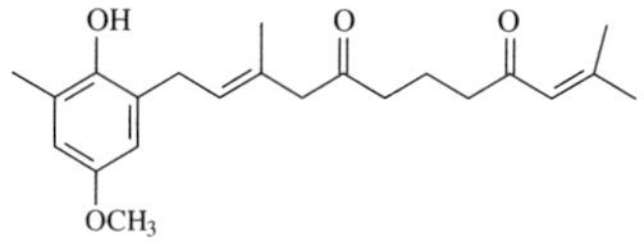

B-9

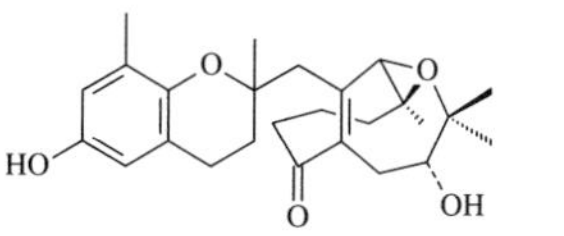

B-10

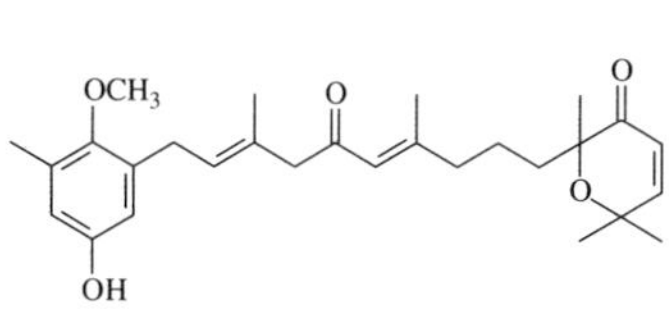

B-11

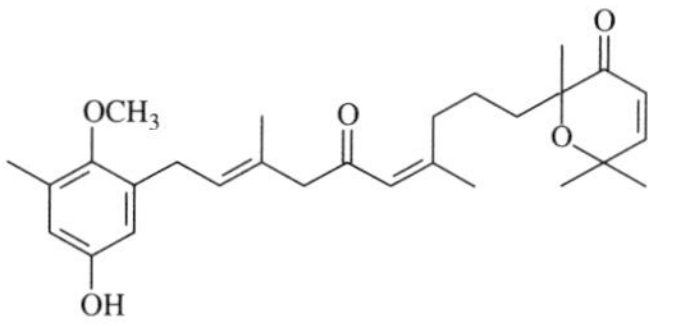

B-12

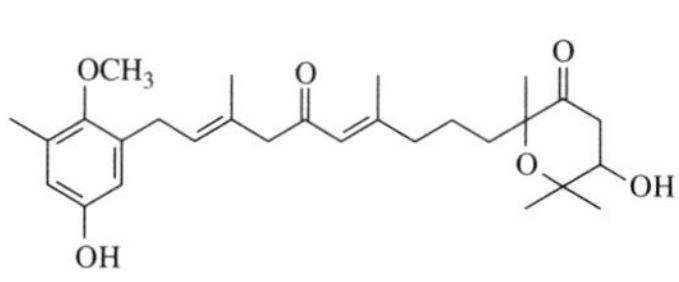

B-13

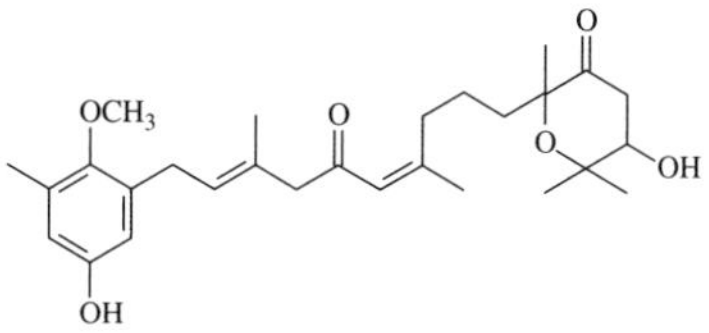

B-14

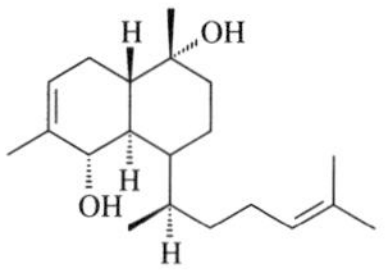

B-15

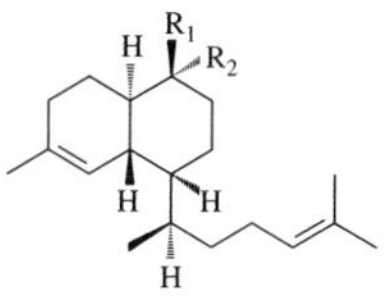

R_1 = Me, R_2 = OH **B-16**

R_1 = OH, R_2 = Me **B-17**

R_1 = Me, R_2 = OMe **B-32**

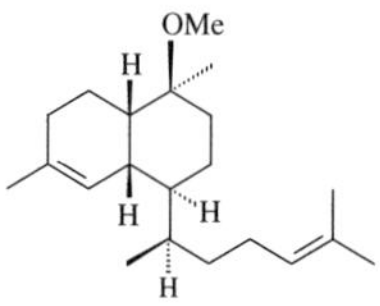

B-33

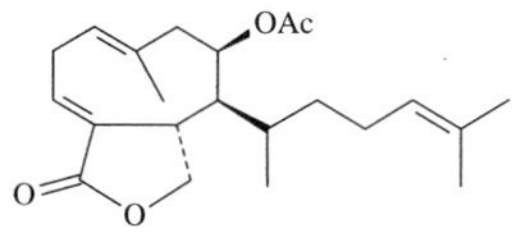

B-18

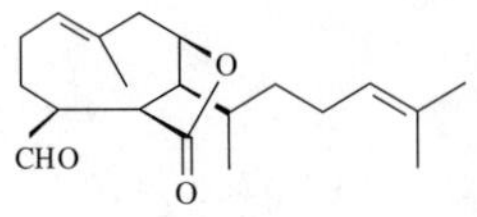

B-19

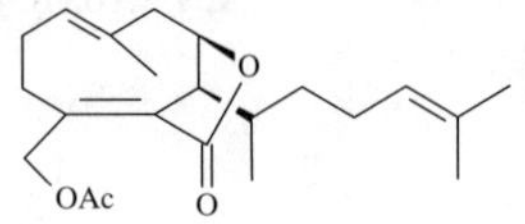

B-20

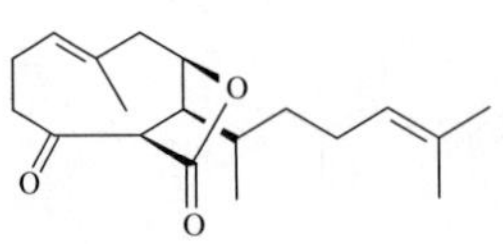

B-21

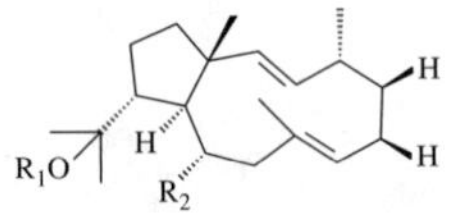

	R_1	R_2
B-22	Ac	OH
B-25	Ac	H

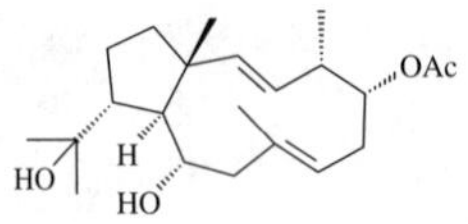

B-23

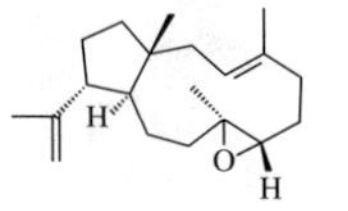

B-24

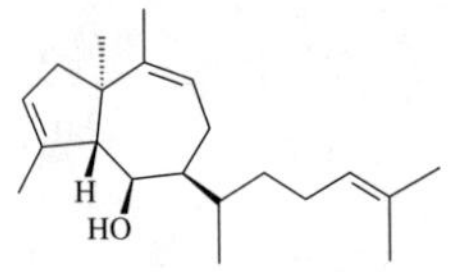

B-26

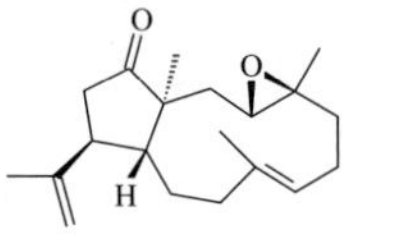

B-27

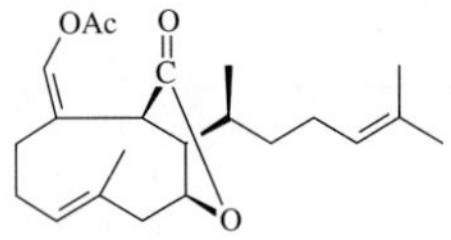

B-28

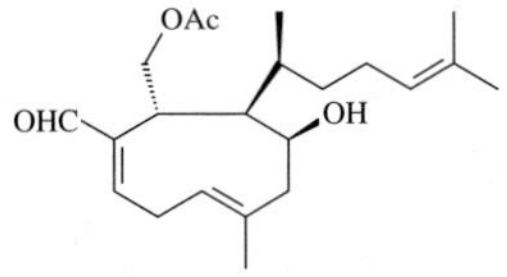

B-29

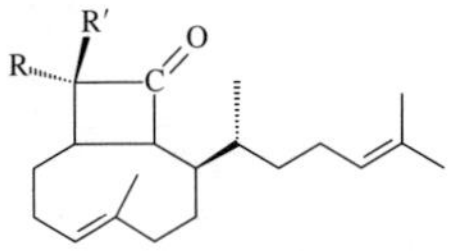

	R	R′
B-30	OAc	H
B-31	H	OAc

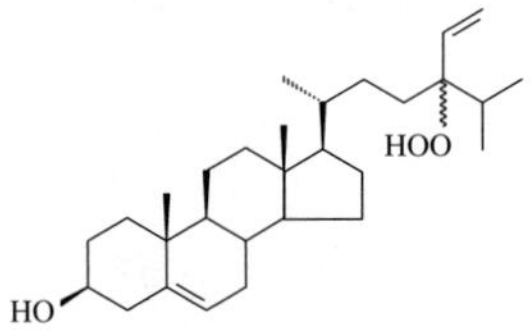

B-34

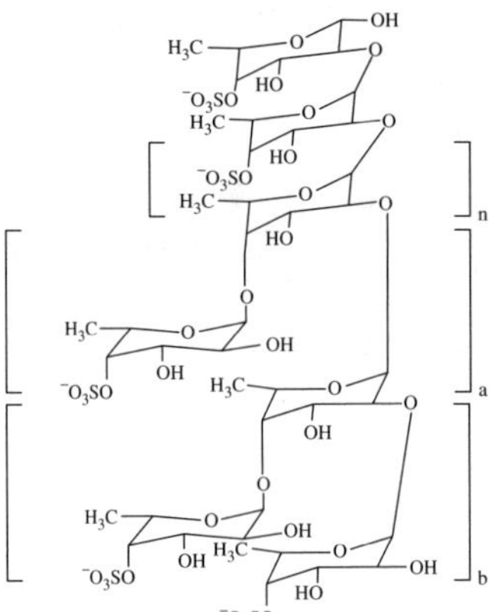

B-35

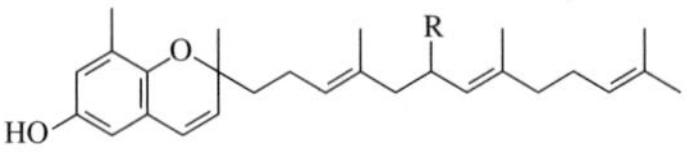

R: H **B-36**

R: OH **B-37**

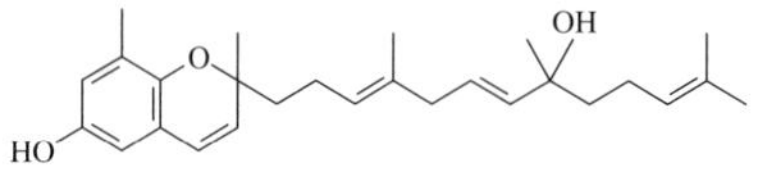

B-38

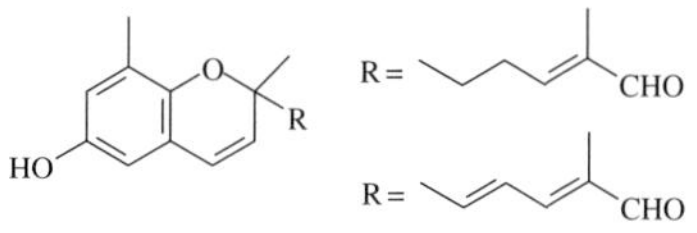

B-39

B-40

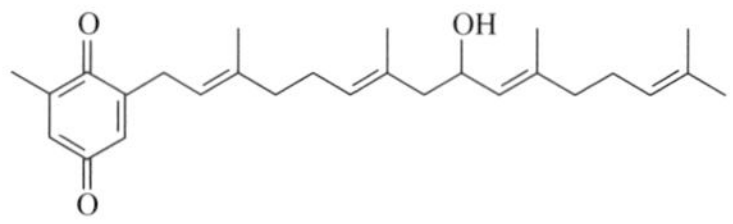

B-41

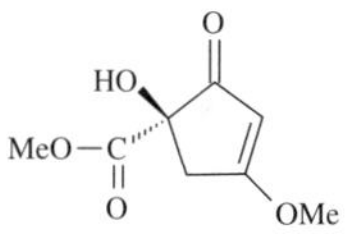

B-42

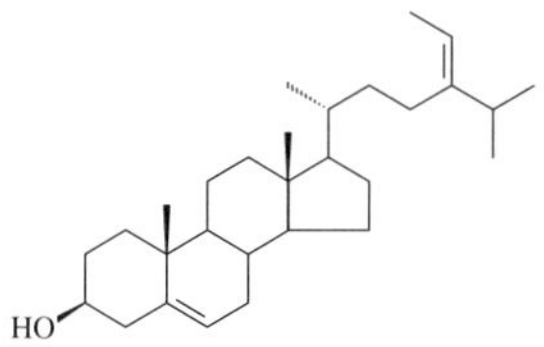

B-43

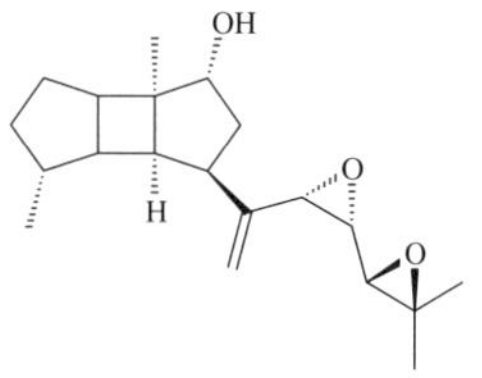

B-44

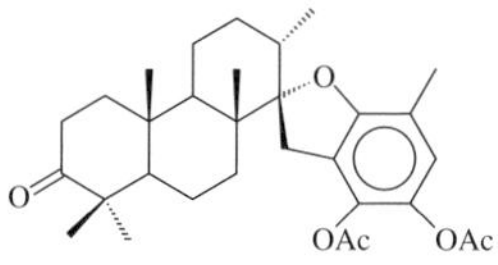

B-45

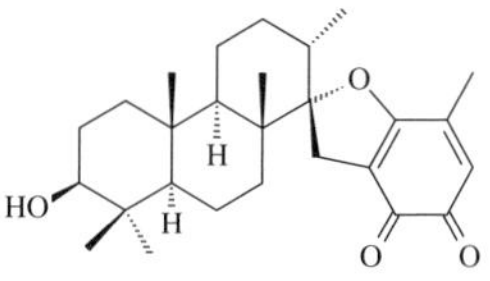

B-46

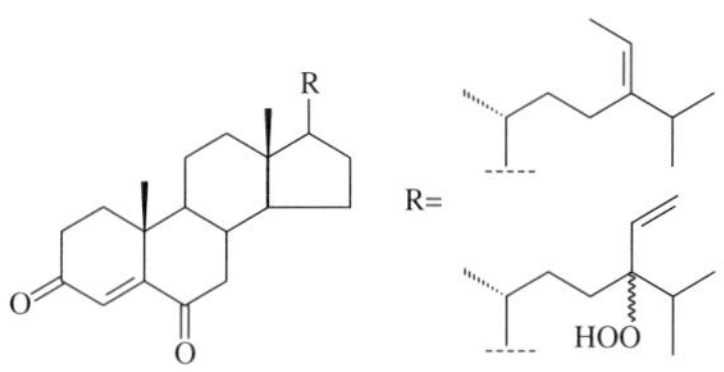

B-47

B-48

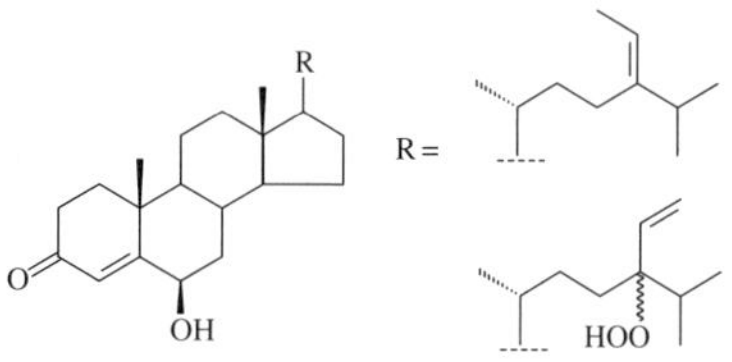

B-49

B-50

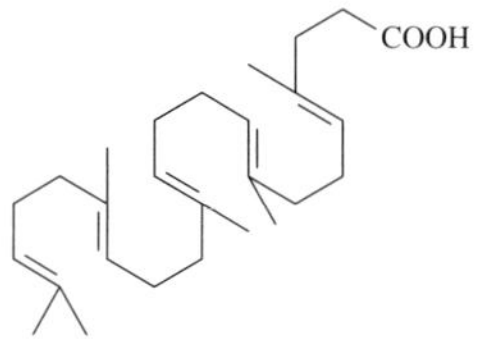

B-51

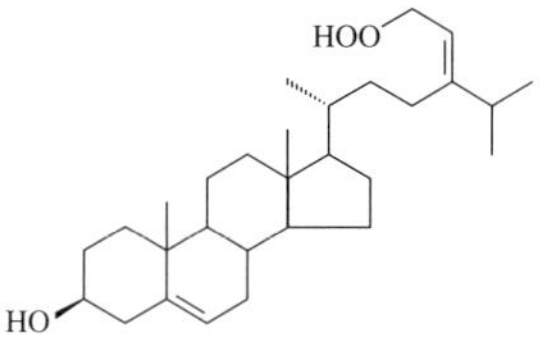

B-52

4.5 Cytotoxic metabolites from microalgae

Amphidinolides A (**M-1**), B (**M-2**), C (**M-3**) and D (**M-4**) have been isolated from the cultured cells of the marine dinoflagellate *Amphidinium sp.*, a symbiotic microalga. These potent cytotoxic 25-membered macrolides exhibited strong antineoplastic activity against L-1210 murine leukemia cells *in vitro* with IC_{50} values of 2.4, 0.00014, 0.0058 and 0.019 $\mu g\,mL^{-1}$, respectively.

Amphidinolide B is the most active and 10,000 times more potent than Amphidinolide A. It is worth noting that these macrolides isolated from the same dinoflagellate are quite different in substitution patterns and activities.

Amphidinolide R (**M-5**) and S (**M-6**), isolated from the cultured dinoflagellate *Amphidinium* sp., showed cytotoxicity against murine lymphoma L-1210 (IC_{50}: 1,4 and 4.0 $\mu g\,mL^{-1}$) and human epidermoid carcinoma KB cells (IC_{50}: 0.67 and 6.5 $\mu g\,mL^{-1}$) *in vitro*, respectively. Amphidinolide V (**M-7**) exhibited cytotoxicity against murine lymphoma L-1210 (IC_{50}: 3.2 $\mu g\,mL^{-1}$) and epidermoid carcinoma KB cells (IC_{50}: 7 $\mu g\,mL^{-1}$) *in vitro*.

Carbenolide (**M-8**) isolated from *Amhidinium* sp. was assessed against the human colon carcinoma cell line HCT-116 by XTT assay and the IC_{50} found to be 1.6 nM. Further, *in vivo* studies found that when P-388 mouse leukemia was implanted intraperitoneally, a dose of 0.03 $mg\,kg^{-1}$ day produced a 50% increase in life span.

A cytotoxic carbohydrate-conjugated ergosterol (Astasin) (**M-9**) was found in cells of the colorless euglenoid *Astasia longa*. When cells of HL 60, human lymphoma, were cultured with Astasin, 50% of the cell growth was inhibited at 5.0 μg Astasin mL^{-1} medium. With 10.0 μg Astasin mL^{-1} medium the cell growth was inhibited completely and 50% of the initial cells were killed.

Cell extracts from photoautrophic cultures of two cyanobacterial *Calothrix* isolates inhibited the growth *in vitro* of a chloroquine-resistant strain of the malaria parasite, *Plasmodium falciparum*, and of human HeLa cancer cells, in a dose-dependent manner. Bioassay-directed fractionation of the extracts led to the isolation and structural characterization of calothrixins A (**M-10**) and B (**M-11**), pentacyclic metabolites with an indolo[3,2-j]phenanthridine ring system unique amongst natural products. Calothrixins exert their growth-inhibitory effects at nanomolar concentrations moreover **M-10** and **M-11** inhibited *in vitro* the growth of HeLa human servical cancer cells with IC_{50} 40 nM and 350 nM, respectively.

Two antitumor promoters, monogalactosyl diacylglycerols (**M-12, M-13**) were isolated from the freshwater green alga, *Chlorella vulgaris*, along with three other monogalactosyl diacylglycerols (**M-14** to **M-16**) and two digalactosyl diacylglycerols (**M-17, M-18**). The monogalactosyl diacylglycerol containing (7Z,10Z)-hexadecadienoic acid (**M-13**) showed a more potent inhibitory effect toward tumor promotion [on the Epstein–Barr virus-associated early antigen (EBV-EA) activation on Raji cells induced by 12-O-tetradecanoylphorbol-13-acetate (TPA)], than the other metabolites.

Increases in the cytotoxic activity of peritoneal macrophages has been attributed to the action of β-Carotene (**M-19**) which has also been reported to increase the number of tumor necrosis factor positive cells considered by many to be endogenous antineoplastic agents. β-Carotene (**M-19**) has been isolated from *Dunaliella* sp. as well as from cyanobacteria such as *Spirulina* sp. In addition carotenoids, which were detected in cyanobacterial extracts, have been found to be mitogenic and to enhance the cytotoxic action of thymus derived cells.

β-Carotene-rich alga *Dunaliella bardawil* has been found to inhibit spontaneous mammary tumorigenesis of mice and the results strongly suggest that this is performed by increasing the homeostatic potential of the host animals as well as by the well-known antioxidant function of β-Carotene.

Welwitindolinones are a family of novel alkaloids recently isolated from the blue-green alga *Hapalosiphon witschii*. Incubation of SK-OV-3 human ovarian carcinoma cells and A-10 vascular smooth muscle cells with welwistatin (**M-20**), results in dose-dependent inhibition of cell proliferation, which is correlated with increases in the percentage of cells in mitosis. Treatment of A-10 cells with welwistatin resulted in reversible depletion of cellular microtubules but did not affect microfilaments. Pretreatment of A-10 cells with paclitaxel prevented microtubule depolymerization in response to welwistatin. Welwistatin (**M-20**), inhibited the polymerization of purified tubulin *in vitro* but did not alter the ability of tubulin to bind [3H]colchicine or to hydrolyze GTP. Also, welwistatin (**M-20**) did not induce the formation of topoisomerase/DNA complexes. These results indicate that welwistatin is a new antimicrotubule compound that circumvents multiple drug resistance and so may be useful in the treatment of drug-resistant tumors.

Hormothamnione (**M-21**) is a cytotoxin isolated from the marine cyanophyte *Hormothamnion enteromorphoides*. This metabolite was found to be a potent cytotoxic agent to P-388 lymphocytic leukemia ($ID_{50} = 4.6\,ng\,mL^{-1}$) and HL-60 human promyelocytic leukemia cell lines ($ID_{50} = 0.1\,ng\,mL^{-1}$) and appears to be a selective inhibitor of RNA synthesis.

Debromoaplysiatoxin (**M-22**) isolated from *Lynbya gracilis*, *Oscillatoria nigroviridis*, *Schizothrix calcicola* and *Symploca muscorum*, as well as from deep and shallow specimens of *Lynbya majuscula* exhibited T/C (Ratio of the survival time of Treated compared to Control diseased mice) 186 and 140 with $1.8\,\mu g\,kg^{-1}$ and $0.6\,mg\,kg^{-1}$ doses, respectively. From the same organism was Aplysiatoxin (**M-23**) originally isolated.

Curacins A, B and C were isolated from the marine cyanobacterium *Lyngbya majuscula*. Curacin A (**M-24**) is an extremely potent antimitotic agent, which is under examination for its potential anticancer utility. Also Curacin B (**M-25**) and C (**M-26**) are both toxic to brine shrimp, demonstrate strong cytotoxicity against murine L-1210 leukemia and human CA46 Burkitt lymphoma cell lines, inhibit the polymerization of purified tubulin *in vitro*, and the NCI *in vitro* 60-cell line assay, show potent antiproliferative activity to many cancer-derived cell lines in a manner characteristic of antimitotic agents. Even though Curacin D (**M-27**) was found to be comparable active to Curacin A (**M-24**) as a potent inhibitor of colchicine binding, it was 7-fold less active than Curacin A in its ability to inhibit tubulin polymerization, 10-fold less active in inhibiting MCF-7 breast cancer cell growth and 13-fold less active as a brine shrimp toxin.

The marine cyanobacterium *Lyngbya majuscula* has yielded also two toxic natural products Hermitamides A (**M-28**) and B (**M-29**). Metabolites **M-28** and **M-29** exhibited LD_{50} values of $5\,\mu M$ and $18\,\mu M$ in the brine shrimp bioassay, and IC_{50} values of $2.2\,\mu M$ and $5.5\,\mu M$ to Neuro-2a neuroblastoma cells in tissue culture, respectively.

Dolastatin 3 (**M-30**) previously reported from the sea hare *Dolabella auricularia* was isolated from an extract of the macroscopic cyanophyte *Lyngbya majuscula*.

Dolastatin 12 (**M-31**) and Lyngbyastatin 1 (**M-32**), a new cytotoxic analogue of Dolastatin12, were isolated as inseparable mixtures with their C-15 epimers from extracts of *Lyngbya majuscula/Schizothrix calcicola* assemblages collected near Guam. Both metabolites proved toxic with only marginal or no antitumor activity when tested against colon adenocarcinoma #38 or mammary adenocarcinoma #16/C. Both compounds were shown to be potent disrupters of cellular microfilament networks.

The lipopeptide Microcolin A (**M-33**) was also isolated from the marine blue green alga *L. majuscula*. Microcolin A suppressed concavalin A, phytohemagglutinin and lipopolysaccharide-induced proliferation of murine splenocytes. Mixed lymphocyte reaction, anti-IgM, and phorbol 12-myristate 13-acetate plus ionomycin stimulation of murine splenocytes were all similarly suppressed by Microcolin A. The inhibitory activity of Microcolin A was time-dependent and reversible and was not associated with a reduction in cell viability. These results indicated that Microcolin A is a potent immunosuppressive and antiproliferative agent.

Apratoxin A (**M-34**) a potent cytotoxin with a novel skeleton has been isolated from *L. majuscula*. This cyclodepsipeptide of mixed peptide–polyketide biogenesis bares a thiazoline ring flanked by polyketide portions, one of which possesses an unusual methylation pattern. Apratoxin A possesses IC_{50} values for *in vitro* cytotoxicity against human tumor cell lines ranging from 0.36 to 0.52 nM; however it was only marginally active *in vivo* against a colon tumor and ineffective against a mammary tumor.

The cytotoxic depsipeptides Lyngbyabellin A (**M-35**) and Lyngbyabellin B (**M-36**) were isolated from a Guamanian strain of *L. majuscula*. Both metabolites found to be cytotoxic with **M-36** being slightly less active *in vitro* than **M-35**. The IC_{50} values for **M-35** and **M-36** were 0.03 $\mu g\,mL^{-1}$ and 0.10 $\mu g\,mL^{-1}$ against KB cells and 0.5 $\mu g\,mL^{-1}$ and 0.83 $\mu g\,mL^{-1}$ against LoVo cells, respectively. Lyngbyabellin A was proved to be potent microfilament-disrupting agent and the same mode of action is speculated for Lyngbyabellin B.

From extracts of the same cyanobacterium two new lipopeptides; Malyngamides D (**M-37**) and Malyngamide H (**M-38**) were isolated by bioassay-guided fractionations. Malyngamide D was mildly cytotoxic with an $ID_{50} < 30\,\mu g\,mL^{-1}$ to KB cells in tissue culture, while Malyngamide H exhibited an ichthyotoxic effect with an $LC_{50} = 5\,\mu g\,mL^{-1}$ and $EC_{50} = 2\,\mu g\,mL^{-1}$. End points in this assay were death and inability to swim against a manually induced current.

The novel lipopeptides Laxaphycin A (**M-39**) and Laxaphycin B (**M-40**) were isolated from *L. majuscula* extracts during screening against three cell lines. The cytotoxicity of Laxaphycins were evaluated for the parent drug-sensitive CCRF-CEM human leukemic lymphoblasts, CEM/VLB100 vinblastine-resistant subline which presents a MDR phenotype7 and CEM/VM-1 subline usually referred to as atypical MDR cells8. Laxaphycin A was not active when tested at a concentration of 20 mM. Laxaphycin B showed pronounced cytotoxic activities on the drug-sensitive cells with IC_{50} of 1.1 mM and was practically equally active against the drug-sensitive cells and the drug-resistant cells. Both sublines showed no resistance to Laxaphycin B whereas those lines showed a 62- and 9-fold resistance to adriamycin. So, unlike the clinically used antitumor antibiotic adriamycin, Laxaphycin B preserved equal cytotoxicity on Pgp-MDR cells and altered DNA-topoisomerase II-associated MDR cells.

Yanucamide A (**M-41**) and Yanucamide B (**M-42**) were isolated from the lipid extract of *L. majuscula* and *Schizothrix* sp. assemblage collected at Yanuca island, Fiji. Both Yanucamides exhibited strong brine shrimp toxicity with a $LD_{50} = 5$ ppm.

Grenadadiene (**M-43**) and grenadamide are structurally unique cyclopropyl-containing metabolites isolated from the organic extract of a Grenada collection of *Lyngbya majuscula*. These were the first reported cyclopropyl-containing fatty acid derivatives from a *Lyngbya* sp. Grenadadiene (**M-43**) has an interesting profile of cytotoxicity in the NCI 60 cell line assay, while grenadamide exhibited modest brine shrimp toxicity ($LD_{50} = 5\,\mu g\,mL^{-1}$).

Kalkipyrone (**M-44**), a novel α-methoxy-β,β'-dimethyl-γ-pyrone possessing an alkyl side chain, was isolated from an assemblage of *Lyngbya majuscula* and *Tolypothrix* sp. Kalkipyrone (**M-44**) is toxic to brine shrimp (LD_{50} 1 $\mu g\,mL^{-1}$) and gold fish (LD_{50} 2 $\mu g\,mL^{-1}$) and is structurally related to the actinopyrones that were previously isolated from *Streptomyces* sp.

Microcystilide A (**M-45**) was isolated from the methanolic extract of the cyanobacterium *Microcystis aeruginosa* NO-15-1840. The compound was found to be only weakly cytotoxic against HCT116 and HCTVP35 cell lines (IC_{50} 0.5 $mg\,mL^{-1}$), but found to be active in the cell differentiation assay using HL-60 cells at a concentration of 0.5 $mg\,mL^{-1}$.

The lipophilic extract of a marine strain of *Nostoc linckia* was found to display appreciable cytotoxicity against LoVo (MIC 0.066 $\mu g\,mL^{-1}$) and KB (MIC 3.3 $\mu g\,mL^{-1}$). This algal extract was among the most LoVo-cytotoxic found in screening extracts of 665 blue-green algae. Bioassay-directed chromatography led to the isolation of Borophycin (**M-46**).

Cryptophycin A (**M-47**) was initially isolated from cyanobacterium *Nostoc* sp. ATCC 53789 and demonstrated antitumor activity. *In vitro* testing showed tumor selective cytotoxicity, that is, higher cytotoxicity for tumor cells (leukemia and solid tumor cells) compared to a low malignant potential fibroplast cell line. The *in vitro* cytotoxicity spectrum of the Cryptophycins included tumors of non-human (L-1210 and P-388 leukemias, colon adenocarcinoma 38, pancreatic ductual adenocarcinoma 03, mammary adenocarcinoma 16/C) and human (colon adenocarcinomas: LoVo, CX-1, HCT-8 and H-116; mammary adenocarcinomas: MX-1 and MCF-7; lung adenosquamous carcinoma: H-125; ovarian adenocarcinoma: SKOV-3; and nasopharyngeal carcinoma: KB) origin.

Six other Cryprophycins were isolated from the same species in minor amounts and structure–activity relationship studies were conducted. The cytotoxicities of epoxides Cryptophycin A (**M-47**) and Cryptophycin B (**M-48**) were the two strongest and were surprisingly identical in potency, implying that the chloro substituent on the *O*-methyltyrosine was unnecessary for exhibiting cytotoxicity. Removal of the epoxide oxygen or hydroxy groups from C-7 and C-8 of unit A as in Cryptophycin C (**M-49**) and Cryptophycin D (**M-50**) resulted in 100-fold decrease in cytotoxicity. The leucic acid unit was clearly required for the potent activity, since Cryptophycin F methyl ester (**M-52**) and Cryptophycin G (**M-53**) were only weakly cytotoxic. The ester bond connecting 3-amino-2-methylpropionic acid and leucic acid was also clearly necessary for optimal activity. Cryptophycin E methyl ester (**M-51**) was 1000-fold less cytotoxic than **M-47** and **M-48**.

In addition, potent *in vitro* cytotoxicity was demonstrated against cells that were known to have multiple drug resistance (mammary 17/C/ADR, MCF-7/ADR, SKVLB1). Thus, Cryptophycin A (**M-47**) belongs to a class of compounds with a broad spectrum of *in vitro* antitumor activity, which is clearly maintained when administered *in vivo* by a route different from the tumor inoculation. Growth of L-1210 cells was inhibited by 85% upon exposure to Cryptophycin A. Cryptophycin A binds strongly with tubulin and disrupts the assembly of microtubules especially needed for mitotic spindle formation and cell proliferation. Because of the impressive *in vitro* and *in vivo* activities exhibited by Cryptophycin A, a number of analogues were synthesized by Eli Lilly & Co. The synthetic derivative Cryptophycin-145 (**M-54**) had an IC_{50} of 0.015 pM against the GC3 human colon carcinoma cell line.

From the extract of *Oscillatoria acutissima* Acutiphycin (**M-55**) was isolated and showed antineoplastic activity with a T/C = 186 with a dose treatment of $50\,\mu g\,kg^{-1}$. Acutiphycin and the 20, 21didehydroacutiphycin (**M-56**) showed $ED_{50} < 1\,\mu g\,mL^{-1}$ against KB and N1H/3T3 cell lines, respectively.

From a mixed culture of *Oscillatoria nigroviridis* and *Schizothrix calcicola*, metabolite Oscillatoxin A (**M-57**) was isolated, and showed antineoplastic activity level, T/C = 140 with a dose $0.2\,\mu g\,kg^{-1}$.

The nucleoside Tubercidin–5-α-D glucopyranoside (**M-58**) was isolated from the cyanophyceae *Plectonema radiosum* and *Tolypothrix distorta* and showed cytotoxicity on KB cells with MIC $3\,\mu g\,mL^{-1}$.

The cytotoxic macrolide Prorocentrolide (**M-59**) was isolated from the dinoflagellate *Prorocentrum lima* and exhibited cytotoxicity against L-1210 with an $IC_{50} = 20\,\mu g\,mL^{-1}$. The structurally related macrolide Prorocentrolide B (**M-60**) was isolated from *Prorocentrum macolosum* and the pharmacological evaluation is under investigation.

Tolytoxin (**M-61**), the most potent of the Scytophycin compounds, has been shown to inhibit cell proliferation, induce morphological changes, and disrupt stress fiber organization in cultured mammalian cells. These effects are manifested rapidly (less than 15 min) and at concentrations significantly lower than other F-actin disrupting agents such as Cytochalasins B or D, Latrunculin A, or Swinholide A. Tolytoxin also inhibits G-actin polymerization and induces F-actin depolymerization *in vitro*. Tolytoxin has been also isolated from the cyanophyta *Tolypothrix conglutinata, Scytonema mirabile* and *S. ocellatum*. The Scytophycins (**M-62** to **M-65**) are antifungal, cytotoxic macrolides produced by cyanobacteria of the genera *Tolypothrix* and *Scytonema*.

The nucleoside Tubercidin (**M-66**), isolated from *Tolypothrix byssoidea* and *Scytonema saleyeriense*, was tested on KB and N1H/3T3 cell lines *in vivo* and the levels of toxicity were found to be high. The MIC on KB cells was found to be $2\,\mu g\,mL^{-1}$. Tubercidin is an inhibitor of DNA, RNA and protein synthesis in growing KB cells, acting by disruption of nucleic acid structure following incorporation. Synthesis of messenger RNA was found to be particularly susceptible.

Symbioramide (**M-67**), a sphingosine derivative, isolated from the cultured dinoflagellate *Symbiodinium* sp. exhibits antileukemic activity against L-1210 murine leukemia cells *in vitro* with an IC_{50} value of $9.5\,\mu g\,mL^{-1}$. The α-hydroxy-β-γ-dehydro fatty acid contained in Symbioramide is seldom found from natural sources.

A new solid tumor selective cytotoxic analogue of Dolastatin 10, Symplostatin 1 (**M-68**) has been isolated from the marine cyanobacterium *Symploca hydnoides*, collected near Guam. Symplostatin 1 exhibited a cytotoxicity IC_{50} value of $0.3\,ng\,mL^{-1}$ against KB cells (an epidermoid carcinoma line), as opposed to $<0.1\,ng\,mL^{-1}$ for Dolastatin 10. Since **M-68** induced 80% microtubule loss at $1\,ng\,mL^{-1}$ when tested on A-10 cells, its mechanism of action must be similar, if not identical, to that of Dolastatin 10. Dolastatin 10 appears to be one of the most potent antineoplastic compounds known to date and is in phase I trials as an anticancer agent. A second metabolite Symplostatin 2 (**M-69**) an analogue to Dolastatin 13 was also isolated from the same cyanobacterium. It has been suggested that Dolastatins isolated from *Dolabella auricularia*, probably have a cyanobacterial dietary origin. The sequestration of algal metabolites by sea hares is well documented in the ecological literature.

Tolyporphin (**M-70**), a porphyrin extracted from the cyanobacteria *Tolypothrix nodosa*, was found to be a very potent photosensitizer of EMT-6 tumor cells grown both *in vitro* as suspensions or monolayers and *in vivo* in tumors implanted on the backs of C.B17/Icr severe combined immunodeficient mice. Thus, during photodynamic treatment (PDT) of EMT-6 tumor cells *in vitro*, the photokilling effectiveness of TP measured as the product of the reciprocal of D_{50} (the light dose necessary to kill 50% of cells) and the concentration of TP is ~5000 times higher than that of Photofrin II (PII), the only PDT photosensitizer thus far approved for clinical trials. The outstanding PDT activity of TP observed *in vivo* may be due to its unique biodistribution properties, in particular low concentration in the liver, resulting in a higher delivery to the other tissues, including tumor.

Tolyporphins J and K (**M-71** and **M-72**) were tested for biological activity in MDR reversal and [^{3}H]vinblastine accumulation assays alone with Tolyporphin as a comparison. In the MDR reversal assay Tolyporphin J (**M-71**) exhibited virtual identical activity to **M-70**. Both compounds sensitized MCF-7/ADR cells to actinomycin D, reversing MDR and verifying their abilities to enhance drug accumulation. Tolyporphin K (**M-72**) exhibited little activity. In contrast to **M-70** and **M-71**, Tolyporphin K promoted only modest increases in [^{3}H]vinblastine accumulation, consistent with its poor ability to sensitize these cells to cytotoxic drugs.

Cyano nucleoside Toyocamycin-5-α-D-glucopyranoside (**M-73**), closely related to Tubercidin-5-D glucopyranose (**M-58**), was isolated from *Tolypothrix tenuis* and was assayed on KB and HL-60 cell lines showing MICs 12 and $6\,\mu g\,mL^{-1}$, respectively.

Table 4.8 Cytotoxic metabolites from microalgae

Source	*Metabolite*	*Code*	*Literature*
Amphidinium sp.	Amphidinolide A	M-1	Ishibashi *et al.*, 1987; Ishiyama *et al.*, 1996;
	Amphidinolide B	M-2	Kobayashi, 1989
	Amphidinolide C	M-3	Kobayashi *et al.*, 1986, 1988a, 1989a
	Amphidinolide D	M-4	
	Amphidinolide R, S	M-5, M-6	Ishibashi *et al.*, 1997
	Amphidinolide V	M-7	Kubota *et al.*, 2000
	Carbenolide	M-8	Shimizu, 1996
Astasia longa	Astasin	M-9	Kaya *et al.*, 1995
Calothrix sp.	Calothrixins A, B	M-10, M-11	Rickards *et al.*, 1999
Chlorella vulgaris	Glyceroglycolipids	M-12 to M-18	Morimoto *et al.*, 1995, Soeder, 1976
Dunaliella sp.	B-Carotene and other carotenoids	M-19	Nagasawa *et al.*, 1989, 1991; Schwartz *et al.*, 1986; 1993; Schwartz and Shklar, 1989; Shklar and Schwartz, 1988; Tomita *et al.*, 1987
Hapalosiphon witschii	Welwistatin	M-20	Zhang and Smith, 1996
Hormothamnione enteromorphoides	Hormothamnione	M-21	Gerwick *et al.*, 1986, 1989
Lyngbya gracilis	Debromoaplysiatoxin	M-22	Mynderse *et al.*, 1977; Mynderse and Moore, 1978
Lyngbya majuscula	Debromoaplysiatoxin	M-22	Moore, 1982
	Aplysiatoxin	M-23	
	Curacin A	M-24	Blokhin *et al.*, 1995; Bonnard *et al.*, 1997, Gerwick *et al.*,
	Curacin B	M-25	1987, 1994; Graber and Gerwick, 1998;
	Curacin C	M-26	Harrigan *et al.*, 1998a; Luesch *et al.*, 2000a,b,
	Curacin D	M-27	2001; Marquez *et al.*, 1998; Mitchell *et al.*, 2000;
	Hermitamides A	M-28	Nagle *et al.*, 1995; Orjala *et al.*, 1995; Pettit *et al.*, 1987;
	Hermitamides B	M-29	Sitachitta and Gerwick, 1998; Sitachitta *et al.*, 2000;
	Dolastatin 3	M-30	Tan *et al.*, 2000b; Verdier-Pinard *et al.*, 1998;
	Dolastatin 12	M-31	Yoo and Gerwick, 1995; Zhang *et al.*, 1997
	Lyngbyastatin 1	M-32	
	Microcolin A	M-33	
	Apratoxin A	M-34	
	Lyngbyabellin A	M-35	
	Lyngbyabellin B	M-36	
	Malyngamide D	M-37	

(continued)

Table 4.8 (Continued)

Source	*Metabolite*	*Code*	*Literature*
	Malyngamide H	M-38	
	Laxaphycin A	M-39	
	Laxaphycin B	M-40	
	Yanucamide A	M-41	
	Yanucamide B	M-42	
	Grenadadiene	M-43	
	Kalkipyrone	M-44	
Microcystis aeruginosa NO-15-1840	Microcystilide A	M-45	Tsukamoto *et al.*, 1993
Nostoc sp. *ATCC 53789*	Borophycin	M-46	Foster *et al.*, 1999
	Cryptophycin A	M-47	Golakoti *et al.*, 1994, 1995
	Cryptophycin B	M-48	Hemscheidt *et al.*, 1994
	Cryptophycin C	M-49	Valeriote *et al.*, 1995
	Cryptophycin D	M-50	Smith *et al.*, 1994a
	Cryptophycin E methyl ester	M-51	
	Cryptophycin F methyl ester	M-52	
	Cryptophycin G	M-53	
	Cryptophycin-145	M-54	Eli Lilly & Co *et al.*, 1998
Oscillatoria acutissima	Acutiphycin	M-55	Barchi *et al.*, 1984
	20,21 Didehydroacutiphycin	M-56	
Oscillatoria nigroviridis	Oscillatoxin A	M-57	Moore, 1982
	Debromoaplysiatoxin	M-22	Mynderse and Moore, 1978; Mynderse *et al.*, 1977
Plectonema radiosum	Tubercidin-5-D glucopyranose	M-58	Stewart *et al.*, 1988
Prorocentrium lima	Prorocentrolide	M-59	Torigoe *et al.*, 1988
Prorocentrium maculosum	Prorocentrolide B	M-60	Hu *et al.*, 1996
Schizothrix calcicola	Oscillatoxin A	M-57	Harrigan *et al.*, 1998a; Moore, 1982
	Debromoaplysiatoxin	M-22	Mynderse and Moore 1978; Mynderse *et al.*, 1977
	Dolastatin 12	M-31	Sitachitta *et al.*, 2000
	Lyngbyastatin 1	M-36	
	Yanucamide A and B	M-41 and M-42	

Scytonema conglutinata	Tolytoxin	M-61	Stewart *et al.*, 1988
Scytonema mirabile	Tolytoxin	M-61	Carmeli *et al.*, 1990; Stewart *et al.*, 1988
Scytonema ocellatum	Tolytoxin	M-61	Stewart *et al.*, 1988
Scytonema pseudohofmanni	Scytophycins A – D	M-62 to M-65	Barchi *et al.*, 1984; Patterson *et al.*, 1993 Smith *et al.*, 1993
Scytonema saleyeriense	Tubercidin	M-66	Stewart *et al.*, 1988
Symbiodinium sp.	Symbioramide	M-67	Kobayashi *et al.*, 1988b
Symploca hydnoides	Symplostatin 1	M-68	Harrigan *et al.*, 1998b, 1999; Poncet, 1999
	Symplostatin 2	M-69	
Symploca muscorum	Debromoaplysiatoxin	M-22	Mynderse *et al.*, 1977
Tolypothrix nodosa	Tolyporphin	M-70	Mayer, 1998; Minehan *et al.*, 1999
	Tolyporphin J	M-71	Morlière *et al.*, 1998; Prinsep *et al.*, 1992, 1995, 1998
	Tolyporphin K	M-72	Smith *et al.*, 1994b
Tolypothrix tenuis	Toyocamycin-5-D glucopyranose	M-73	Renau *et al.*, 1994; Stewart *et al.*, 1988
Tolypothrix byssoidea	Tubercidin	M-66	Barchi *et al.*, 1983; Furusawa *et al.*, 1983
Tolypothrix conglutinata	Tolytoxin	M-61	Moore, 1981
Tolypothrix distorta	Tubercidin-5-D glucopyranose	M-60	Stewart *et al.*, 1988

Cytotoxic metabolites from microalgae

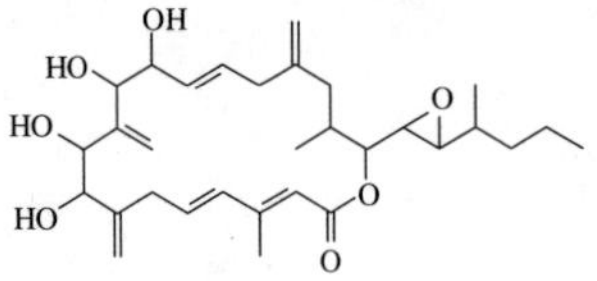

M-1

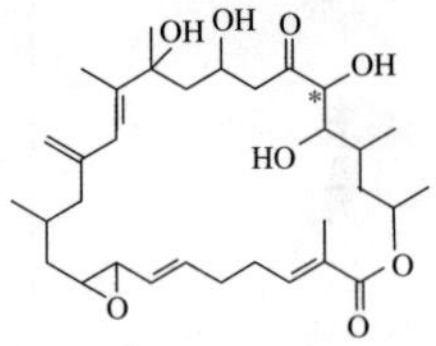

Srereoisomers at *C21

M-2
M-4

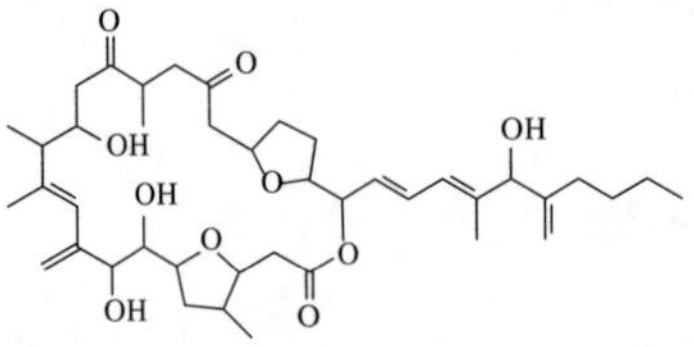

M-3

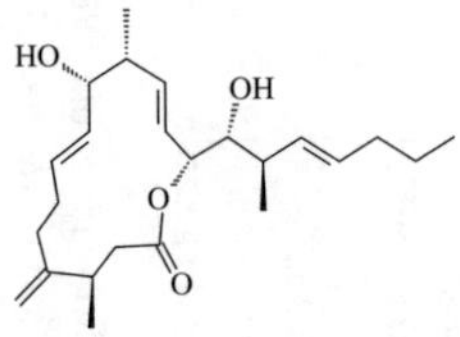

M-5

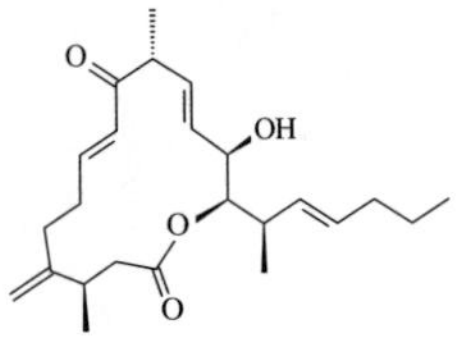

M-6

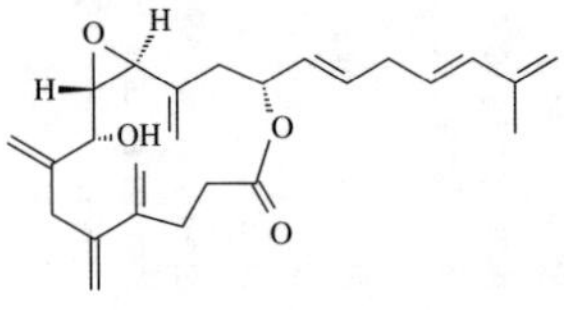

M-7

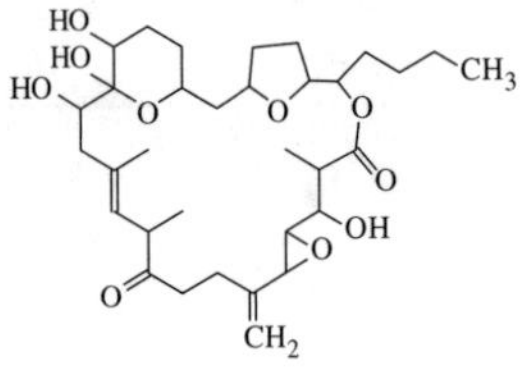

M-8

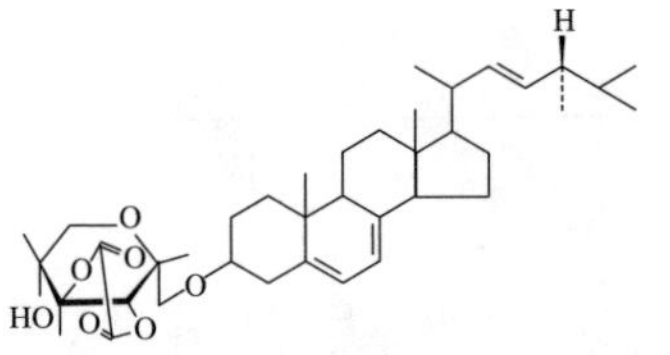

M-9

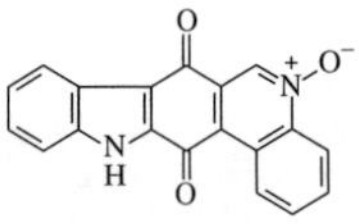

M-10

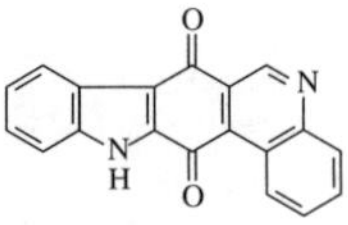

M-11

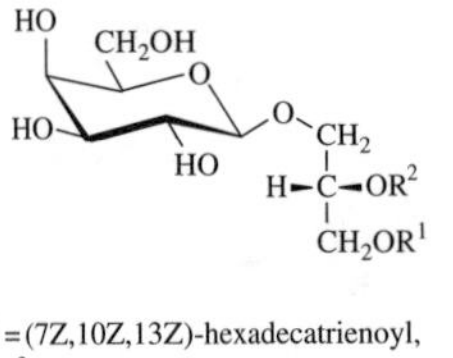

R^1 = (7Z,10Z,13Z)-hexadecatrienoyl, R^2 = (7Z,10Z)-hexadecadienoyl — M-12

R^1, R^2 = (7Z,10Z)-hexadecadienoyl — M-13

R^1 = linolenoyl, R^2 = (7Z,10Z,13Z)hexadecatrienoyl — M-14

R^1 = linolenoyl, R^2 = (7Z,10Z)-hexadecadienoyl — M-15

R^1, R^2 = linoleoyl — M-16

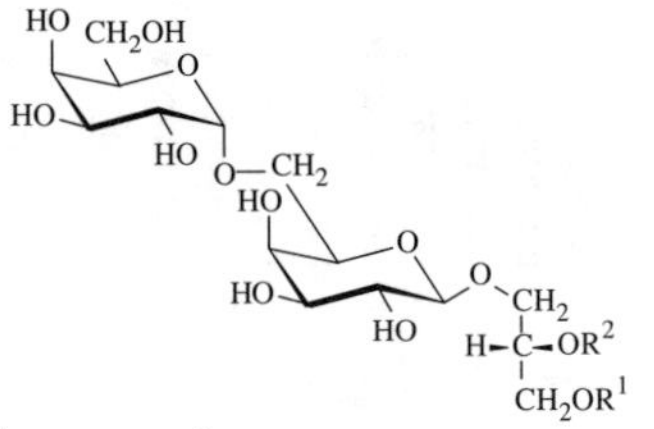

R^1 = linolenoyl, R^2 = (7Z,10Z)-hexadecadienoyl — M-17

R^1 = linolenoyl, R^2 = (7Z,10Z,13Z)-hexadecatrienoyl — M-18

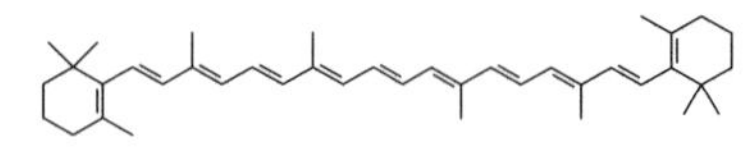

M-19

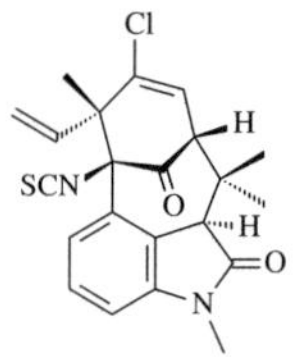

M-20

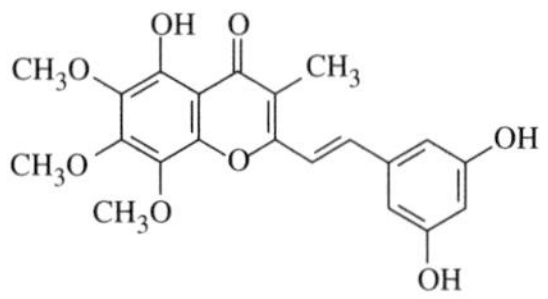

M-21

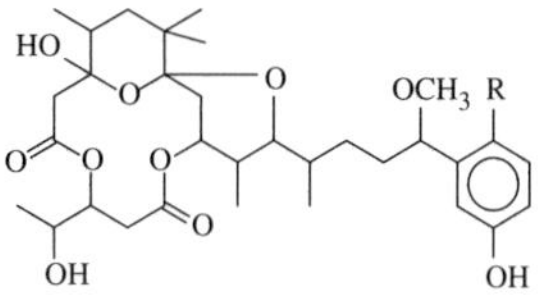

R: H **M-22**

R: Br **M-23**

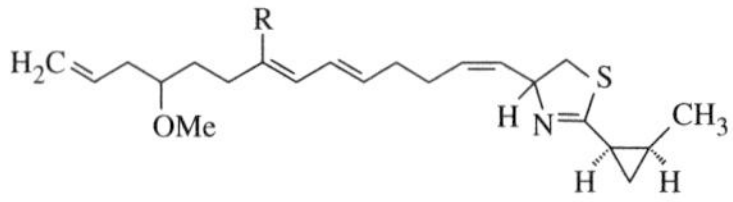

R: Me **M-24**

R: H **M-27**

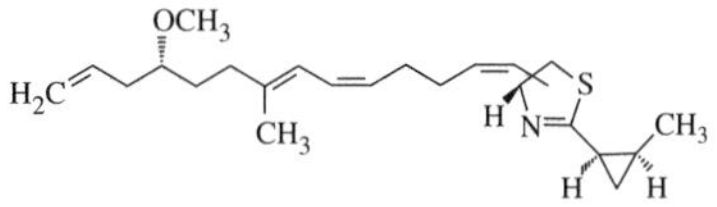

M-25

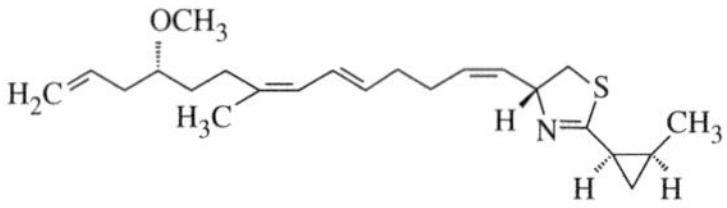

M-26

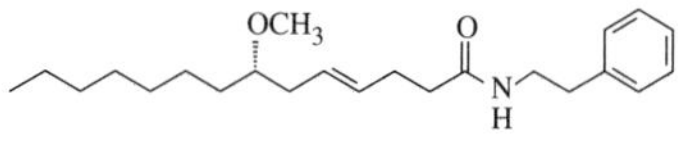

M-28

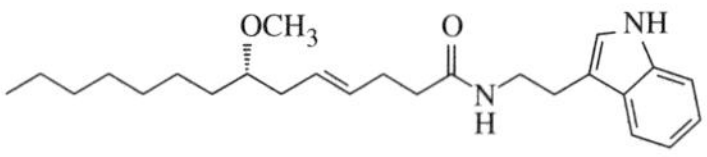

M-29

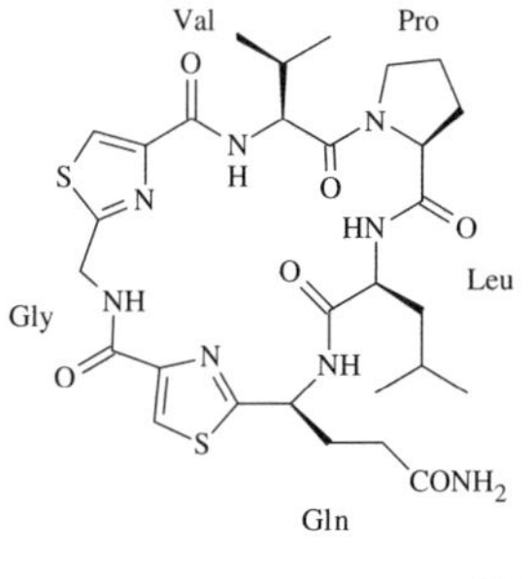

M-30

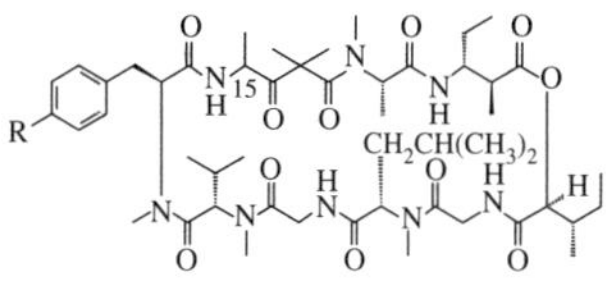

R	
H	**M-31**
OCH3	**M-32**

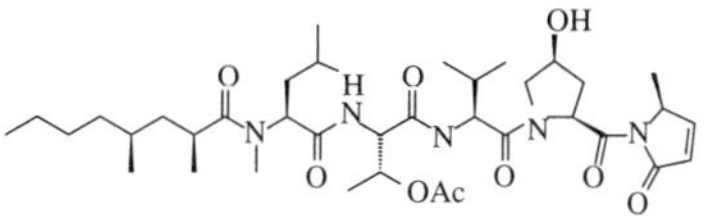

M-33

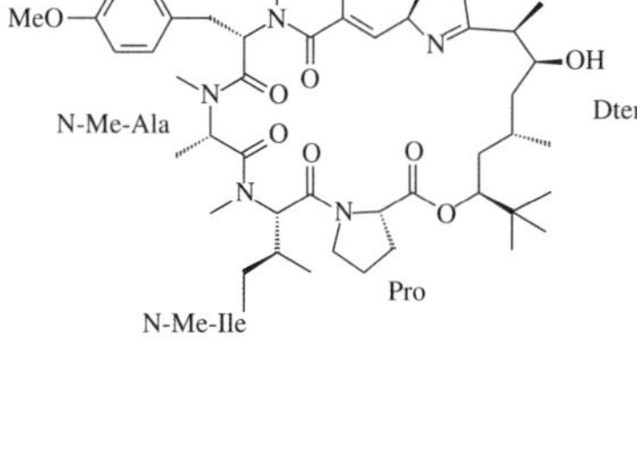

M-34

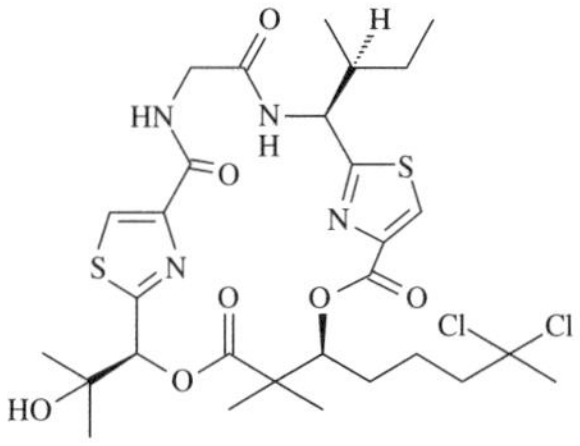

M-35

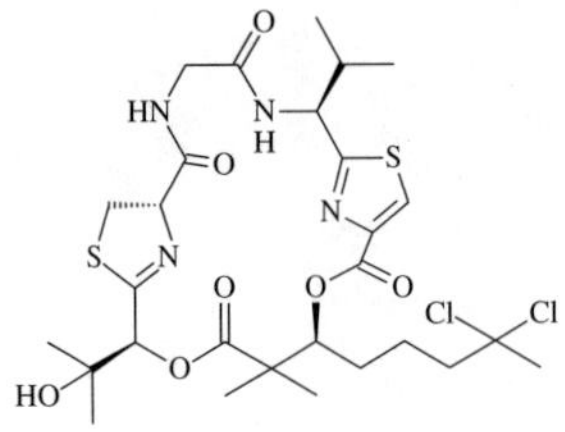

M-36

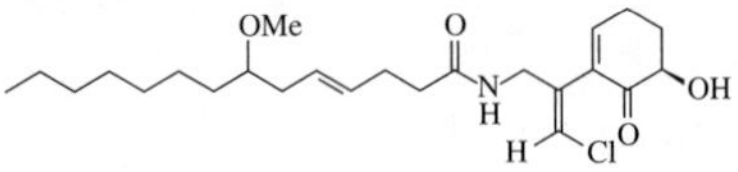

M-37

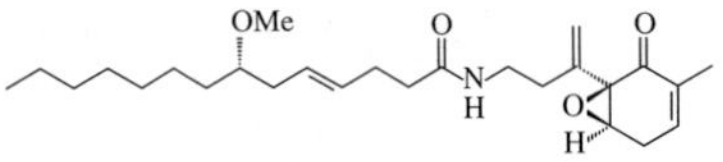

M-38

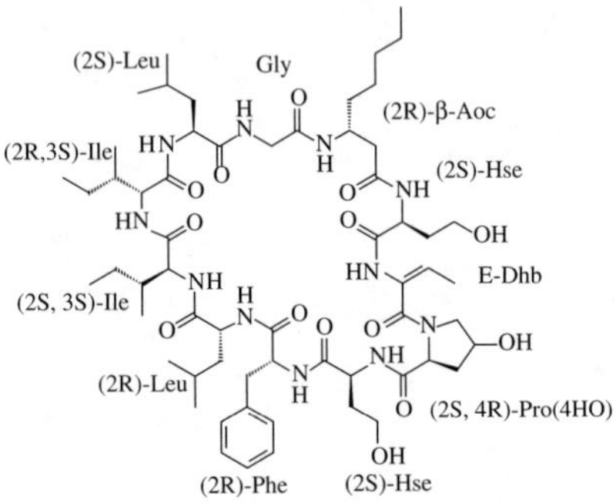

M-39

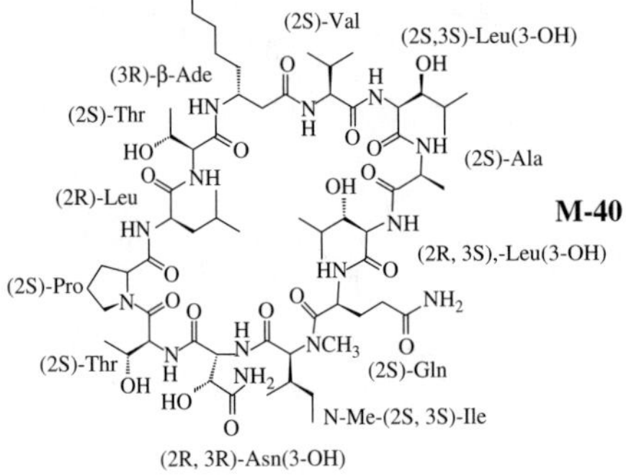

M-40

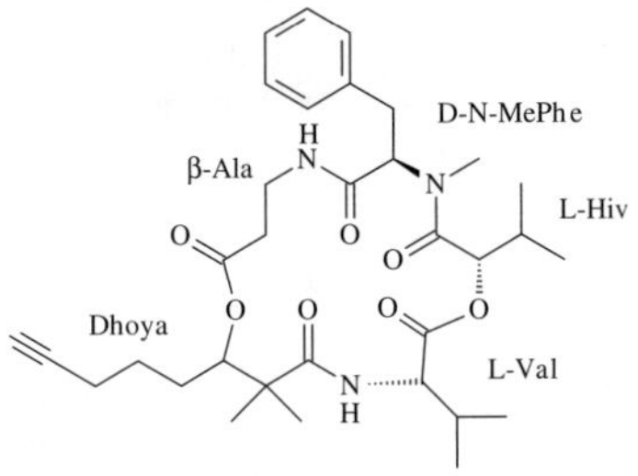

M-41

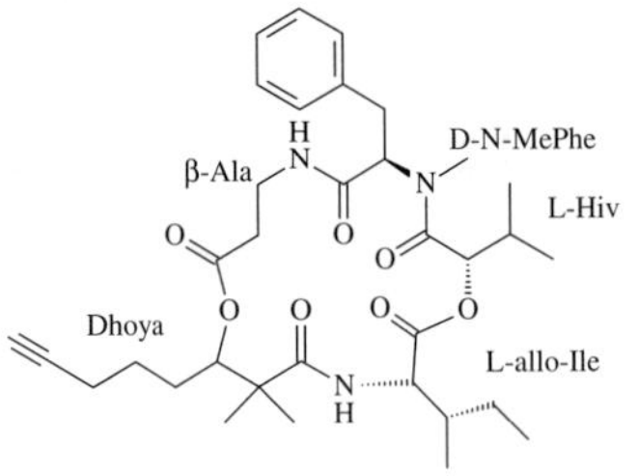

M-42

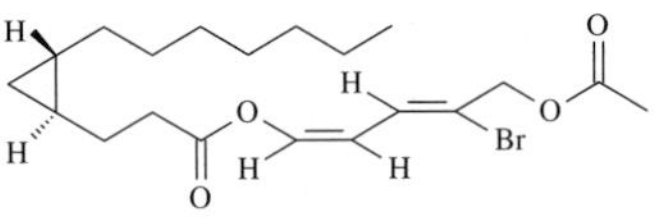

M-43

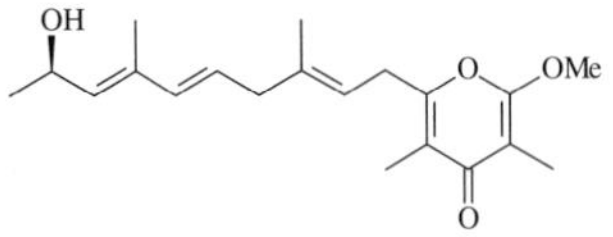

M-44

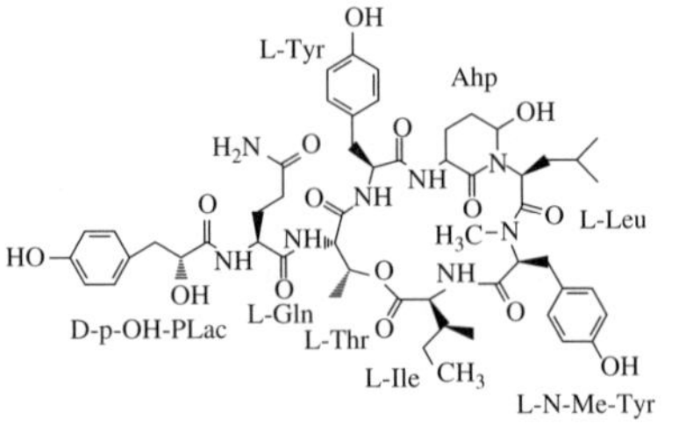

M-45

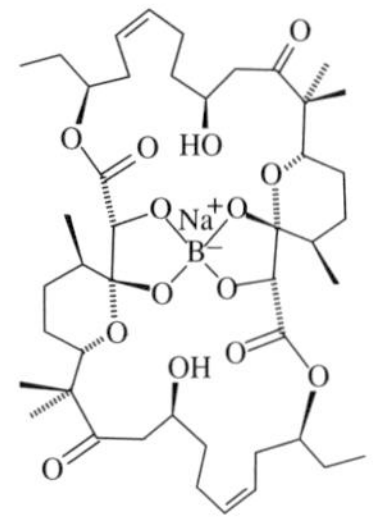

M-46

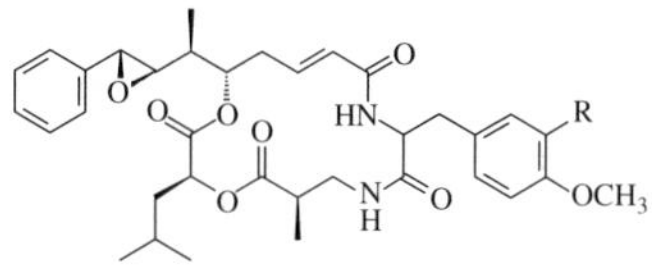

R: Cl **M-47**
R: H **M-48**

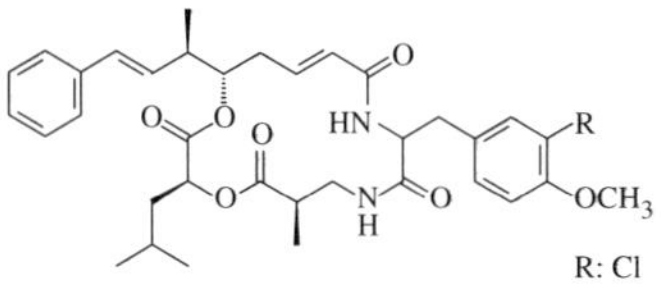

R: Cl **M-49**
R: H **M-50**

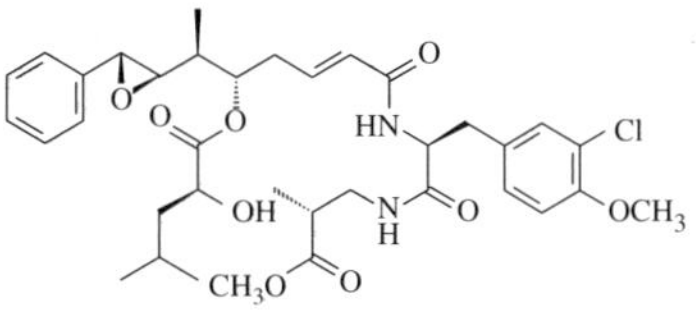

M-51

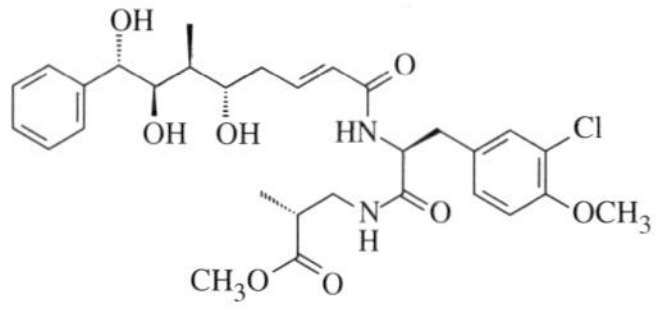

M-52

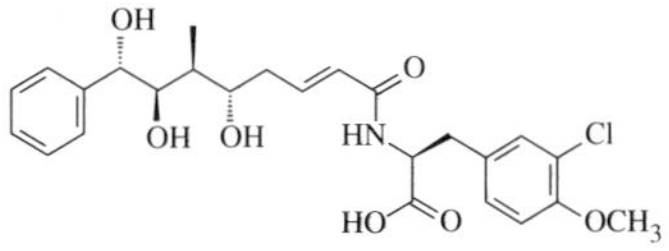

M-53

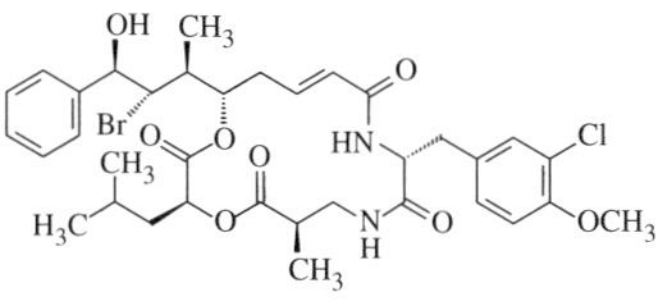

M-54

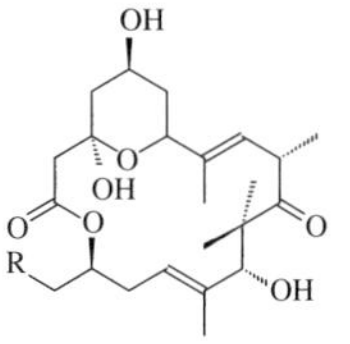

R: Bu **M-55**
R: $CH_2CH = CHCH_3$ **M-56**

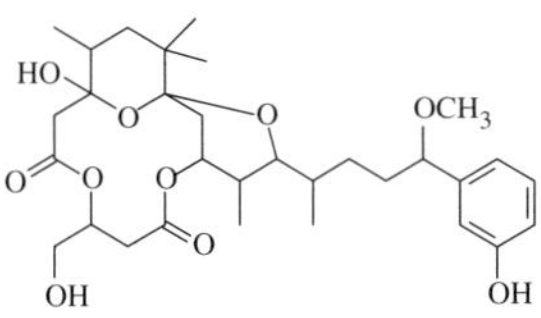

M-57

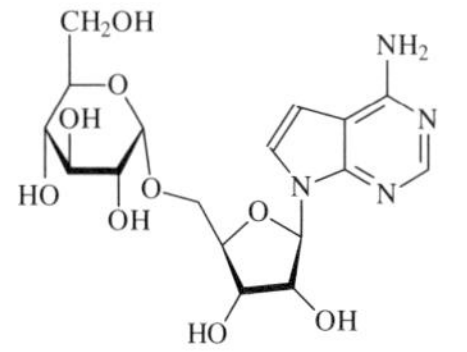

M-58

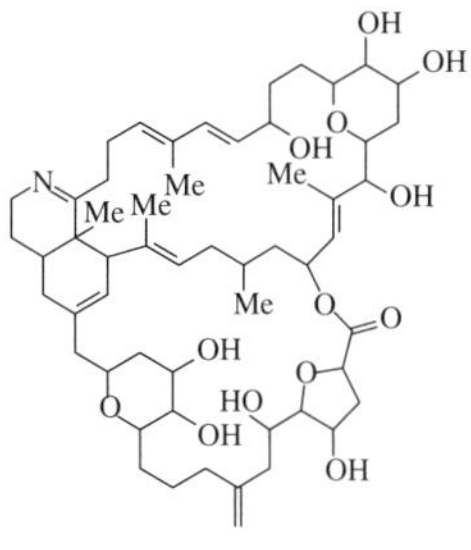

M-59

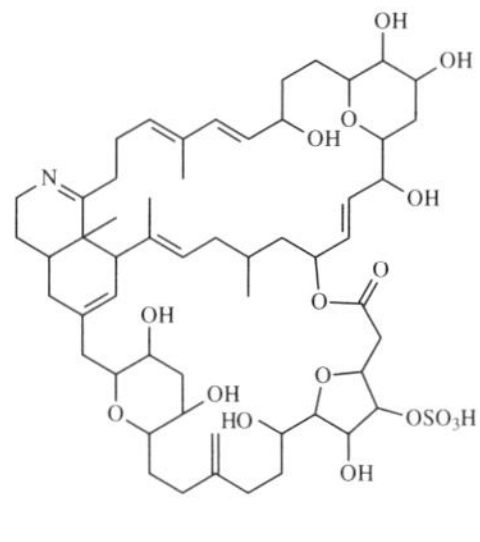

M-60

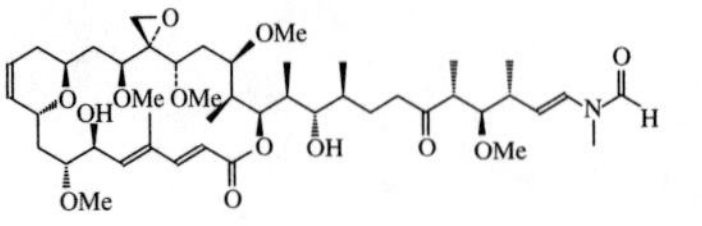

M-61

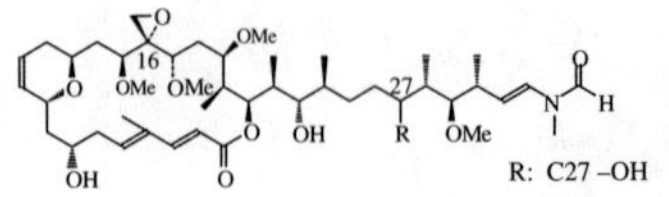

R: C27 –OH **M-62**

R: C27 = O **M-63**

R: C16 –Me **M-64**

R: C16 –Me, –OH **M-65**

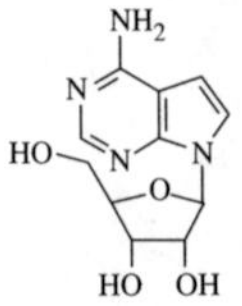

M-66

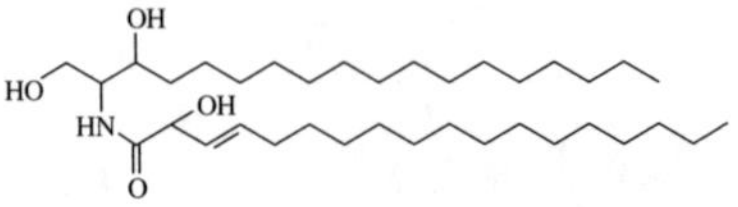

M-67

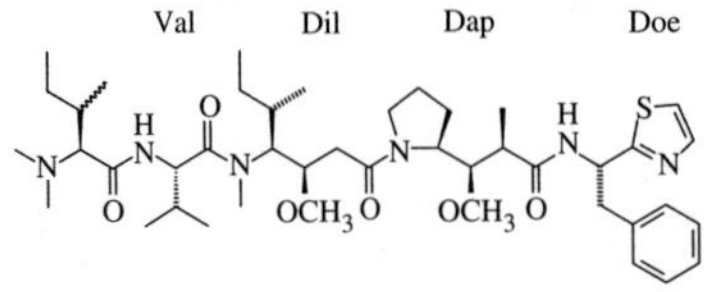

M-68

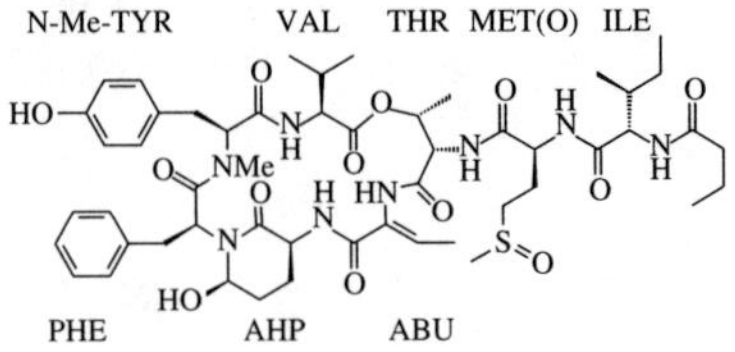

M-69

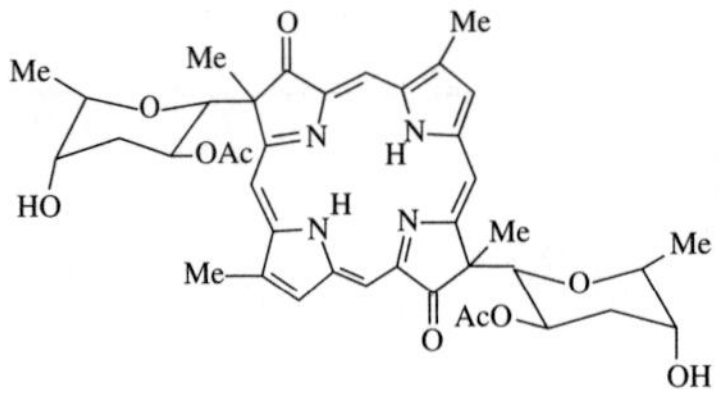

M-70

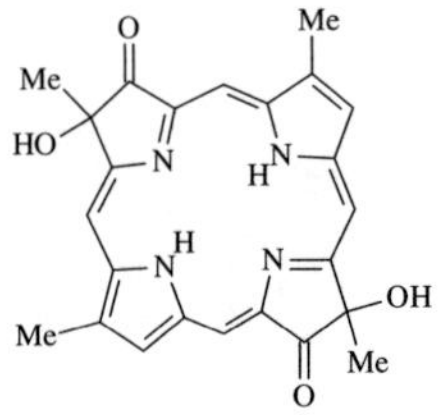

M-71

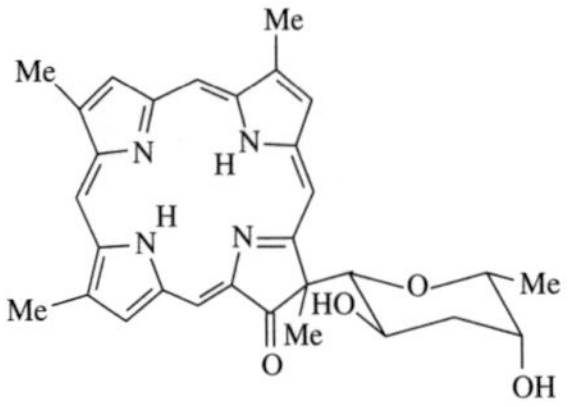

M-72

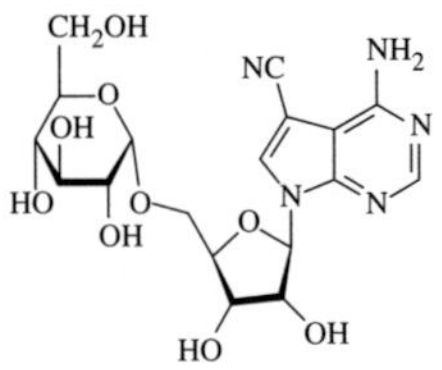

M-73

Conclusions

In the last 25 years, marine organisms (algae, invertebrates and microbes) have provided key structures and compounds that proved their potential in several fields, particularly as new therapeutic agents for a variety of diseases. The interest in the field is reflected by the number of scientific publications, the variety of new structures and the wide scope of the organisms investigated. As indicated in a review (Bongiorni and Pietra, 1996) covering the patents on different aspects of marine natural products applications, filed during the last 25 years, human health, health food and cosmetics account for more than 80% of the applications. As reported by Bongiorni and Pietra (1996), approximately 200 patents on marine natural products had been recorded between 1969 and 1995. In the period from 1996 until April 1999, close to 100 new patents had been issued in this area.

As yet, no compound isolated from a marine source has been approved for commercial use as a chemotherapeutic agent, though, Ziconotide® which is conotoxin VII from *Conus magnus* is awaiting final approval from the US FDA as a non-narcotic analgesic. In the antitumour area, several compounds are in the various phases of clinical development as potential agents.

Seaweeds have afforded, to date the highest number of compounds within a single group of marine organisms. A high percentage of recent reports concern bioactive metabolites with interesting biological properties. The reported pharmacological activities in this review have focused on the cytotoxicity against tumoral cells.

Algae were some of the first marine organisms that were investigated and proven to be rich sources of extraordinary chemical structures. Up to date only a small percentage of algae has been studied and the fact that many species exhibit geographic variation in their chemical composition shows the huge potential algae still hold as sources of interesting bioactive metabolites. Also since some of the investigations on the algal chemistry preceded the development of many of the current pharmacological bioassays it is well profitable to reexamine the pharmacological potential of these algal metabolites as well.

On the basis of the reviewed literature, it can be predicted that further intense research on bioactive algal metabolites will be stimulated from the advancement of sophisticated NMR techniques and the development of new faster and more efficient pharmacological evaluation assays.

Appendix

Chemical structures of selected compounds

List of compounds

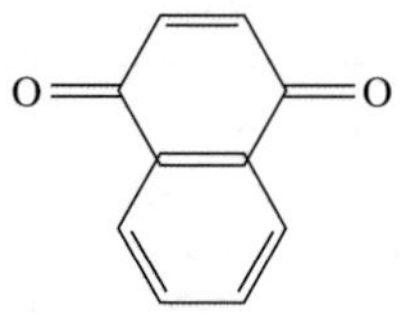

1,4-Naphthoquinone

Synonyms: 1,4-Naphthalenedione; 1,4-dihydro-1,4-diketonaphthalene; α-naphthoquinone.

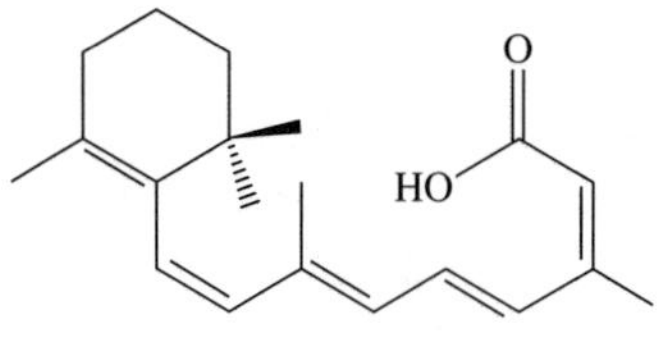

Isoretinoin

Synonyms: Accutane; 13-*cis*-Vitamin A acid; 13-*cis*-Retinoic acid; *cis*-retinoic acid; neovitamin A acid; 13-RA; ro-4-3780; retinoic acid, 9*Z* form; 3,7-Dimethyl-9-(2,6,6-trimethyl-1-cyclohexen-1-yl)-2,(*E*),4,6,8(*Z,Z,Z*)-nonatetraenoic acid; Isotretinoin; Accure; IsotrexGel; Roaccutane; Isotrex; Teriosal; 3,7-Dimethyl-9-(2,6,6-trimethyl-1-cyclohexen-1-yl)2-*cis*-4-*trans*-6-*trans*-8-*trans*-nonatetraenoic acid; Tasmar.

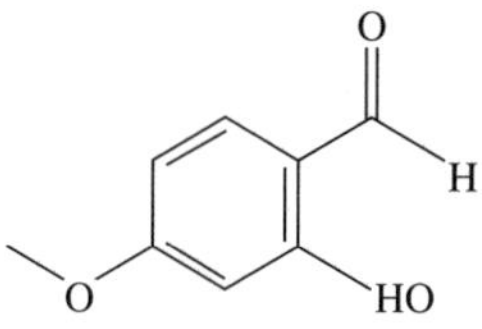

2-Hydroxy-4-methoxybenzaldehyde

Synonyms: 4-Methoxysalicylaldehyde.

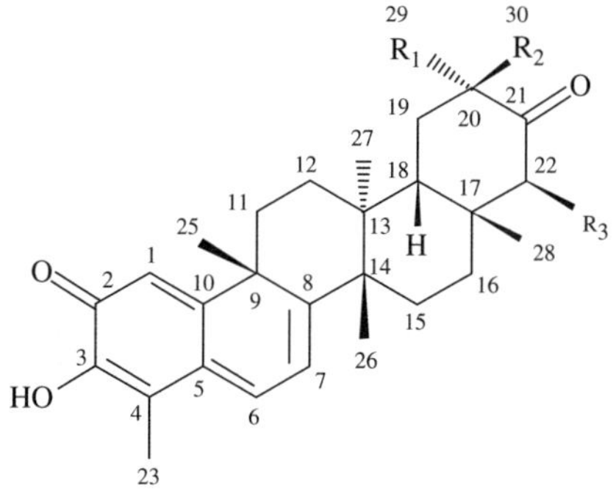

	R_1	R_2	R_3
1	H	CH_3	OH

22-Hydroxytingenone

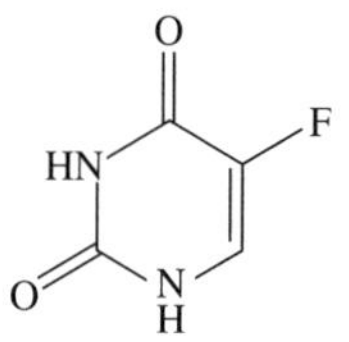

5-Fluorouracil

Synonyms: Fluorouracil, FU; 5-FU; 5-fluoro-2,4(1H,3H)-Pyrimidinedione; Adrucil; Efudex; Fluoroplex; Ro 2-9757; Arumel; Carzonal; Effluderm (free base); Efudix; Fluoroblastin; Fluracil; Fluri; Fluril; Kecimeton; Timazin; U-8953; Ulup; 5-Fluoro-2,4-pyrimidinedione; 5-Fluoropyrimidine-2,4-dione; 5-Ftouracyl; efurix; fluracilum; ftoruracil; queroplex; 50fluoro uracil; Fluorouracil (Topical); Fluroblastin.

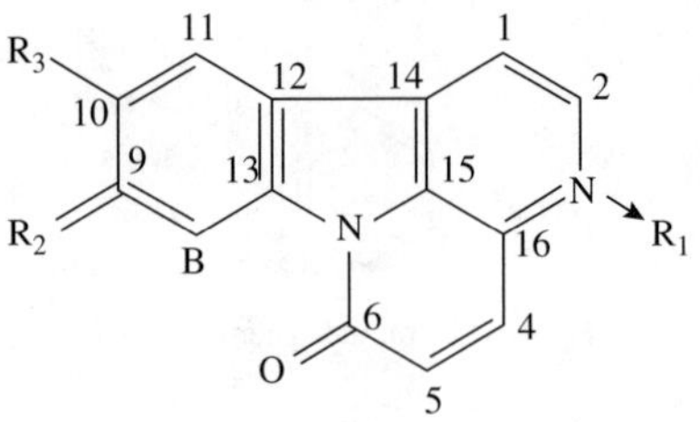

R_1	R_2	R_3
–	OMe	H

9-Methoxycanthin-6-one

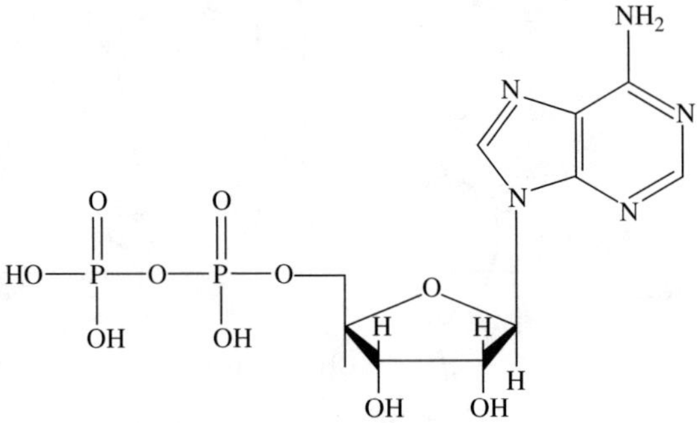

Adenosine diphosphate

Synonyms: ADP; adenosine 5′-diphosphate.

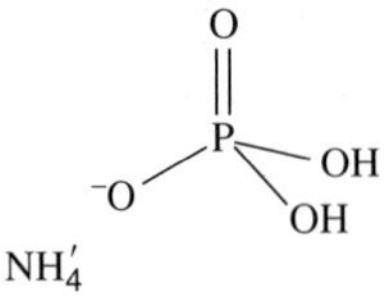

Ammonium phosphate monobasic

Synonyms: Ammonium biphosphate; Ammonium Dihydrogen Phosphate; ADP; Ammonium phosphate; Phosphoric acid, monoammonium salt; Monoammonium phosphate.

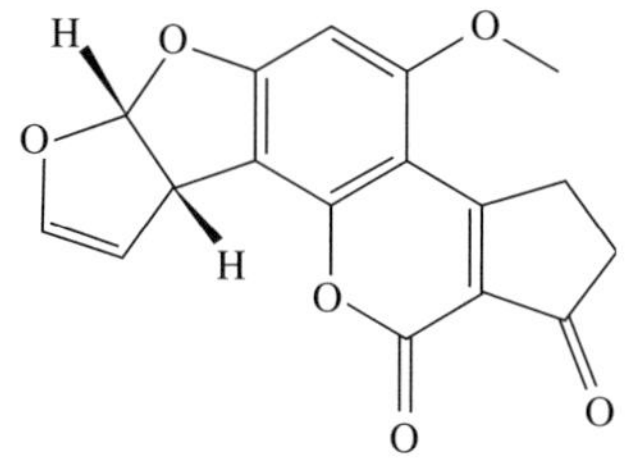

Aflatoxin B1

Synonyms: AFB1; AFBI aflatoxin b; 2,3,6a,9a-tetrahydro-4-methoxycyclopenta(c)furo(3′,2′:4,5)furo-(2,3-h)(1)benzopyran-1,11-dione; Aflatoxin B1, crystalline.

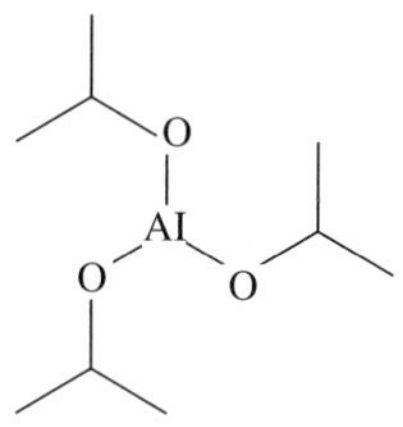

Aluminum isopropoxide

Synonyms: AIP; 2-Propanol, aluminum salt; Aluminum(III)isopropoxide.

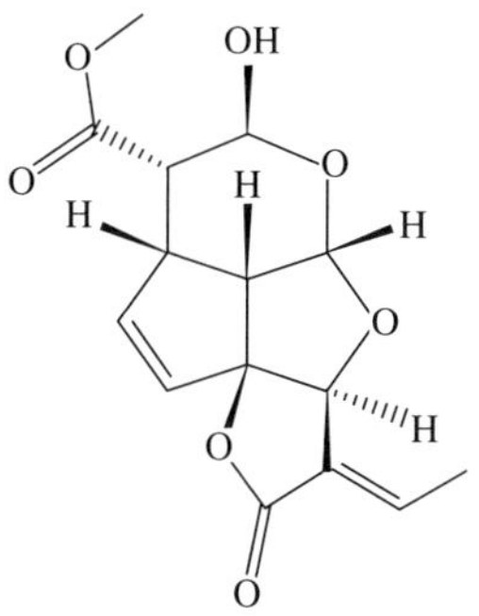

Allamandin

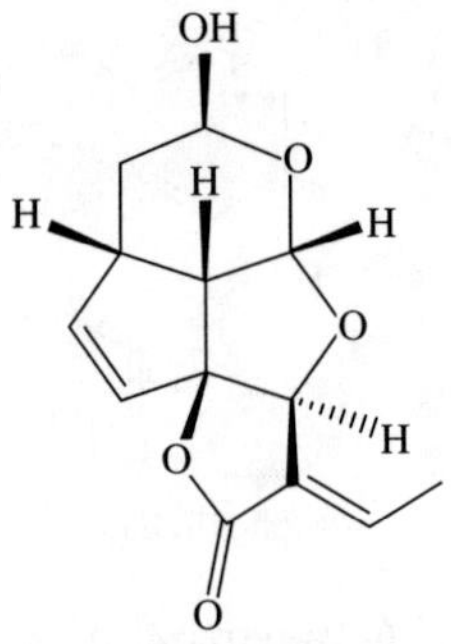

Allamcin

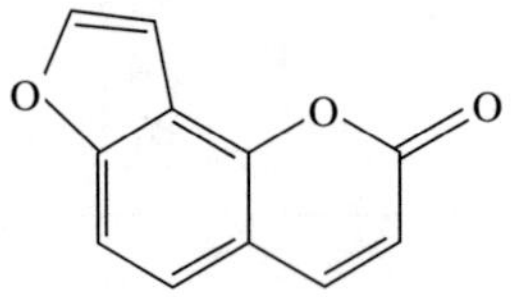

Angelicin

Synonyms: Furo(2,3-h)coumarin.

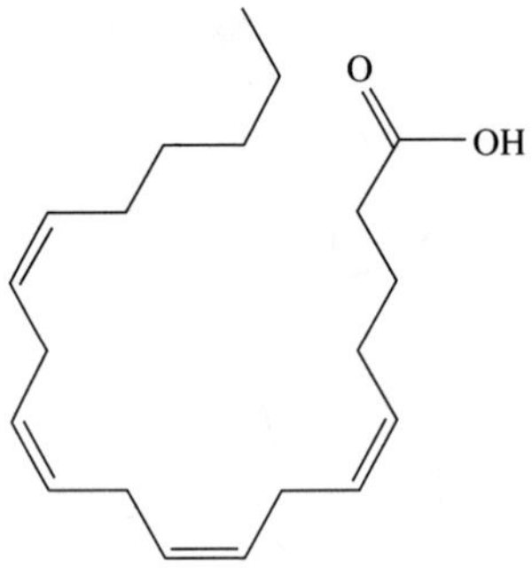

Arachidonic acid

Synonyms: 5,8,11,14-icosatetraenoic acid; 5,8,11,14-Eicosatetraenoic acid, (all-*Z*)-; Eicosa-5*Z*,8*Z*,11*Z*,14*Z*-tetraenoic acid.

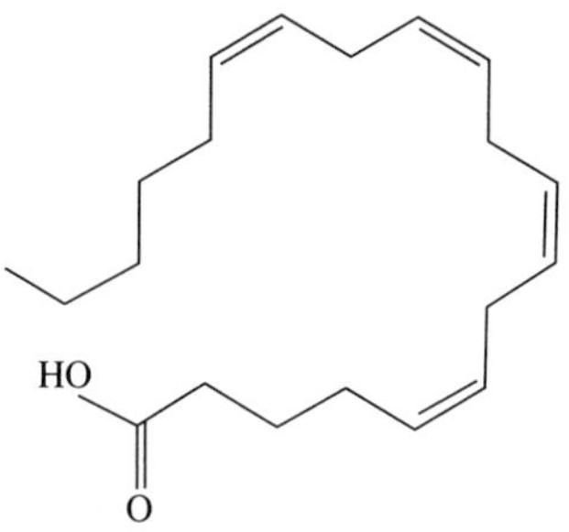

Arachidonic acid

Synonyms: all *cis*-Delta-5,8,11,14-icosatetraenoate.

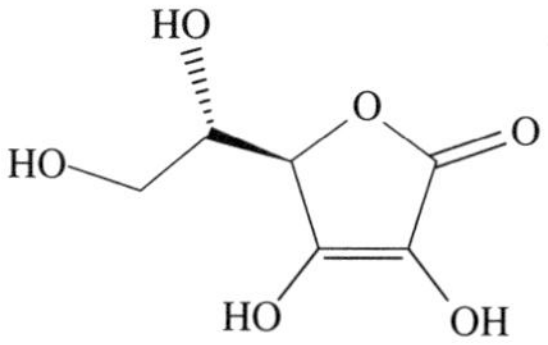

Vitamin C

Synonyms: L-ascorbic acid; L-3-ketothreohexuronic acid; Ascorbicap; Cebid; Cecon; Cevalin; Cemill; Sunkist; L-(+)-Ascorbic Acid; Acid Ascorbic; antiscorbic vitamin; antiscorbutic vitamin; cevitamic acid; 3-keto-L-gulofuranolactone; L-3-ketothreohexuronic acid lactone; laroscorbine; L-lyxoascorbic acid; 3-oxo-L-gulofuranolactone; L-xyloascorbic acid; adenex; allercorb; cantan; proscorbin; vitacin; AA; arco-cee; ascoltin; ascorb; ascorbajen; ascorbicab; ascor-b.i.d.; ascorbutina; ascorin; ascorteal; ascorvit; cantaxin; catavin c; cebicure; cebion; cee-caps td; cee-vite; cegiolan; ceglion; celaskon; ce lent; Celin; cemagyl; ce-mi-lin; cenetone; cereon; cergona; cescorbat; cetamid; cetemican; cevatine; Cevex; cevibid; cevimin; ce-vi-sol; cevital; cevitamin; cevitan; cevitex; Cewin; ciamin; Cipca; citriscorb; c-level; C-Long; colascor; concemin; C-Quin; C-Span; c-vimin; dora-c-500; davitamon c; duoscorb; L-*threo*-hex-2-enonic acid, γ-lactone; Hicee; hybrin; IDO-C; lemascorb; liqui-cee; Meri-c; natrascorb injectable; 3-oxo-L-gulofuranolactone (enol form); planavit c; redoxon; ribena; roscorbic; scorbacid; scorbu-c; secorbate; testascorbic; vicelat; Vicin; vicomin c; viforcit; viscorin; vitace; vitacee; vitacimin; vitamisin; vitascorbol; Xitix; Ascorbic Acid.

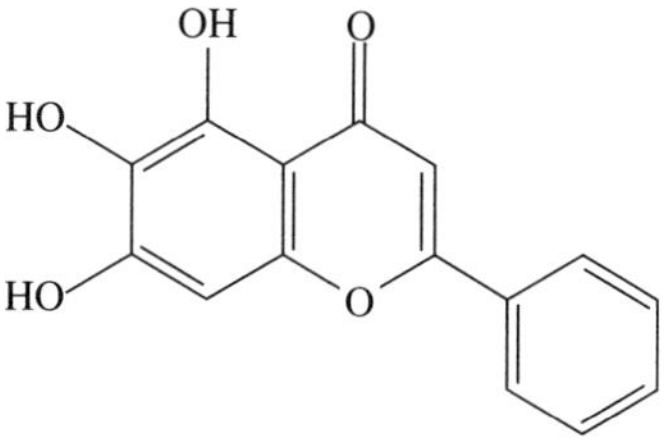

Baicalein

Synonyms: 5,6,7-Trihydroxyflavone.

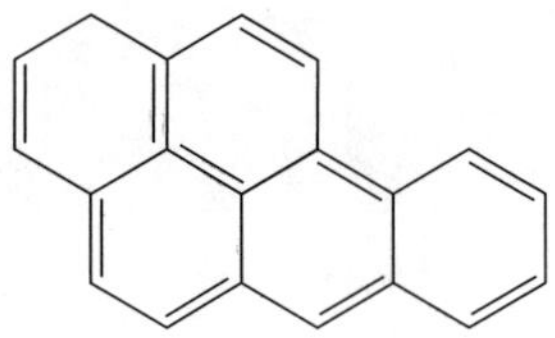

Benzo[*a*]pyrene

Synonyms: 6,7-Benzopyrene; B[A]P; BP; 3,4-Benzopyrene; Benzo[d,e,f]chrysene; 3,4-Benzpyrene; Benzpyrene; 3,4-benzylpyrene; 3,4-benz[*a*]pyrene; 3,4-BP; Benzo[*a*]pyrene.

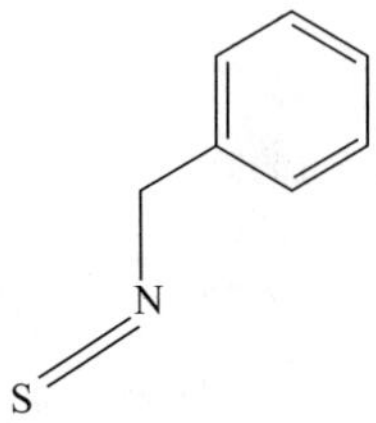

Benzylisothiocyanate

Synonyms: Benzene, (isothiocyanatomethyl)-.

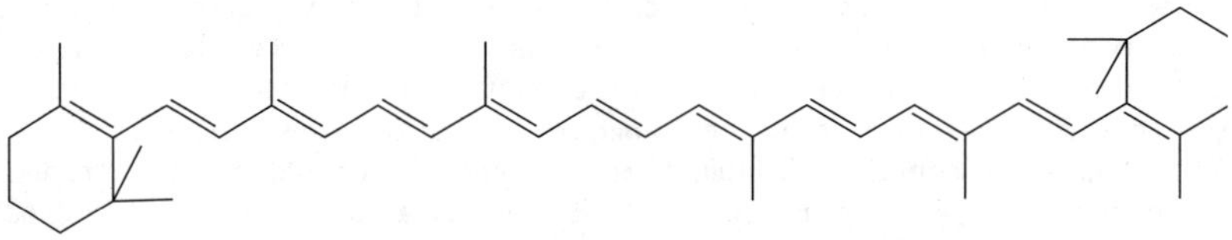

β-Carotene

Synonyms: Solatene; *trans*-*β*-Carotene; Carotene; *β*,*β*-Carotene.

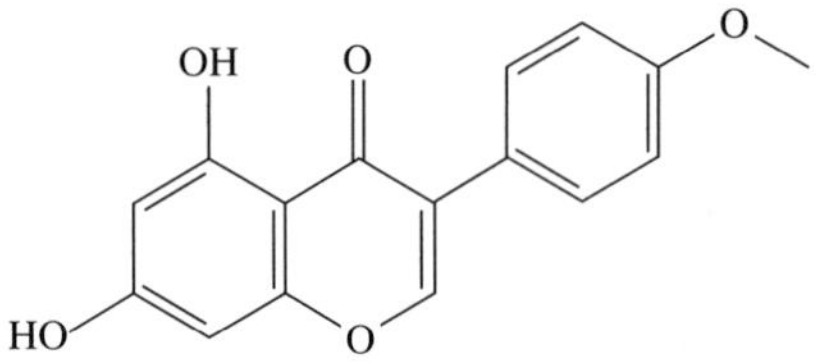

Biochanin A

Synonyms: 5,7-dihydroxy-4′-methoxyiso-flavone;olmelin.

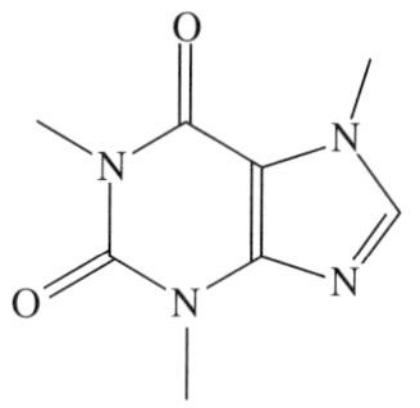

Caffeine

Synonyms: 1,3,7-Trimethylxanthine; 3,7-dihydro-1,3,7-trimethyl-1*H*-Purine-2,6-dione; 1,3,7-Trimethyl-2,6-dioxopurine; 7-Methyltheophylline; Alert-Pep; Cafeina; Cafipel; Guaranine; Koffein; Mateina; Methyltheobromine; No-Doz; Refresh'n; Stim; Theine; 1-methyltheobromine; methyltheobromide; eldiatric c; organex; 1,3,7-trimethyl-2,6-dioxo-1,2,3,6-tetrahydropurine; caffenium.

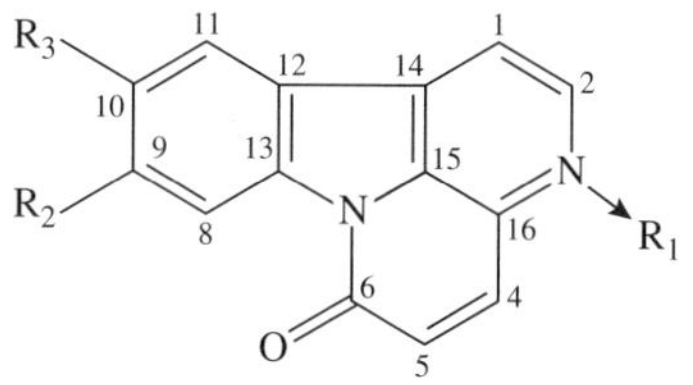

R_1	R_2	R_3
O	H	H

Canthin-6-one

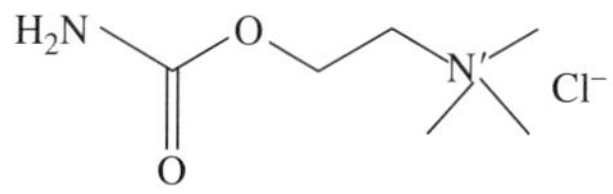

Carbamylcholine chloride

Synonyms: Carbachol chloride; 2-[(aminocarbonyl)oxy]-*N,N,N*-trimethyl-Ethanaminium chloride; Carbachol; (2-Hydroxyethyl)trimethylammonium chloride carbamate; (2-Carbamoyloxyethyl)trimethylammonium chloride.

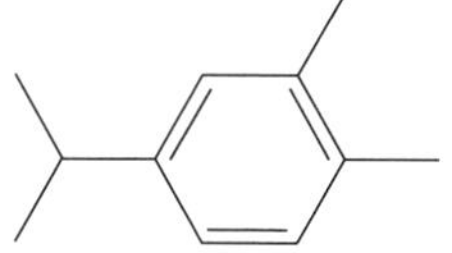

5-Isopropyl-2-methyl-phenol

Synonyms: carvacrol; Phenol, 2-methyl-5-(1-methylethyl)-; Cymenol; Hydroxy-*p*-cymene; Isopropyl-*o*-cresol; Isothymol; Methyl-5-(1-methylethyl)phenol.

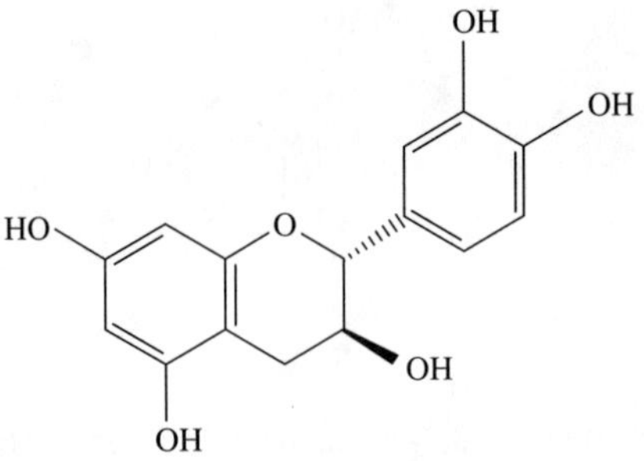

Catechin

Synonyms: Cianidanol; (+)-CATECHIN.

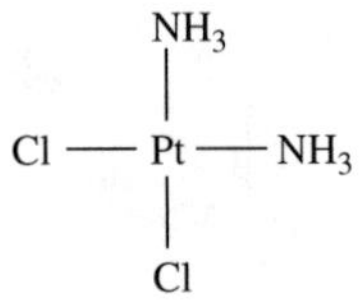

Cisplatin

Synonyms: cis-Diaminedichloroplatinum(II); *cis*-Platinous Diamine Dichloroplatin; CACP; CDDP; CPDD; Platinol; *cis*-Platinous diamine dichloride; dCDP; *cis* Pt II; *cis*-Diaminedichloroplatinum; DDP; DDPt; Platiblastin; *cis*-Dichlorodiamineplatinum(II); (SP-4-2)-diaminedichloroplatinum; *cis*-diaminodichloroplatinum(II); *cis*-platinum(II) diamine dichloride; cisplatyl; CPDC; *cis*-ddp; neoplatin; peyrone's chloride; platinex; PT-01; diaminedichloroplatinum; *cis*-dichlorodiaminoplatinum(II); *cis*-dichlorodiamineplatinum; *cis*-platinous diaminodichloride; 2′-Deoxycytidine diphosphate; *cis*-Diammine dichloroplatinum(II); *cis*-Dichlorodiammine platinum (II); CISPLATIN (CIS-DIAMINEDICHLOROPLATIUM (II)).

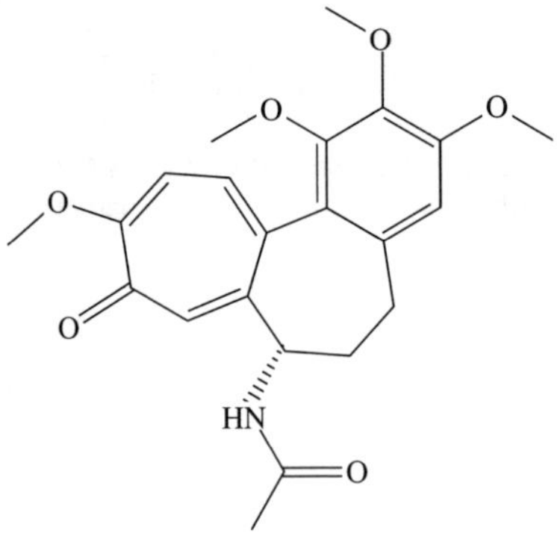

Colchicine

Synonyms: (*S*)-*N*-(5,6,7,9-tetrahydro-1,2,3,10-tetramethoxy-9-oxobenzo[*a*]heptalen-7-yl)acetamide; *N*-(5,6,7,9-tetrahydro-1,2,3,10-tetramethoxy-9-oxobenzo[*a*]heptalen-7-yl)-acetamide; *N*-acetyltrimethylcolchicinic acid methyl ether; 7-acetamido-6,7-dihydro-1,2,3,10-tetramethoxy-benzo[*a*]heptalen-9(5H)-one; 7-α-*H*-colchicine; colchineos; colchisol; colcin; colsaloid; condylon; colchiceine methyl ether; Colgout; COLCHICINE CRYSTALLINE.

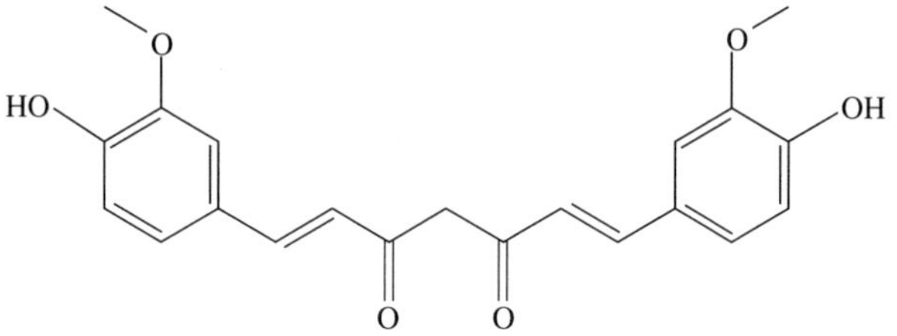

Curcumin

Synonyms: C.I. 75300; 1,6-Heptadiene-3,5-dione, 1,7-bis(4-hydroxy-3-methoxyphenyl)-, (*E*,*E*)-; tumeric yellow; 1,7-bis(4-hydroxy-3-methoxyphenyl)-1,6-heptadiene-3,5-dione; C.I. natural yellow 3; curouma; diferuloylmethane; gelbwurz; Haidr; Halad; haldar; Halud; indian saffron; kachs haldi; merita earth; safra d'inde; souchet; terra merita; yellow ginger; yellow root; YO-KIN; Natural Yellow 3; E-100.

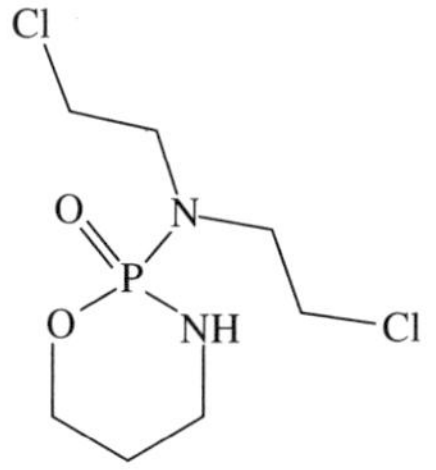

Cyclophosphamide

Synonyms: *N*, *N*-Bis(2-Chloroethyl)tetrahydro-2*H*-1,3,2-Oxazaphosphorin-2-Amine, 2-Oxide; Cytoxan; Cyclophosphane; B 518; Procytox; Neosar; Cyclophosphamides; Cyclophosphoramide; Sendoxan; bis(2-Chloroethyl)phosphamide cyclic propanolamide ester; bis(2-Chloroethyl)phosphoramide cyclic propanolamide ester; *N*,*N*-bis(beta-Chloroethyl)-*N*′,*O*-propylenephosphoric acid ester diamide; *N*-bis(β-Chloroethyl)-*N*′,*O*,trimethylenephosphoric acid ester diamide; Cytophosphane; 2-(bis(2-Chloroethyl)-amino)tetrahydro-2*H*-1,3,2-oxazaphosphorine 2-oxide; Cycloblastin; Cyclostin; (−)-Cyclophosphamide; Asta B 518; Clafen; Claphene; Cyclophosphamidum; cb 4564; Endoxan R; Endoxan-Asta; Endoxana; Endoxanal; Endoxane; Enduxan; Genoxal; Mitoxan; *N*,*N*-Bis(β-chloroethyl)-*N*′,*O*-trimethylenephosphoric acid ester diamide; *N*,*N*-Bis(2-chloroethyl)-*N*′,*O*-propylenephosphoric acid ester diamide; *N*,*N*-Di(2-chloroethyl)-*N*,*O*-propylene-phosphoric acid ester diamide; Semdoxan; Senduxan; sk 20501; tatrahydro-2-(Bis(2-chloroethyl)amino)-2*H*-1,3,2-oxazaphosphorine 2-oxide; 2-(di(2-chloroethyl)amino)-1-oxa-3-aza-2-phosphacyclohexane 2-oxide; ASTA; *N*,*N*-bis(2-chloroethyl)-*N*′-(3-hydroxypropyl)phosphorodiamidic acid intramol. ester; tetrahydro-*N*,*N*-bis(2-chloroethyl)- 2*H*-1,3,2-oxazaphosphorin-2-amine 2-oxide; 1-(bis(2-chloroethyl)amino)-1-oxo-2-aza-5-oxaphosphoridine.

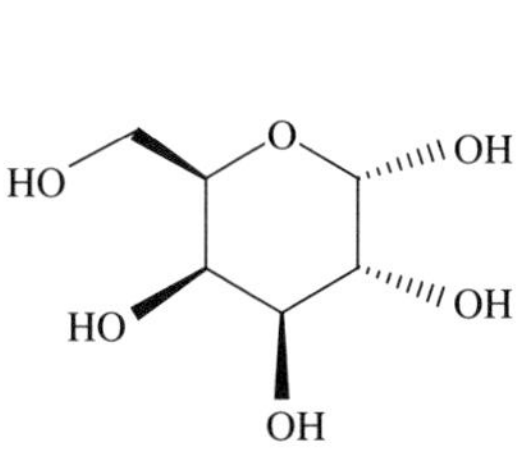

D-Galactose

Synonyms: D-(+)-Galactose; Galactose; Gal; α-Galactose(D); D(+)GALACTOSE SIGMA GRADE.

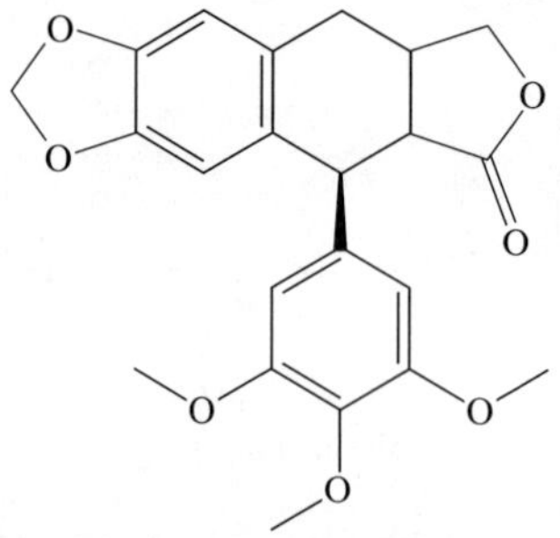

Desoxypodophyllotoxin

Synonyms: (5*S*)-5,8,8a,9-Tetrahydro-5-(3,4,5-trimethoxyphenyl)-6H-furo[3′,4′:6,7]-naphtho[2,3-d]-1,3-dioxol-6-one.

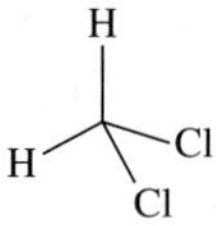

Dichloromethane

Synonyms: Methylene dichloride; Methane dichloride; R 30; Aerothene MM; Refrigerant 30; Freon 30; DCM; narkotil; solaesthin; solmethine; Methylene chloride; Plastisolve; METHYLENE CHLORIDE (DICHLOROMETHANE); Dichloromethane.

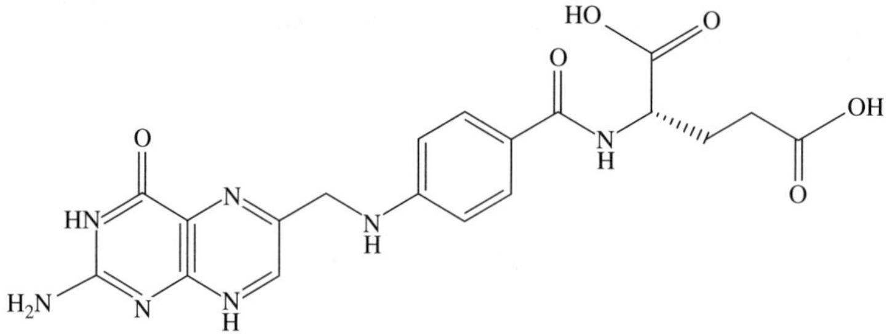

Dihydrofolate

Synonyms: 7,8-dihydrofolic acid.

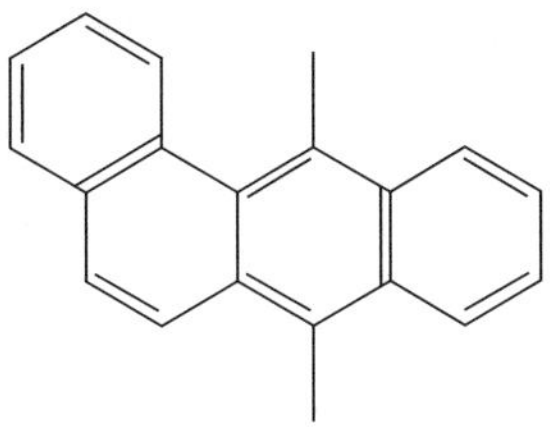

7,12-Dimethylbenz[a]anthracene

Synonyms: 9,10-Dimethyl-1,2-benzanthracene; DMBA; Dimethylbenzanthracene; dimethylbenz[*a*]anthracene; 7,12-dimethylbenzanthracene; 9,10-dimethyl-benzanthracene; 9,10-dimethylbenz[*a*]anthracene; dimethylbenzanthrene; 1,4-dimethyl-2,3-benzophenanthrene; 7,12-dmba; 7,12-dimethyl-1,2-benzanthracene.

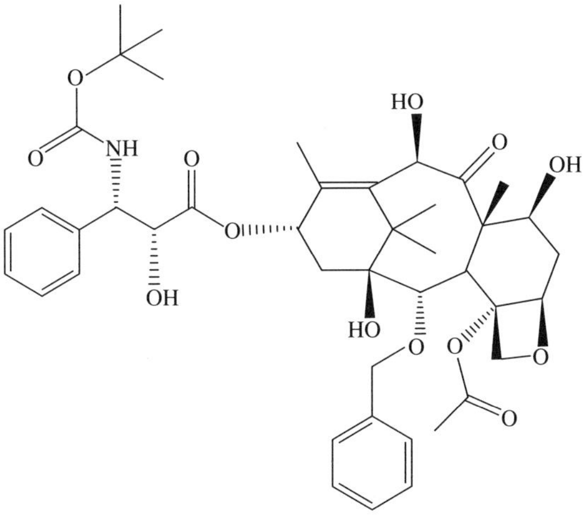

Taxotere

Synonyms: docetaxel; *N*-debenzoyl-*N*-*tert*-butoxycarbonyl-10-deacetyl taxol.

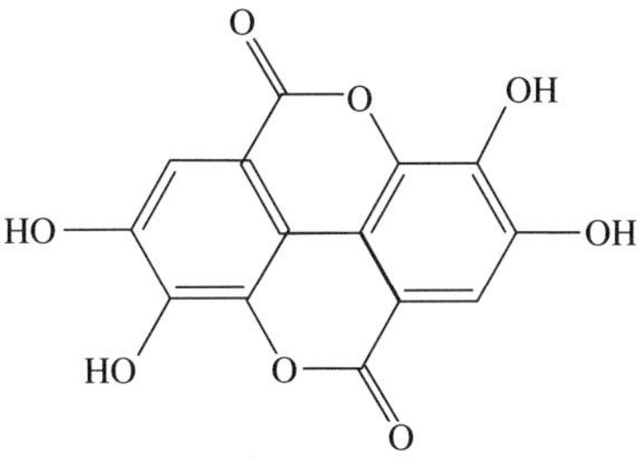

Ellagic acid

Synonyms: 4,4′,5,5′,6,6′-hexahydrodiphenic acid 2,6,2′,6′-dilactone; 2,3,7,8-tetrahydroxy(1)benzopyrano(5,4,3-cde)(1)benzopyran-5,10-dione; alizarine yellow; benzoaric acid; elagostasine; eleagic acid; gallogen; lagistase; C.I. 55005; C.I. 75270; Ellagic acid dihydrate.

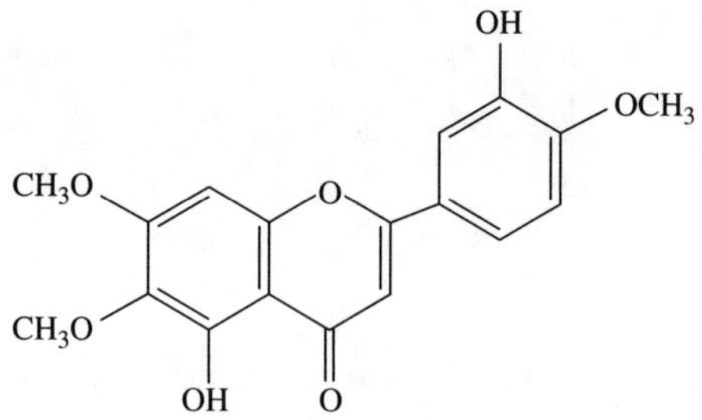

Eupatorin

Synonyms: 5-Hydrohy-2-(3-hydroxy-4-methoxy-phenyl)-6,7-dimethoxy-4H-1-benzopyran-4-one; 3′,5-dihydroxy-4′,6,7-trimethoxyflavone.

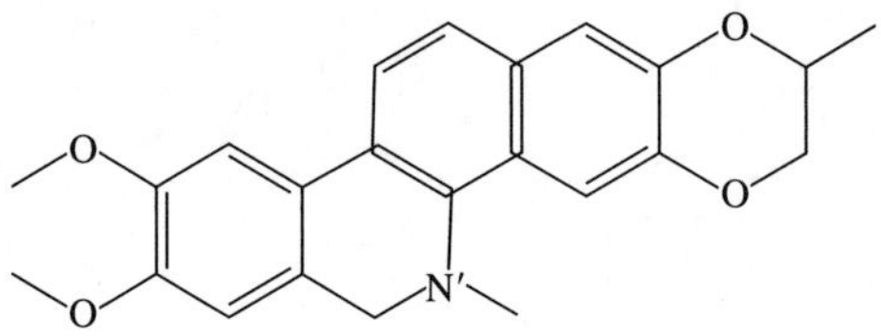

Fagaronine

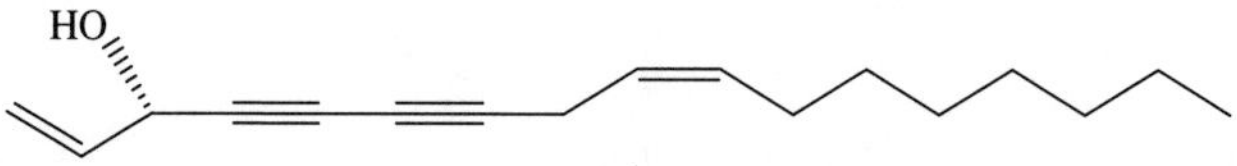

Falcarinol

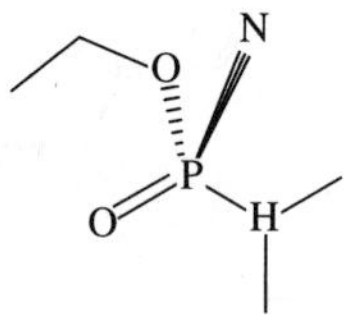

Tabun

Synonyms: Ethyl *N,N*-dimethylphosphoramidocyanidate; Ethyl dimethylphosphoramidocyanidate; Dimethylaminoethoxy-cyanophosphine oxide; Dimethylamidoethoxyphosphoryl cyanide; Ethyldimethylaminocyanophosphonate; Ethyl ester of dimethylphosphoroamidocyanidic acid; Ethylphosphorodimethylamidocyanidate; GA; EA1205; *O*-Ethyl *N,N*-dimethyl phosphoramido-cyanidate; dimethylphosphoramidocyanidic acid ethyl ester; *O*-ethyl dimethylamidophosphoryl-cyanide.

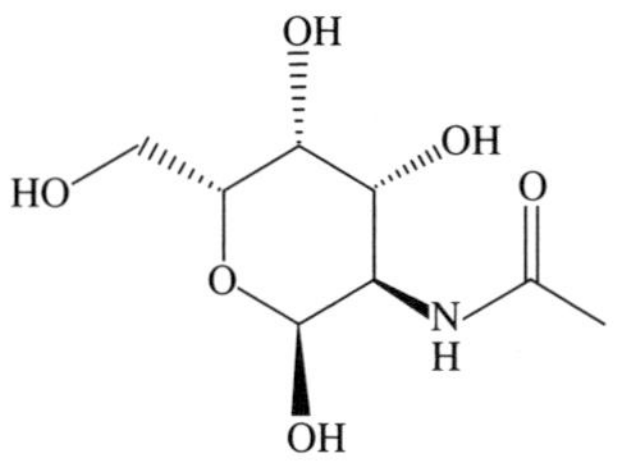

N-acetyl-D-Galactosamine

Synonyms: 2-Acetamido-2-deoxy-D-galactopyranose; *N*-Acetyl-D-chondrosamine; 2-Acetamido-2-deoxy-D-galactose; GalNAc.

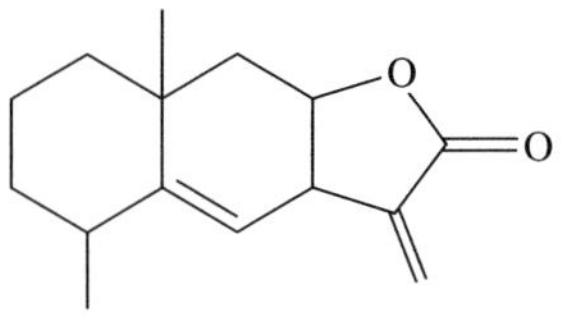

Alantolactone

Synonyms: [3a*R*-(3aa,5b,8ab,9aa)]-3a,5,6,7,8,8a,9,9a-Octahydro-5,8a-dimethyl-3-methylenenaphtho-[2,3-b]furan-2(3H)-one; 8b-hydroxy-4aH-eudesm-5-en-12-oic acid; γ-lactone; Helenin; Alant camphor; Elecampane camphor; Inula camphor; Eupatal.

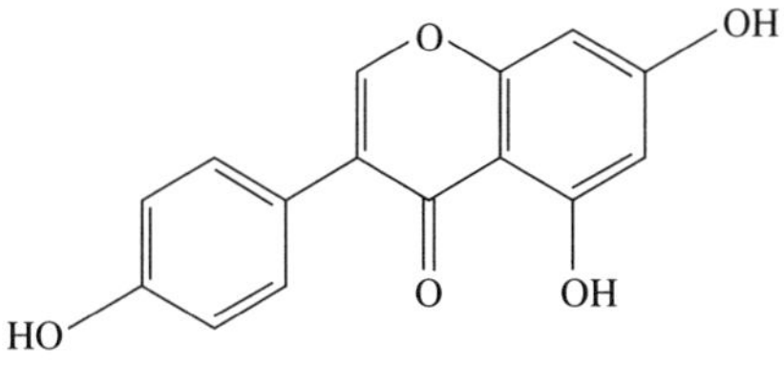

Genistein

Synonyms: 4′,5,7-Trihydroxyisoflavone.

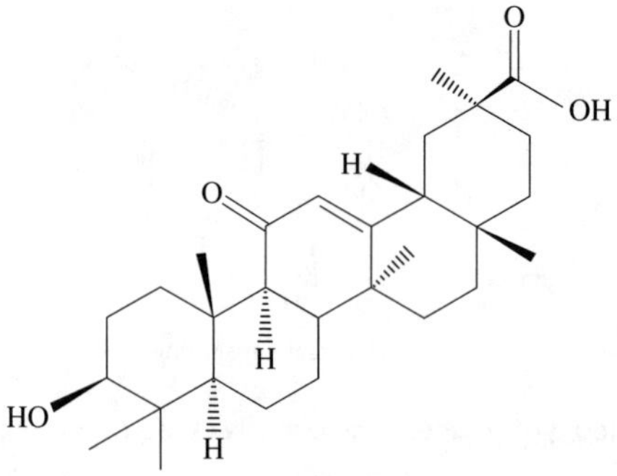

Glycyrrhetinic acid

Synonyms: 18beta-Glycyrrhetinic acid; Enoxolone; 18-beta-Glycyrrhetinic acid, (Titr., on the anhydrous basis).

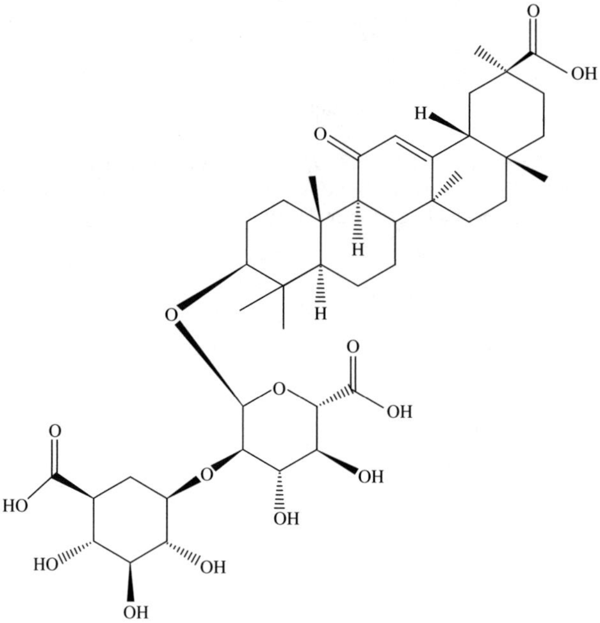

Glycyrrhizic acid

Synonyms: Glycyrrhizinate; Glycyrrhizin.

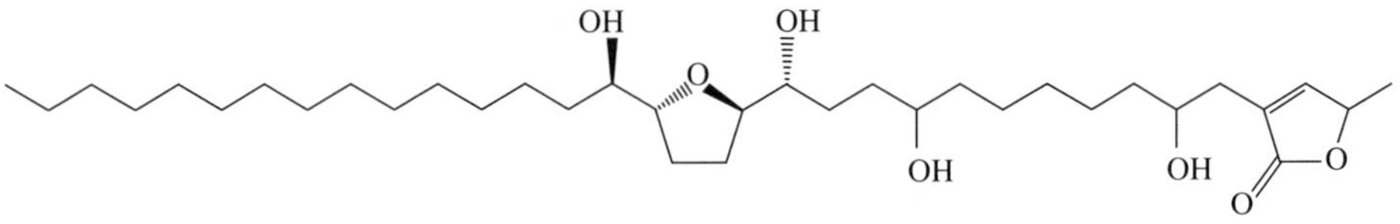

Goniothalamicin

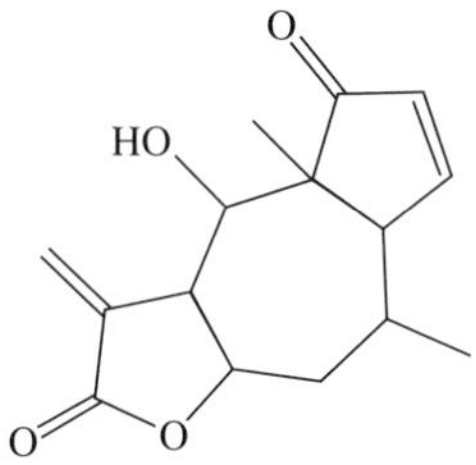

Helenalin

Synonyms: 3,3a,4,4a,7a,8,9,9a-Octahydro-4-hydroxy-4a,8-dimethyl-3-methyleneazuleno[6,5-b]furan-2,5dione; 6α,8β-dihydroxy-4-oxoambrosa-2,11(13)-dien-12-oic acid 12,8-lactone.

Hexane

Synonyms: Normal hexane; Hexyl hydride; n-Hexane; skellysolve B; dipropyl; gettysolve-b; Hex; n-Hexane.

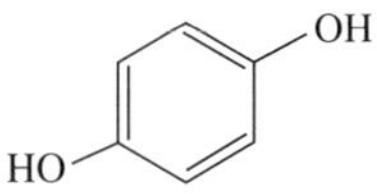

Hydroquinone

Synonyms: Dihydroquinone; 1,4-Dihydroxybenzene; Quinol; 1,4-benzenediol; *p*-Benzendiol; Benzoquinol; *para*-Hydroxyphenol; Dihydroxybenzene; 1,4-Hydroxybenzene; *p*-Hydroquinone; *p*-Dihydroxybenzene; 1,4-Benzendil; Aida; Black and White Bleaching Cream; Eldoquin; Elopaque; quinnone; Tecquinol; Hydroquinol; *p*-Diphenol; Hydrochinon; hydrokinone; *p*-benzenediol; p-dioxobenzene; α-hydroquinone; benzohydroquinone; β-quinol; arctuvin; eldopaque; tenox hq; tequinol; Benzene-1,4-diol; HYDROQUINONE BAKER; Hydroquinone.

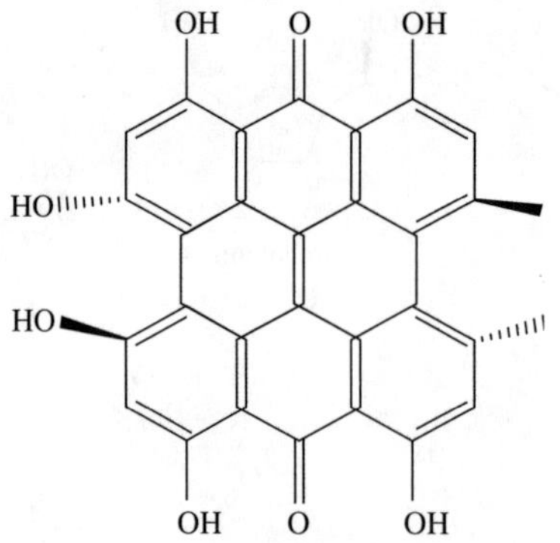

Hypericin

Synonyms: 1,3,4,6,8,13-Hexahydroxy-10,11-di-methylphenanthro(1,10,9,8,opqra)perylene-7,14-dione; Hypericum red; Cyclo-Werrol; Cyclosan; Vimrxyn.

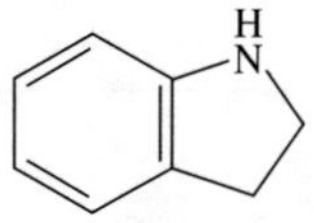

Indole

Synonyms: 2,3-Benzopyrrole; 1-Benzazole; Benzopyrrole; 1H-indole; Indoles; 1-Benzol beta pyrrol.

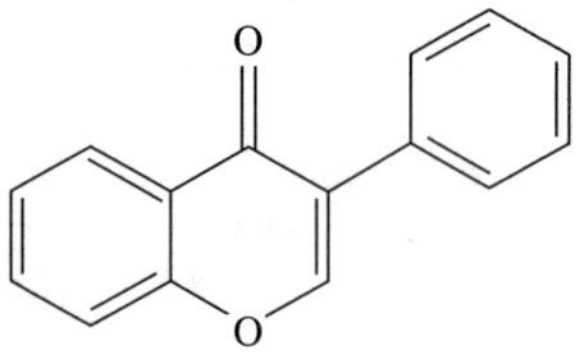

Isoflavone

Synonyms: 3-Phenylchromone.

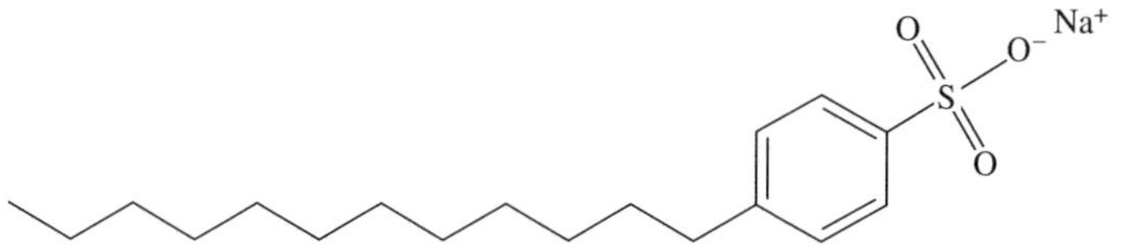

Dodecyl benzenesulfonic acid, sodium salt

Synonyms: Sodium Laurylbenzenesulfonate; sodium dodecyl benzenesulfonate; sodium dodecylphenylsulfonate; AA-9; AA-10; abeson nam; bio-soft D-40; bio-soft D-60; bio-soft D-62; bio-soft D-35x; calsoft f-90; calsoft L-40; calsoft L-60; conco aas-35; conco aas-40; conco aas-65; conco aas-90; conoco c-50; conoco c-60; conoco sd 40; detergent hd-90; mercol 25; mercol 30; naccanol nr; naccanol sw; nacconol 40f; nacconol 90f; nacconol 35SL; neccanol sw; pilot hd-90; pilot sf-40; pilot sf-60; pilot sf-96; pilot sf-40b; pilot sf-40fg; pilot SP-60; richonate 1850; richonate 45b; richonate 60b; santomerse 3; santomerse no. 1; santomerse no. 85; solar 40; solar 90; sulfapol; sulframin 85; sulframin 90; sulframin 40; sulframin 40ra; sulframin 1238 slurry; sulframin 1250 slurry; ultrawet k; ultrawet 60k; ultrawet kx; ultrawet sk; stepan ds 60; ultrawet 1t; marlon a 350; marlon a; maranil; marlon a 375; siponate ds 10; trepolate f 40; conoco c 550; kb (surfactant); nansa sl; santomerse me; merpisap ap 90p; nansa ss; trepolate f 95; nansa hs 80; deterlon; ultrawet 99ls; sulfuril 50; F 90; elfan wa; sandet 60; steinaryl nks 50; sinnozon; NANSA HS 85S; C 550; KB; HS 85S; nansa hf 80; arylan sbc; marlon 375a; X 2073; conco aas 35H; neopelex 05; richonate 40b; DS 60; pelopon a; sulframin 1240; 35SL; SDBS; Nacconol; Santomerse; Sulframin 1238; Ultrawet XK; Arylsulfonat; SODIUM DODECYLBENZENE SULFONATE.

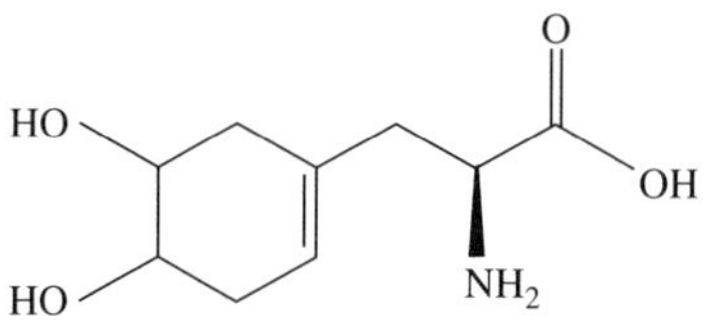

Levodopa

Synonyms: Dopar; Larodopa; Sinemet; [3-(3,4-Dihydroxyphenyl)-L-Alanine]; L-3,4-dihydroxyphenylalanine; 3-hydroxy-L-tyrosine; L-Dihydroxyphenyl-L-alanine; 3,4-Dihydroxy-L-phenylalanine; L-β-(3,4-Dihydroxyphenyl)alanine; L-3-(3,4-Dihydroxyphenyl)alanine.

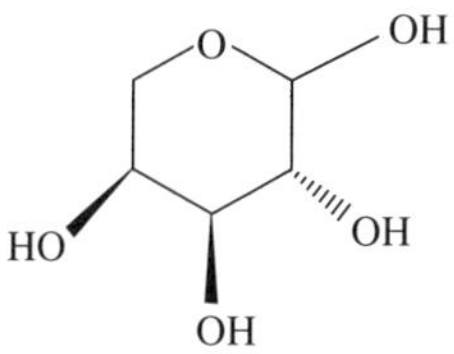

L-(+)-Arabinose

Synonyms: L-Arabinopyranose; Arabinose(L); L-(+)-ARABINOSE CRYSTALLINE.

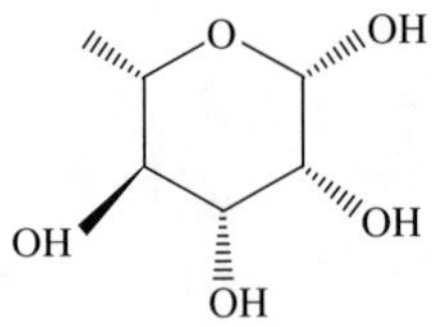

α-L-Rhamnose

Synonyms: α-L-rhamnopyranose; 6-deoxy-L-mannose; L-rhamnose; L-Mannomethylose; α-6-Deoxy-L-mannose; α-L-Mannomethylose; rhamnose.

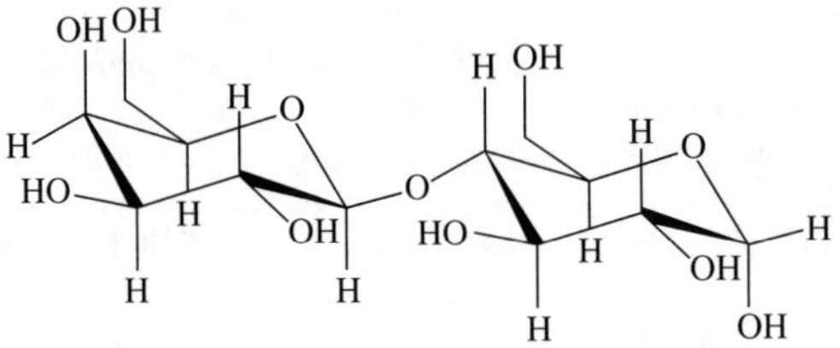

D-(+)-Lactose

Synonyms: Milk sugar; 4-*O*-β-D-galactopyranosyl-D-glucose; β-lactose; β-D-Lactose; Lactose; Lac; lactin; 4-(β-D-galactosido)-D-glucose; lactobiose; saccharum lactin; (+)-β-D-lactose.

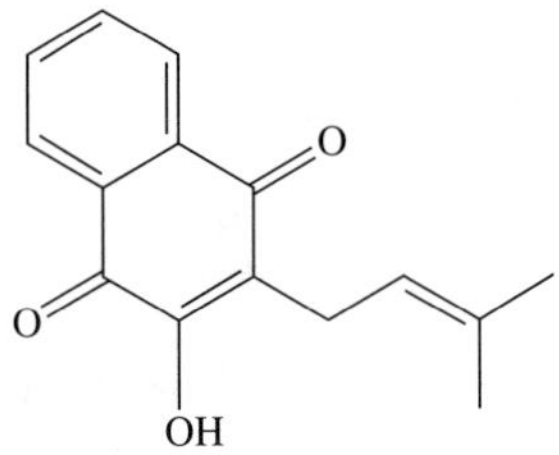

Lapachol

Synonyms: 2-Hydroxy-3-(3-methyl-2-butenyl)-1,4-naphthalenedione.

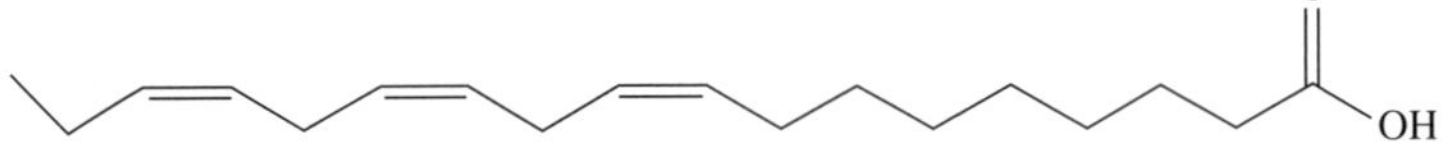

All *cis*-δ-9,12,15-Octadecatrienoate

Synonyms: Linolenic acid; α-linolenic acid; all-*cis*-9,12,15-octadecatrienoic acid; *cis,cis,cis*-9,12,15-octadecatrienoic acid; *cis*-δ-9,12,15-octadecatrienoic acid; *cis*-9,12,15-octadecatrienoic acid; 9,12,15-all-*cis*-octadecatrienoic acid; (*Z,Z,Z*)-9,12,15-Octadecatrienoic acid; (9*Z*,12*Z*,15*Z*)-9,12, 15-Octadecatrienoic acid.

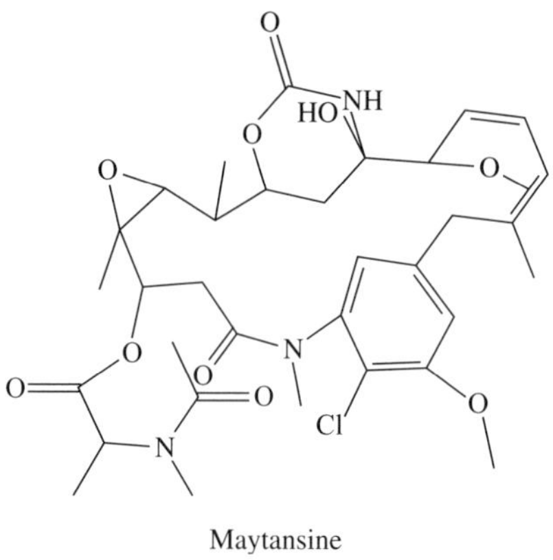

Maytansine

Synonyms: Alanine, *N*-acetyl-*N*-methyl-, 6-ester with 11-chloro-6,21-dihydroxy-12,20-dimethoxy-2,5,9,16-tetramethyl-4,24-dioxa-9,22-diazatetracyclo[19.3.1.1(10,24).0(3,5)]hexacosa-10,12,14[26],16,18-pentaene-8,23-dione; *N*-acetyl-*N*-methyl-L-Alanine, [1*S*-(1*R**,2*S**,3*R**,5*R**,6*R**,16*E*,18*E*,20*S**,21*R**)]-11-chloro-21-hydroxy-12,20-dimethoxy-2,5,9,16-tetramethyl-8,23-dioxo-4,24-dioxa-9,22-diazatetracyclo[19.3.1.110,14.03,5]hexacosa-10,12,14(26),16,18-pentaen-6-yl ester; Maitansine; Maysanine; MTS.

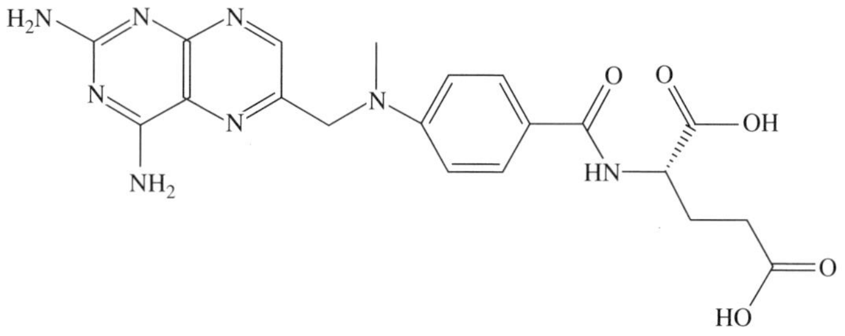

Methotrexate

Synonyms: N-4-((2,4-Diamino-6-Pteridinyl) Methyl Methylamino Benzoyl)-L-Glutamic Acid; Amethopterin; MTX; Hdmtx; Methyl-aminoopterin; Rheumatrex; 4-Amino-*N*10-methyl-pteroylglutamic acid; 4-Amino-10-methylfolic acid; Methylaminopterin; Emtexate; *N*-(p(((2,4-Diamino-6-pteridinyl)methyl)-methylamino)-benzoyl)-L-glutamic acid; cl-14377; emt 25,299; Metatrexan; Methopterin; R 9985; L-(+)-amethopterin dihydrate; 4-amino-4-deoxy-*N*(sup 10)-methylpteroylglutamate; *N*-bismethylpteroylglutamic acid; *N*-(p-(((2,4-diamino-6-pteridyl)methyl)methylamino) benzoyl)glutamic acid; 4-amino-4-deoxy-*N*(sup 10)-methylptero ylglutamic acid; 4-amino-*N*(sup 10)-methylpteroylglutamic acid; methotextrate; antifolan; L-(+)-*N*-(p-(((2,4-diamino-6-pteridinyl)methyl)methylamino)benzoyl)glutamic acid; ledertrexate; methylaminopterinum; Methotrexate dihydrate; MTX dihydrate; L-(+)-4-Amino-*N*10-methylpteroylglutamic acid dihydrate; Amethopetrin; Folex; Folex PFS; Methoblastin; Mexate; (+)-4-Amino-10-methylfolic acid; Mexate (disodium salt of Methotrexate); Folex (disodium salt of Methotrexate); L-(+)-*N*-; Abitrexate; Brimexate; Emthexate; Farmitrexat; Maxtrex; Methotrexato; Metotrexato; Neotrexate; Tremetex.

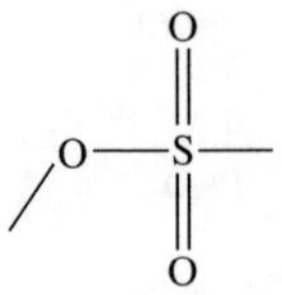

Methyl methanesulfonate

Synonyms: MMS; Methanesulfonic acid methyl ester; Methyl mesylate; as-dimethyl sulfite; methyl ester of methanesulfonic acid; methyl methansulfonate; Methylsulfonic acid, methyl ester.

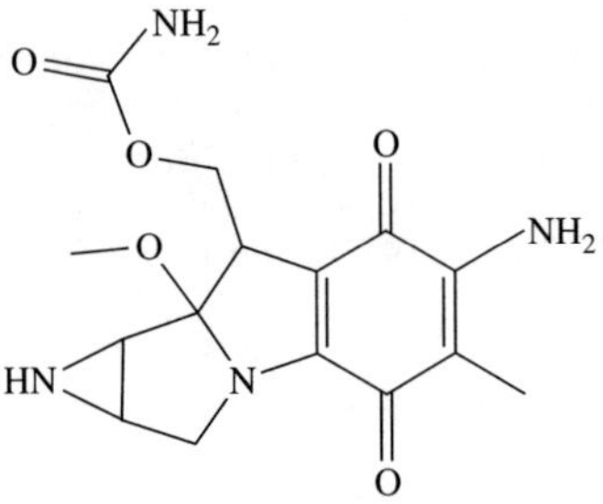

Mitomycin C

Synonyms: MMC; Mitomycin; Mutamycin; 6-amino-8-[[(aminocarbonyl)oxy]methyl]-1,1a,2,8,8a,8b-hexahydro -8a-methoxy-5-methyl, [1aS-(1aα,8β,8aα,8bα)]-azirino[2′,3′:3,4]pyrrolo[1,2a]indole-4,7-dione; [1a*R*-(1aα,8β,8aα,8bα)]-6-amino-8-[[(aminocarbonyl)oxy]methyl]-1,1a,2,8,8a,8b-hexahydro-8a-methoxy-5-methylazirino[2′,3′:3,4]pyrrolo[1,2-α]indole-4,7-dione; Ametycin; Mit-C; Mito-C; Mitocin-C; Mitomycinum; Mytomycin; 7-Amino-9α-methoxymitosane; Azirino[2′,3′:3,4]pyrrolo[1,2-a]indole-4,7-dione, 6-amino-8-[[(aminocarbonyl)oxy]methyl]-1,1a,2,8,8a,8b-hexahydro-8a-methoxy-5-methyl-, [1aS-(1aα,8β,8aα,8bα)]-; 6-amino-1,1a,2,8,8a,8b-hexahydro-8-(hydroxymethyl)-8a-methoxy-5-methylazirino[2′,3′:3,4]pyrrolo[1,2-a]indole-4,7-dione, carbamate (ester); (1ar)-6-amino-8-(((aminocarbonyl) oxy)methyl)-1,1a,2,8,8a,8b-hexahydro-8a-methoxy-5-methylazirino[2′,3′:3,4]pyrrolo[1,2-a]indole-4,7-dione.

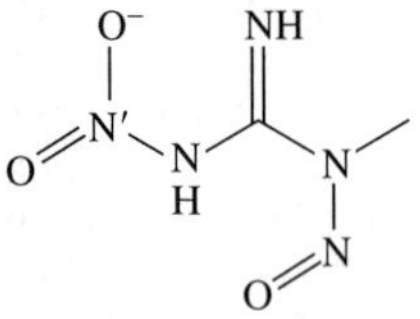

N-methyl-*N*′-nitro-*N*-nitrosoguanidine

Synonyms: MNNG; *N*-methyl-*N*-nitroso-*N*′-nitroguanidine; *N*′-nitro-*N*-nitroso-*N*-methylguanidine; 1-methyl-3-nitro-1-nitrosoguanidine; methylnitronitrosoguanidine; *N*-nitroso-*N*-methylnitroguanidine; 1-methyl-1-nitroso-*N*-methylguanidine; 1-nitro-*N*-nitroso-*N*-methylguanidine; MNG; Methyl-*N*′-nitro-*N*-nitrosoguanidine; *N*-Methyl-*N*-Nitroso-*N*′-Nitroguanidine, 97% - Carc.

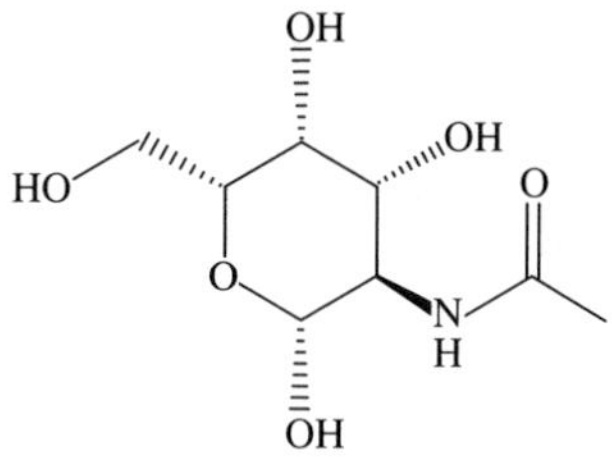

N-Acetylgalactosamine

Synonyms: *N*-Acetylchondrosamine; 2-Acetamido-2-deoxygalactose; *N*-Acetyl-β-D-galactosamine; *N*-Acetyl-D-galactosamine.

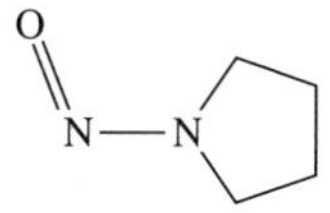

N-Nitrosopyrrolidine

Synonyms: 1-nitroso-Pyrrolidine; Nitrosopyrrolidine; NPYR; *N*-*N*-PYR; No-pyr; Pyrrole, tetrahydro-*N*-nitroso-; *N*-Nitrosopyrrolidine.

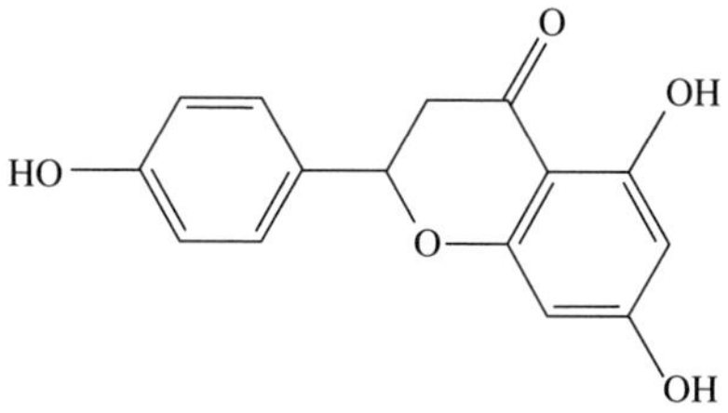

Naringenin

Synonyms: 5,7-Dihydroxy-2-(4-hydroxyphenyl)chroman-4-one; 4[,5,7-Trihydroxyflavanone.

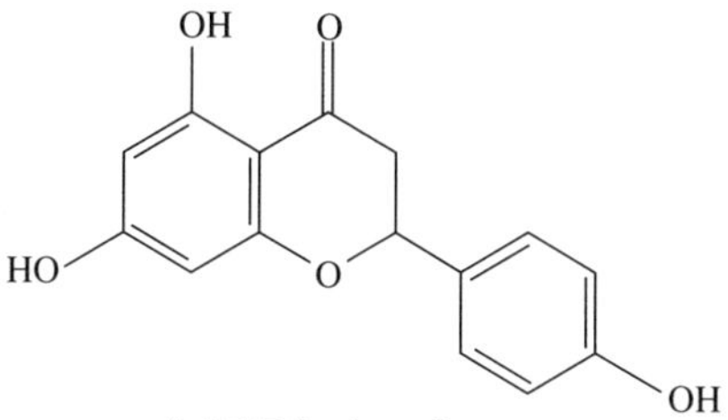

4′,5,7-Trihydroxyflavanone

Synonyms: Naringenin.

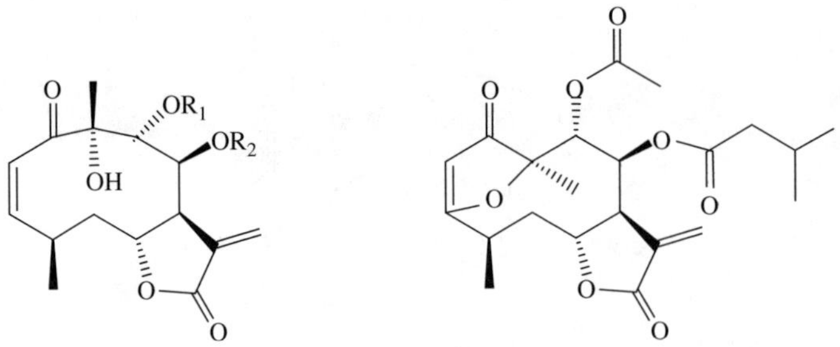

6

	R[1]	R[1]
1	Ac	i-Val
2	i-Val	H
3	H	i-Val
4	Ac	i-But
5	Ac	2-Me-but

Neurolanin

Ni'^{2}

Cl^{-} Cl^{-}

Nickel chloride

Synonyms: Nickel (II) Chloride; Nickelous Chloride; Nickel dichloride; Nickel (II) chloride, ultra dry, anhydrous, 99.9% (metals basis); Nickel chloride.

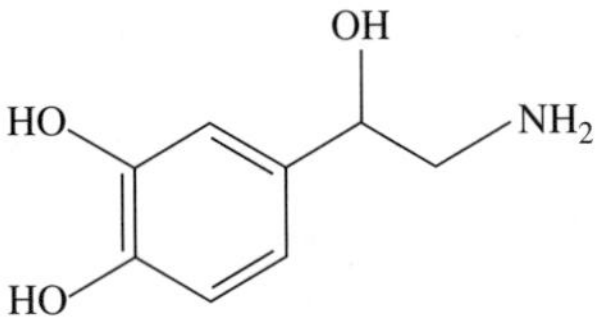

Norepinephrine

Synonyms: NE; NA; noradrenalin; Arterenol; Levophed.

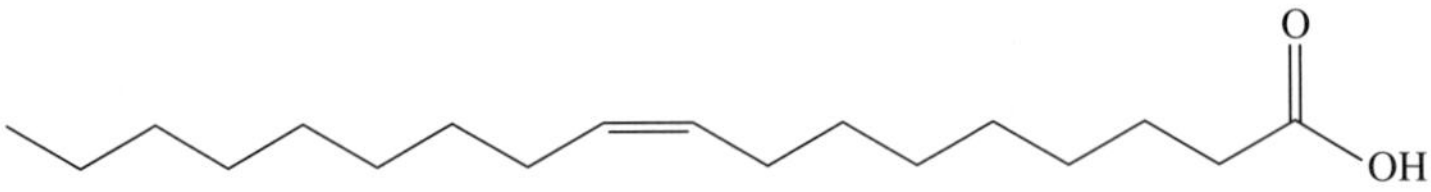

Oleic acid

Synonyms: cis-δ-9-octadecanoate; 9-Octadecenoic acid (*Z*)-; *cis*-9-Octadecenoic acid; *cis*-octadec-9-enoic acid; century cd fatty acid; emersol 210; emersol 213; emersol 6321; emersol 233ll; glycon ro; glycon wo; *cis*-δ(sup 9)-octadecanoic acid; 9-octadecenoic acid; wecoline oo; tego-oleic 130; vopcolene 27; groco 2; groco 4; groco 6; groco 5l; hy-phi 1055; hy-phi 1088; hy-phi 2066; K 52; neo-fat 90-04; neo-fat 92-04; hy-phi 2088; hy-phi 2102; Metaupon; red oil; (*Z*)-9-Octadecenoic acid; Octadecenoic acid; oleate.

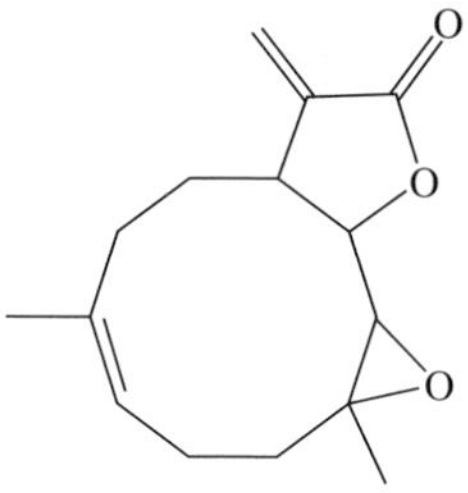

Parthenolide

Synonyms: [1a*R*-(1a*R**,4[E].7aS*,10a*S**,-10b*R**)]-2,3,6,7,7a,8,10a,10b-Octahydro-1a,5-dimethyl-8-methyleneoxireno[9,10]cyclodeca[1,2-b]furan-9(1aH)-one; 4,5α-epoxy-6β-hydrohy-germacra-1(10),11(13)-dien-12-oic acid,γ-lactone.

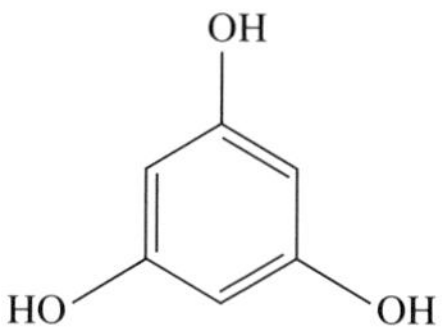

Phloroglucinol

Synonyms: 1,3,5-Benzenetriol; 1,3,5-trihydroxybenzene; 1,3,5-THB; 1,3-Trihydroxybenzene.

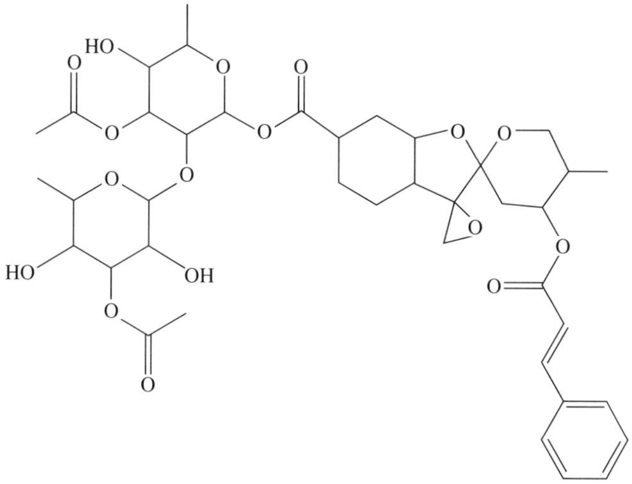

Phyllanthoside

Synonyms: Phyllantoside.

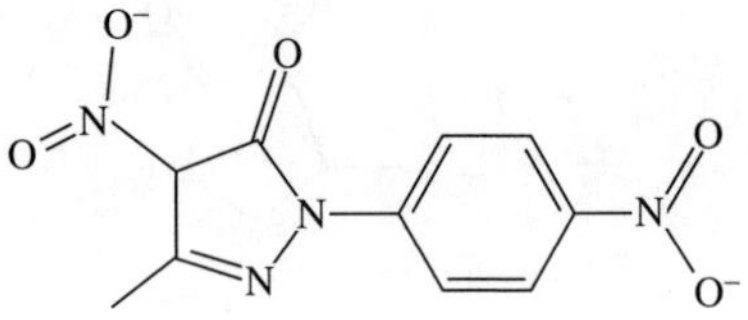

Picrolonic Acid

Synonyms: 3*H*-Pyrazol-3-one-2,4-dihydro-5-methyl-4-nitro-2-(4-nitrophenyl).

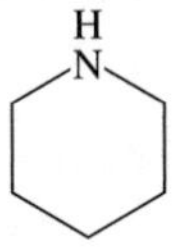

Piperidine

Synonyms: Hexahydropyridine; Pentamethyleneimine; Azacyclohexane; cyclopentimine; cypentil; hexazane.

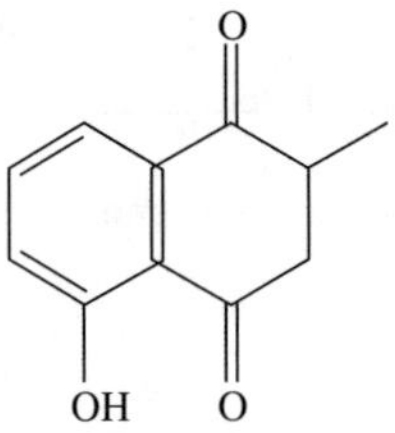

Plumbagin

Synonyms: 5-Hydroxy-2-methyl-1,4-naphthoquinone.

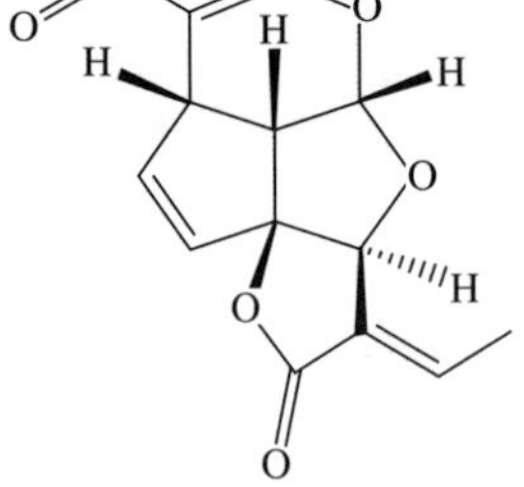

Plumericin

Synonyms: [3a*S*-(3*E*,3aα,4aβ;,7aβ,9a*R**,9bβ)]-3-Ethylidene-3,3a,7a,9b-tetrahydro-2-oxo-2*H*,4a*H*-1,4,5-trioxadicyclopent[a,hi]indene-7-carboxylic acid methyl ester.

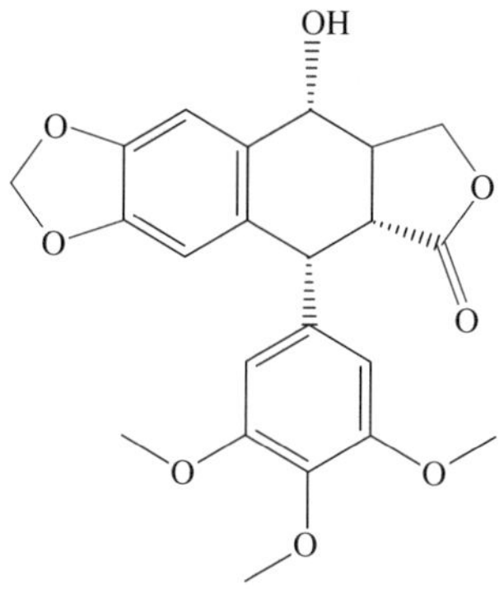

Podophyllotoxin

Synonyms: 5,8,8a,9-Tetrahydro-9-hydroxy-5-(3,4,5-trimethoxyphenyl)furo[3′,4′:6,7]naphthol[2,3-d]-1,3-dioxol-6(5aH)-one; 1-hydroxy-2-hydroxymethyl-6,7-methylenedioxy-4-(3′,4′,5′-trimethoxyphenyl)-1,2,3,4-tetrahydronaphthalene-3-carboxylic acid lactone; podophyllinic acid lactone; podofilox; Condyline; Condylox; Martec; Warticon

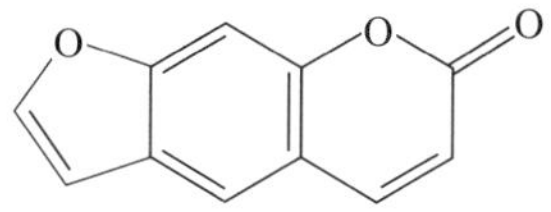

Psoralen

Synonyms: 7H-Furo[3,2-g][1]benzopyran-7-one; 6-hydroxy-5-benzofuranacrylic acid δ-lactone; furo[3,2-]-coumarin; ficusin.

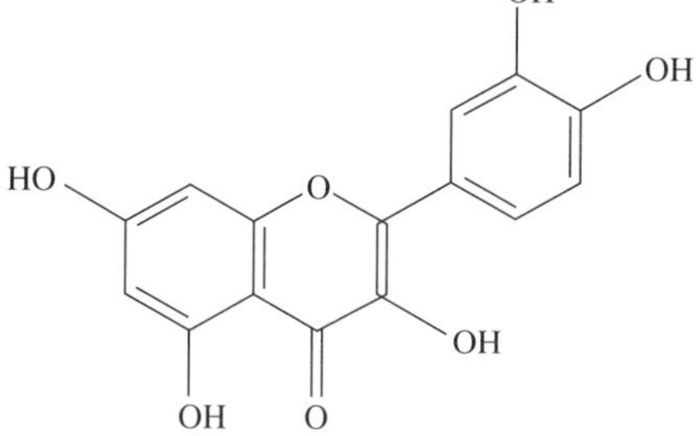

Quercetin

Synonyms: 3,3′,4′,5,7-pentahydroxyflavone; 2-(3,4-dihydroxyphenyl)-3,5,7-trihydroxy-4H-1-benzopyran-4-one; 3,5,7,3′,4′-pentahydroxyflavone; 3′,4′,5,7-tetrahydroxyflavon-3-ol; cyanidelonon 1522; C.I. natural yellow 10; C.I. natural yellow 10 & 13; C.I. natural red 1; C.I. 75670; meletin; quercetol; quertine; sophoretin; t-gelb bzw. grun 1; xanthaurine.

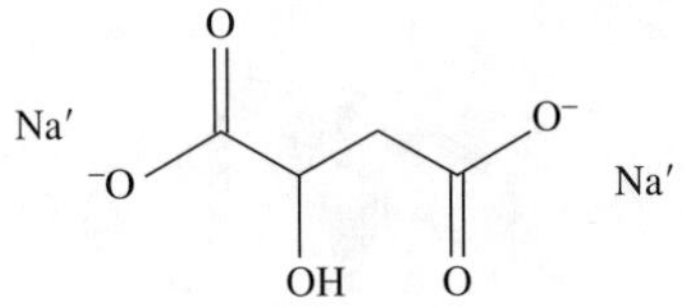

L-Malic acid, sodium salt

Synonyms: Hydroxybutanedioic acid; hydroxy-succinic acid.

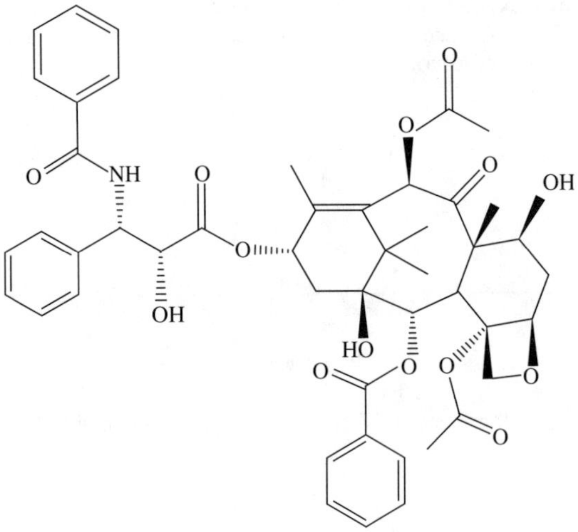

Paclitaxel

Synonyms: Taxol; Taxal; Taxol A; 7,11-Methano-5H-cyclodeca[3,4]benz[1,2-b]oxete,benzene-propanoic acid deriv.; TAX; 5-β,20-epoxy-1,2-α,4,7-β,10-β,13-α-hexahydroxy-tax-11-en-9-one 4,10-diacetate 2-benzoate 13-ester with (2*R*,3*S*)-*N*-benzoyl-3-phenyl-isoserine.

Tetrahydrofuran

Synonyms: THF; 1,4-Epoxybutane; Butylene oxide; Cyclotetramethylene; tetramethylene oxide; oxacyclopentane; Cyclotetramethylene oxide; Furanidine; Hydrofuran; oxolane.

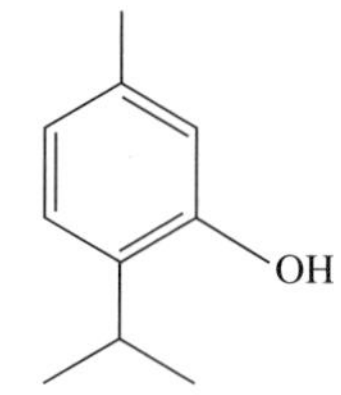

Thymol

Synonyms: 6-Isopropyl-*m*-cresol; 3-Hydroxy-*p*-cymene; Isopropyl cresol; 5-Methyl-2-(1-methylethyl)phenol; 5-Methyl-2-isopropyl-1-phenol; 3-*p*-Cymenol; 2-Isopropyl-5-methyl phenol; THYMOL CRYSTALS USP.

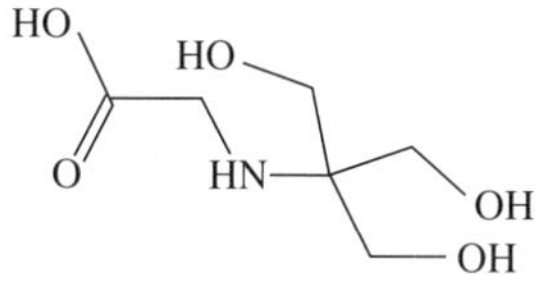

Tricine

Synonyms: *N*-(tris(hydroxymethyl)methyl)glycine; Glycine, *N*-[2-hydroxy-1,1-bis(hydroxymethyl)ethyl]-.

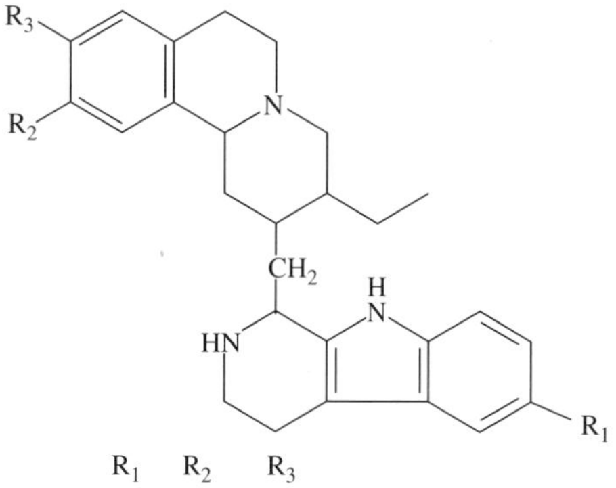

R_1	R_2	R_3	
OH	OCH_3	OCH_3	tubulosine

Tubulosine

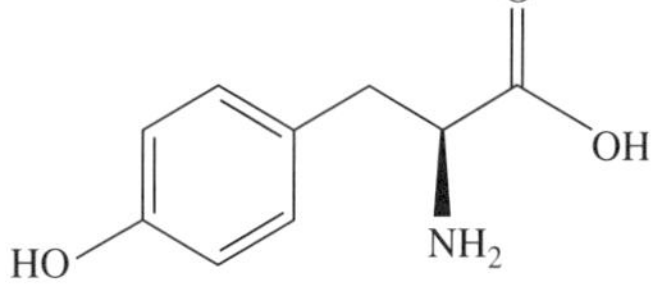

(*S*)-(−)-Tyrosine

Synonyms: *p*-tyrosine; Tyr; Y; Tyrosine; L-Tyrosine; L-(-)-tyrosine; 2-Amino-3-(4-hydroxyphenyl)-propanoic acid; 3-(4-Hydroxyphenyl)-L-alanine; 3-(p-hydroxyphenyl)alanine; 2-amino-3-(*p*-hydroxyphenyl)propionic acid; L-TYROSINE FREE BASE.

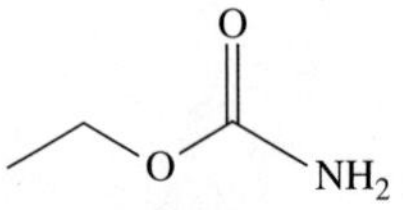

Urethane

Synonyms: Ethyl carbamate; Carbamic acid ethyl ester; Ethyl urethane; *o*-ethylurethane; ethyl ester of carbamic acid; leucethane; leucothane; pracarbamin; a 11032; u-compound; X 41; *o*-Ethyl carbamate; Ethyl carbamate.

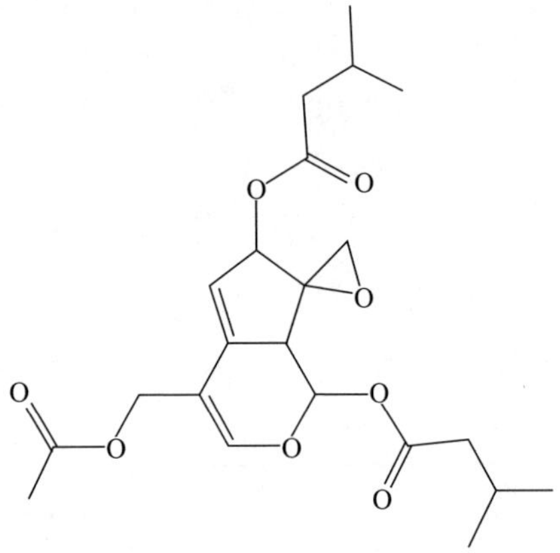

Valtrate

Synonyms: Valtrate.

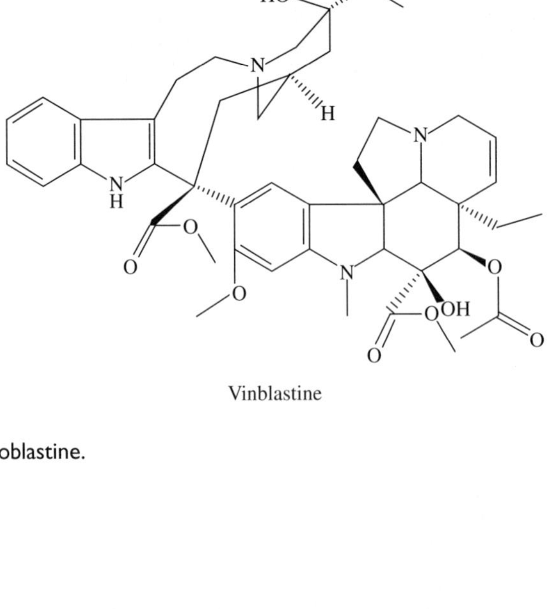

Vinblastine

Synonyms: Vincaleukoblastine.

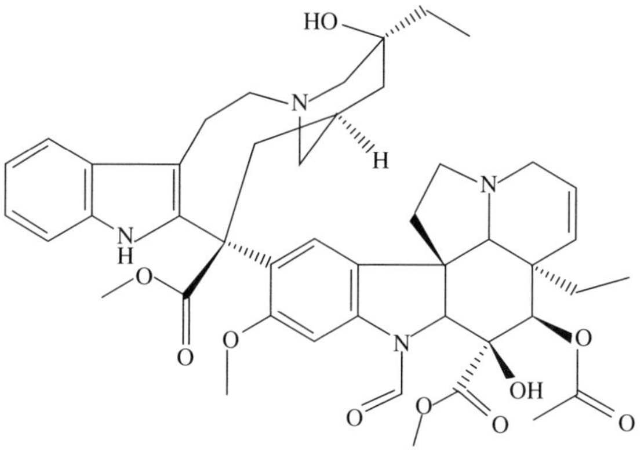

22-Oxovincaleukoblastine

Synonyms: Vincristine; Oncovin; Vincasar; Vincrex; Leurocristine; VCR; LCR; Kyocristine; PES; Vincosid; Vincasar PES; Vincasar (Vincristne sulfate); Oncovin (Vincristne sulfate); Kyocristine (Vincristine sulfate); Vincrex (Vincristine sulfate).

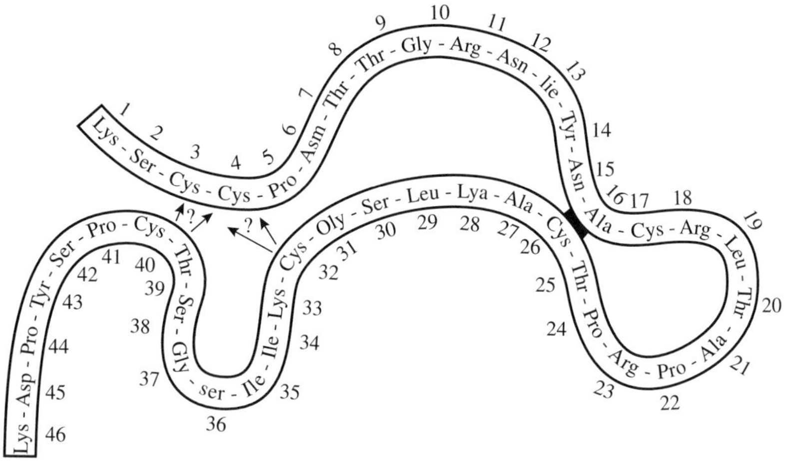

Viscotoxin A3

References

Alali, F.Q., Liu, X.X. and McLaughin, J.L. (1999) Annonaceous acetogenins: recent progress. *J. Nat. Prod.* 62, 504–40.

Anderson, W.J. and Beardall, J. (1991) Molecular Activities of Plant Cells. Blackwell Scientific Publications, Oxford.

Anonymous (2000) Loss of loved one affects cancer risk and survival. Healthcentral News.

Asari, F., Kusumi, T. and Kakisawa, H. (1989) Turbinaric acid, a cytotoxic secosqualene carboxylic acid from the brown alga *Turbinaria ornata*. *J. Nat. Prod.* 52, 1167–9.

Aulas, J.J. (1996) Alternative cancer treatments. *Sci. American* 275, 126–7.

Barbounaki-Konstantakou, E. (1989) *Chemotherapy*. Beta Medical Arts, Athens.

Barchi, J.J., Moore, R.E. and Patterson, G.M.L. (1984) Acutiphycin and 20,21-Didehydroacutiphycin, new antineoplastic agents from the cyanoplyte *Oscillatoria acutissima*. *J. Am. Chem. Soc.* **106**, 8193–7.

Barchi, Jr J.J., Norton, T.R., Furusawa, E., Patterson, G.M.L. and Moore, R.E. (1983) Identification of a cytotoxin from *Tolypothrix byssoidea* as tubercidin. *Phytochemistry* 22, 2851–2.

Barnekow, D.E., Cardellina II, J.H., Zektzer, A.S. and Martin, G.E. (1989) Novel cytotoxic and phytotoxic halogenated sesquiterpenes from the green alga *Neomeris annulata*. *J. Am. Chem. Soc.* 111, 3511–17.

Baslow, M.H. (1969) *Marine Pharmacology*. The Williams & Wilkins Co, Baltimore.

Becker, H. and Schwarz, G. (1971) Callus cultures from *Viscum album*. A possible source of raw materials for gaining therapeutic interesting extracts. *Planta Med.* 20(4), 357–62.

Beljanski, M. and Beljanski, M.S. (1982) Selective inhibition of *in vitro* synthesis of cancer DNA by alkaloids of beta-carboline class. *Exp. Cell Biol.* 50(2), 79–87.

Beljanski, M., Crochet, S. and Beljanski, M.S. (1993) PB-100: a potent and selective inhibitor of human BCNU resistant glioblastoma cell multiplication. *Anticancer Res.* 13(6A), 2301–8.

Berge, J.P., Bourgougnon, N., Carbonnelle, D., Le Bert, V., Tomasoni, C., Durand, P. and Roussakis, C. (1997) Antiproliferative effects of an organic extract from the marine diatom *Skeletonema costatum* (Grev.) Cleve. Against a non-small-cell bronchopulmonary carcinoma line (NSCLC-N6). *Anticancer Res.* 17(3C), 2115–20.

Beuth, J. (2000) Natural versus recombinant mistletoe lectin-1. Market trends. In: *Medicinal and Aromatic Plants-Industrial Approaches: The Genus Viscum*. (Ed. A. Bussing), Harwood Publishers, Amsterdam, pp. 237–46.

Biard, J.F. and Verbist, J.F. (1981) Agents antineoplastiques des algues marines: substances cytotoxiques de *Colpomenia peregrina* Sauvageau (Scytosiphonacees). *Pl. Med. Phytoth.* 15, 167–71.

Blokhin, A.V., Yoo, H.D., Geralds, R.S., Nagle, D.G., Gerwick, W.H. and Hamel, E. (1995) Characterization of the interaction of the marine cyanobacterial natural product curacin A with the colchicine site of tubulin and initial structure–activity studies with analogues. *Mol. Pharmacol.* 48(3), 523–31.

Bongiorni, L. and Pietra, F. (1996) Marine natural products for industrial applications. *Chem. Ind.* 74, 54–8.

Bonnard, I., Rolland, M., Francisco, C. and Banaigs, B. (1997) Total structure and biological properties of laxaphycins A and B, cyclic lipopeptides from the marine cyanobacterium *Lyngbya majuscula*. *Lett. Peptide Sci.* 4, 1–4.

Bouaicha, N., Tringali, C., Pesando, D., Malléa, M. and Verbist, J.F. (1993a) Bioactive diterpenoids isolated from *Dilophus ligulatus*. *Planta Med.* 59, 256.

Bouaicha, N., Pesando, D. and Puel, D. (1993b) Cytotoxic diterpenoids from the brown alga *Dilophus ligulatus*. *J. Nat. Prod.* 56, 1747.

Boyd, M.R. (1997) The NCI *in vitro* anticancer drug discovery screen. In: *Anticancer Drug Development Guide: Clinical Trials, and Approval* (Ed. B. Teicher), Humana Press Inc., Totowa, NJ.

Brower, V. (1999) Tumor-angiogenesis – new drugs on the block. *Nature Biotech.* 17, 963–8.

Budavari, S., O'Neil, M.J., Smith, A. and Heckelman, P.E. (1989) *The Merck Index, an Encyclopedia of Chemicals, Drugs, and Biologicals.* Merck & CO., Inc., Rahway, NJ.

Carbonnelle, D., Pondaven, P., Morancais, M., Masse, G., Bosch, S., Jacquot, C., Briand, G., Robert, J. and Roussakis, C. (1999) Antitumor and antiproliferative effects of an aqueous extract from the marine diatom *Haslea ostrearia* (Simonsen) against solid tumors: lung carcinoma (NSCLC-N6), kidney carcinoma (E39) and melanoma (M96) cell lines. *Anticancer Res.* 19(1A), 621–4.

Carmeli, S., Moore, R.E., Patterson, G.M.L., Mori, Y. and Suzuki, M. (1990) Isonitriles from the blue-green alga *Scytonema mirabile*. *J. Org. Chem.* 55, 4431–8.

Chen, J.L. and Gerwick, W.H. (1994) Isorawsonol and related IMP dehydrogenase inhibitors from the tropical green alga *Avrainvillea rawsonii*. *J. Nat. Prod.* 57, 947–52.

Coll, J.C., Skelton, B.W., White, A.H. and Wright, A.D. (1988) Tropical marine algae II. The structure determination of new halogenated minoterpenes from *Plocamium hamatum* (Rhodophyta, Gigartinales, Plocamiaceae). *Aust. J. Chem.* 41, 1743–53.

Cragg, G.M. (1998) Paclitaxel (Taxol): a success story with valuable lessons for natural product drug discovery and development. *Med. Res. Rev.* 18(5), 315–31.

Cragg, G.M. and Newman, D.J. (2000) Antineoplastic agents from natural sources: achievements and future directions. *Exp. Opin. Invest. Drugs* 9, 2783–97.

Cragg, G.M., Schepartz, S.A., Suffness, M. and Grever, M.R. (1993) The taxol supply crisis. New NCI policies for handling the large-scale production of natural product anticancer and anti-HIV agents. *J. Nat. Prod.* 56, 1657–68.

Debinski, W., Obiri, N.I., Powers, S.K., Pastan, I. and Puri, P.K. (1995) Human glioma cells overexpress receptors for interleukin 13 and are extremely sensitive to a novel chimeric protein composed of interleukin 13 and pseudomonas exotoxin. *Clin. Cancer Res.* 1(11), 1253–8.

Depix, M.S., Martinez, J., Santibanez, F., Rovirosa, J., San Martin, A. and Maccioni, R.B. (1998) The compound 14-keto-stypodiol diacetate from the algae *Stypopodium flabelliforme* inhibits microtubules and cell proliferation in DU-145 human prostatic cells. *Mol. Cell. Biochemistry* 187, 191–9.

Diallo, A.O., Mehri, H., Iouzalen, H. and Plat, M. (1995) Alkaloids from leaves of *Alangium bussyanum*. *Phytochemistry* 40(3), 975–7.

Dimitriou, D. (2001) Capitalizing on the industry challenges in the pharma & biotech sector: a novel approach to drug development. *DyoDelta Biosciences Newsletter*. Oral announcement in the 4th Conference Medicinal Chemistry, Drug Design and Development, University of Patras, Department of Chemistry and Pharmacy. March 13–14, 2003, Patras.

Duran, J.D.G., Guindo, M.C., Delgado, A.V., Gonzalez-Caballero, F. (1997) Surface Chemical Analysis and Electrokinetic Properties of Synthetic Spherical Mixed *Zinc-Cadmium* Sulfides. *J. Colloid Interface Sci.* Sep 15(2), 223–33.

Duran, R., Zubia, E., Ortega, M.J. and Salva, J. (1997) New diterpenoids from the alga *Dictyota dichotoma Tetrahedron* 53, 8675–88.

Egorin, M.J., Rosen, D.M., Benjamin, S.E., Callery, P.S., Sentz, D.L. and Eiseman, J.L. (1997) *In vitro* metabolism by mouse and human liver preparations of halomon, and antitumor halogenated monoterpene. *Cancer Chemother. Pharmacol.* 41, 9–14.

Egorin, M.J., Sentz, D.L., Rosen, D.M., Ballesteros, M.F., Kearns, C.M., Callery, P.S. and Eiseman, J.L. (1996) Plasma pharmacokinetics, bioavailability, and tissue distribution in CD_2F_1 mice of halomon, an antitumor halogenated monoterpene isolated from the red algae *Portieria hornemanii. Cancer Chemother. Pharmacol.* 39, 51–60.

ELI LILLY & CO (1998) patent, WO9808505.

Erickson, K.L., Beutler, J.A., Gray, G.N., Cardellina II, J.H. and Boyd, M.R. (1995) Majapolene-A, a cytotoxic sesquiterpenoid peroxide, and related sesquiterpenoids from the marine alga *Laurencia majuscula. J. Nat. Prod.* 58, 1848–60.

Fadli, M., Aracil, J.M., Jeanty, G., Banaigs, B. and Francisco, C. (1991) Novel meroterpenoids from *Cystoseira mediterranea*: use of the Crown–Gall bioassay as a primary screen for lipophilic antineoplastic agents. *J. Nat. Prod.* 54, 261–4.

Faulkner, D.J. (2001) Marine natural products. *Nat. Prod. Rep.* 18(1), 1–49.

Fernádez, J.J., Souto, M.L. and Norte, M. (1998) Evaluation of the cytotoxic activity of polyethers isolated from *Laurencia. Bioorg. Med. Chem.* 6, 2237–43.

Fischel, J.L., Lemée, R., Formento, P., Caldani, C., Moll, J.L., Pesando, D., Meinesz, A., Grelier, P. and Milano, G. (1994) Mise en évidence d'effets antiprolifèrants de la Caulerpényne (de *Caulerpa taxifolia*). Expérience sur cellules tumorales humaines en culture. *Bulletin du Cancer* 81, 489–92.

Fischel, J.L., Lemée, R., Formento, P., Caldani, C., Moll, J.L., Pesando, D., Meinesz, A., Grelier, P., Pietra, F., Guerriero, A. and Milano, G. (1995) Cell growth inhibitory effects of caulerpenyne, a sesquiterpenoid from the marine algae *Caulerpa taxifolia. Anticancer Res.* 15, 2155–60.

Folkman, J. (1996) Fighting cancer by attacking its blood supply. *Sci. American* 275, 116–20.

Fomina, I.P., Navashin, S.M., Preobrazhenskaia, M.E. and Rozenfel'd, E.L. (1966) Comparative study of the biological effect of polysaccharides glucan and laminarin. *Bul. Exp. Biol. Med.* 61, 79–83.

Foster, B.J., Fortuna, M., Media, J., Wiegand, R.A. and Valeriote, F.A. (1999) Cryptophycin 1 cellular levels and effects *in vitro* using L1210 cells. *Invest. New Drugs* 16, 199–204.

Fugihara, M., Iizima, N., Yamamoto, I. and Nagumo, T. (1984a) Purification and chemical and physical characterisation of an antitumour polysaccharide from the brown seaweed *Sargassum fulvellum. Carbohyd. Res.* 125, 97–106.

Fugihara, M., Komiyama, K., Umezawa, I. and Nagumo, T. (1984b) Antitumour activity and action – mechanisms of sodium alginate isolated from the brown seaweed *Sargassum fulvellum. Chemotherapy* 32, 1004–9.

Fujii, K., Ito, H., Shimura, K. and Horiguchi, Y. (1975) Studies on antitumor activities of marine algae polysaccharides. *Mie Igaku* 18, 245–50.

Fukui, M., Azuma, J. and Okamura, K. (1990) Induction of callus from mistletoe and interaction with its host cells. *Bulletin of the Kyoto University Forests* 62, 261–9.

Fuller, R.W., Cardellina, II, J.H., Kato, Y., Brinen, L.S., Clardy, J., Snader, K.M. and Boyd, M.R. (1992) A pentahalogenated monoterpene from the red algae *Portieria hornemanii* produces a novel cytotoxicity profile against a diverse panel of human tumor cell lines. *J. Med. Chem.* 35, 3007–11.

Fuller, R.W., Cardellina II, J.H., Jurek, J., Scheuer, P.J., Alvarado-Lindner, B., McGuire, M., Gray, G.N., Steiner, J.-R., Clardy, J., Menez, E., Shoemaker, R.H., Newman, D.J., Snader, K.M. and Boyd, M.R. (1994) Isolation and structure/activity features of halomon – related antitumor monoterpenes from the red algae *Portieria hornemanii. J. Med. Chem.* 37, 4407–11.

Funayama, Y., Nishio, K., Wakabayashi, K., Nagao, M., Shimoi, K., Ohira, T., Hasegawa, S. and Saijo, N. (1996) Effects of beta- and gamma-carboline derivatives of DNA topoisomerase activities. *Mutat. Res.*, 349(2), 183–91.

Furusawa, E. and Furusawa, S. (1985) Anticancer activity of a natural product, viva-natural extracted from *Undaria pinnatifida* on intraperitoneally implanted Lewis lung carcinoma. *Oncology* 42, 364–9.

Furusawa, E. and Furusawa, S. (1988) Effect of pretazettine and viva-natural, a dietary seaweed extract, on spontaneous AKR leukemia in comparison with standard drugs. *Oncology* 45(3), 180–6.

Furusawa, E. and Furusawa, S. (1989) Anticancer potential of viva-natural, a dietary seaweed extract, on Lewis lung carcinoma in comparison with chemical immunomodulators and on cyclosporine-accelerated AKR leukemia. *Oncology* 46(5), 343–8.

Furusawa, E. and Furusawa, S. (1990) Antitumor potential of low-dose chemotherapy manifested in combination with immunotherapy of viva-natural, a dietary seaweed extract, on Lewis lung carcinoma. *Cancer Lett.* 50(1), 71–8.

Furusawa, E., Patterson, G.M.L. and Moore, R.E. (1983) Identification of a cytotoxin from *Tolypothrix byssoidea* as tubercidin. *Phytochemistry* 22, 2851–2.

Garcia-Rocha, M., Bonay, P. and Avila, J. (1996) The antitumoral compound Kahalalide F acts on cell lysosomes. *Cancer Lett.* 99, 43–50.

Gebhardt, R. (2000) *In vitro* screening of plant extracts and phytopharmaceuticals: novel approaches for the elucidation of active compounds and their mechanisms. *Plant Med.* 66, 99–105.

Gerwick, W.H. (1989) 6-Desmethoxyhormothamnione, a new cytotoxic styrylchromone from the marine cryptophyte *Chrysophaeum taylori. J. Nat Prod.* 52(2), 252–6.

Gerwick, W.H. and Fenical, W.H. (1981) Ichthyotoxic and cytotoxic metabolites of the tropical brown alga *Stypopodium zonale* (Lamouroux) Papenfuss. *J. Org. Chem.* 46, 22–7.

Gerwick, W., Reyes, S. and Alvarado, B. (1987) Two Malyngamides from the Caribbean cyanobacterium *Lyngbya majuscula. Phytochemistry* 26, 1701–4.

Gerwick, W., Lopez, A., Van Duyne, G., Clardy, J., Ortiz, W. and Baez, A. (1986) Hormothamnione, a novel cytotoxic styrylchromone from the marine cyanophyte *Hormonothamnion enteromorphoides* Grunow. *Tetrahedron Lett.* 27, 1979–82.

Gerwick, W.H., Mrozek, C., Moghaddam, M.F. and Agarwal, S.K. (1989) Novel cytotoxic peptides from the tropical marine cyanobacterium *Hormothamnion enteromorphoides*. I. Discovery, isolation and initial chemical and biological characterization of the hormothamnins from wild and cultured material. *Experientia* 45(2), 115–21.

Gerwick, W.H., Fenical, W., Van Engen, D. and Clardy, J. (1993) Isolation and structure of spatol a potent inhibitor of cell replication from the brown seaweed *Spatoglossum schmittii. J. Am. Chem. Soc.* 102, 7991–3.

Gerwick, W.H., Proteau, P.J., Nagle, D.G., Hamel, E., Blokhin, A. and Slate, D.L. (1994) Structure of Curacin A, a novel antiproliferative, and brine shrimp toxic natural product from the marine cyanobacterium *Lyngbya majuscula. J. Org. Chem.* 59, 1243–5.

Goetz, G., Yoshid, W.Y. and Scheuer, P.J. (1999) The absolute stereochemistry of kahalalide F. *Tetrahedron* 55, 7739–41.

Golakoti, T., Ohtani, I., Patterson, G.M.L., Moore, R.E., Corbett, T.H., Valeriote, F.A. and Demchik, L. (1994) Total structures of cryptophycins, potent antitumor depsipeptides from the blue-green alga *Nostoc* sp. strain GSV 224. *J. Am. Chem. Soc.* 116, 4729–37.

Golakoti, T., Ogino, J., Heltzel, C.E., Lehusebo, T., Jensen, C.M., Larsen, L.K., Patterson, G.M.L., Moore, R.E., Mooberry, S.L., Corbett, T.H. and Valeriote, F.A. (1995) Structure determination, conformational analysis, chemical stability studies, and antitumor evaluation of the cryptophycins. Isolation of 18 new analogs from *Nostoc* sp strain GSV 224. *J. Am. Chem. Soc.* 117, 12030–49.

Govindan, M., Abbas, S.A., Schmitz, F.J., Lee, R.H., Papkoff, J.S. and Slate, D.L. (1994) New cycloartanol sulfates from the alga *Tydemania expeditionis*: inhibitors of the protein tyrosine kinase $pp60^{v\text{-}src}$. *J. Nat. Prod.* 57, 74–8.

Graber, M.A. and Gerwick, W.H. (1998) Kalkipyrone, a toxic γ-pyrone from an assemblage of the marine cyanobacteria *Lyngbya majuscula* and *Tolypothrix* sp. *J. Nat. Prod.* 61, 677–80.

Greenwald, P. (1996) Chemoprevention of cancer. *Sci. American* 275, 64–7.

Grieve M. (1994) *A modern herbal.* Edited and introduced by Mrs C.F. Leyel, Tiger books international, London.

Guven, K.C., Guvener, B. and Guler, E. (1990) Pharmacological properties of marine algae. In: *Introduction to Applied Phycology* (Ed. I. Akatsuka) SPB Academic Publishing, The Hague, The Netherlands, pp. 67–92.

Hamann, M.T. and Scheuer, P.J. (1993) Kahalalide F, a Bioactive Depsipeptide from the Sacoglossan Mollusk *Elysia rufescens* and the Green Alga *Bryopsis* sp. *J. Am. Chem. Soc.* 115, 5825–6.

Hamann, M.T., Otto, C.S., Scheuer, P.J. and Dunbar, D.C. (1996) Kahalalides: bioactive peptides from a marine mollusk *Elysia rufescens* and its algal diet *Bryopsis* sp. *J. Org. Chem.* 61, 6594–600.

Harada, H. and Kamei, Y. (1997) Selective cytotoxicity of marine algae extracts to several human leukemic cell lines. *Cytotechnology* 25, 213–19.

Harada, H. and Kamei, Y. (1998) Dose-dependent selective cytotoxicity of extracts from marine green alga, *Cladophoropsis vaucheriaeformis*, against mouse leukemia L1210 cells. *Biol. Pharm. Bull.* 21(4), 386–9.

Harada, H., Noro, T. and Kamei, Y. (1997) Selective antitumor activity *in vitro* from marine algae from Japan coasts. *Biol. Pharm. Bull.* 20(5), 541–6.

Harrigan, G.G., Yoshida, W., Moore, R.E. Nagle, D.G., Park, P.U., Biggs, J., Paul, V.J., Mooberry, S.L., Corbett, T.H. and Valeriote, F.A. (1998a) Isolation of Dolastatin 12 and Lyngbyastatin 1 from *Lyngbya majuscula/Schizothrix calcicola* algal assemblages and their biological evaluation. *J. Nat. Prod.* 61, 1221–5.

Harrigan, G.G., Luesch, H., Yoshida, W.Y., Moore, R.E., Nagle, D.G., Paul, V.J., Mooberry, S.L., Corbett, T.H. and Valeriote, F.A. (1998b) Symplostatin 1: a Dolastatin 10 analogue from the marine cyanobacterium *Symploca hydnoides*. *J. Nat. Prod.* 61, 1075–7.

Harrigan, G.G., Luesch, H., Yoshida, W.Y., Moore, R.E., Nagle, D.G. and Paul, V.J. (1999) Symplostatin 2: a Dolastatin analogue from the marine cyanobacterium *Symploca hydnoides*. *J. Nat. Prod.* 62, 655–8.

Hayflick, L. and Moorhead, P.S. (1961) The serial cultivation of human diploid cell strains. *Exp. Cell Res.* 25, 585–90.

Hemscheidt, T., Puglisi, M.P., Larsen, L.K., Patterson, G.M.L., Moore, R.E., Rios, J.L. and Clardy, J. (1994) Structure and biosynthesis of borophycin, a new boeseken complex of boric acid from a marine strain of the blue-green alga *Nostoc linckia*. *J. Org. Chem.* 59, 3467–71.

Hodgson, E. (1987) Measurement of toxicity. In *Modern Toxicology* (Eds E. Hodgson and P.E. Levi) Elsevier, Amsterdam, pp. 233–43.

Hodgson, L.M. (1984) Antimicrobial and antineoplastic activity in some South Florida seaweeds. *Bot Mar.* 27, 387–90.

Hoppe, H.A. (1969) Marine algae as raw material. In: *Marine Algae* (Eds H.A. Levring, H.A. Hoppe and O.J. Schmid) Cram de Gruyter & Co, Hamburg, pp. 126–287.

Horgen, F.D., Sakamoto, B. and Scheuer, P.J. (2000) New terpenoid sulfates from the red alga *Tricleocarpa fragilis*. *J. Nat. Prod.* 63, 210–16.

Hori, K., Ikegami, S., Miyazawa, K. and Ito, K. (1988) Mitogenic and antineoplastic isoagglutinins from the red alga *Solieria robusta*. *Phytochemistry* 27, 2063–7.

Hu, T., deFreitas, A.S.W., Curtis, J.M., Oshima, Y., Oshima, Y., Walter, J.A. and Wright, J.L.C. (1996) Isolation and structure of Prorocentrolide B, a fast acting toxin from *Prorocentrum maculosum*. *J. Nat. Prod.* 59, 1010–14.

Ichiba, T., Corgiat, J.M., Scheuer, P.J. and Kelly-Borges, M. (1994) 8-Hydroxymanzamine A, a beta-carboline alkaloid from a sponge, *Pachypellina* sp. *J. Nat. Prod.*, 57(1), 168–70.

Ishibashi, M., Ohizumi, Y., Hamashima, M., Nakamura, H., Hirata, Y., Sasaki, T. and Kobayashi, J. (1987) Amphidinolide-B, a novel macrolide with potent antineoplastic activity from the marine dinoflagellate *Amphidinium* sp. *J. Chem. Soc. Commun.* 1127–9.

Ishibashi, M., Takahashi, M. and Kobayashi, J. (1997) Studies on the macrolides from marine dinoflagellate *Amphidinium* sp.: structures of Amphidinolides R and S and a succinate feeding experiment. *Tetrahedron* 53, 7827–32.

Ishitsuka, M.O., Kusumi, T. and Kakisawa, H. (1988) Antitumor xenicane and norxenicane lactones from the brown alga *Dictyota dichotoma*. *J. Org. Chem.* 53, 5010–13.

Ishitsuka, M.O., Kusumi, T., Ichikawa, A. and Kakisawa, H. (1990a) Bicyclic diterpenes from two species of brown algae of the Dictyotaceae. *Phytochemistry* 29, 2605–10.

Ishitsuka, M.O., Kusumi, T. and Kakisawa, H. (1990b) Antitumor xenicane and norxenicane lactones from the brown algae *Dictyotaceae*. *J. Org. Chem.* 53, 5010–13.

Ishiyama, H., Ishibashi, M. and Kobayashi, J. (1996) Studies on the stereochemistry of Amphidinolides: synthesis of a diastereomer of the C1–C9 fragment of Amphidinolide C. *Chem. Pharm. Bull.* 44, 1819–22.

Ito, H. and Suriura, M. (1976) Antitumor polysaccharide fraction from *Sargassum thunbergii*. *Chem. Pharm. Bull.* 24, 1114–15.

Ito, H., Naruse, S., Sugiura, M. (1976) Studies on antitumor activities of *Basidiomycetes* – antitumor activity of *polysaccharides* and sex factors. *Nippon Yakurigaku Zasshi*. 72(1), 77–94.

Itoh, H., Noda, H., Amano, H. and Ito, H. (1995) Immunological analysis of inhibition of lung metastases by fucoidan (GIV-A) prepared from brown seaweed *Sargassum thunbergii*. *Anticancer Res.* 15(5B), 1937–47.

Itoh, H., Noda, H., Amano, H., Zhuaug, C., Mizuno, T. and Ito, H. (1993) Antitumor activity and immunological properties of marine algal polysaccharides, especially fucoidan, prepared from *Sargassum thunbergii* of phaeophyceae. *Anticancer Res.* 13, 2045–52.

Jakubke, H.O. and Jeschkeit, H. (1975) *Biochemie*. Brockhaus Verlag, Leipzig.

Jiang, G.R., Lin, Z.K. and Chen, J.T. (1986) Studies on antitumor effect of hot water solute from *Sargassum* of China. *J. Mar. Drugs/Haiyang Yaowu* 5, 17–19.

Jolles, B., Remington, M. and Andrews, P.S. (1963) Effects of sulphated degraded laminarin on experimental tumour growth. *Br. J. Cancer* 17, 109–15.

Juagdan, E.D., Kalidindi, R. and Sheuer, P. (1997) Two new chamigranes from an Hawaiian red alga, *Laurencia cartilaginea*. *Tetrahedron* 53, 521–8.

Kaeffer, B., Benard, C., Lahaye, M., Blottiere, H.M. and Cherbut, C. (1999) Biological properties of Ulvan, a new source of green seaweed sulfated polysaccharides, on cultured normal and cancerous colonic epithelial cells. *Planta Med.* 65, 527–31.

Kaiser, J. (2000) Controversial cancer therapy finds political support. *Science* 287, 2139–41.

Kan, Y., Fujita, T., Nagai, H., Sakamoto, B. and Hokama, Y. (1998) Malyngamides M and N from the Hawaiian red alga *Gracilaria coronopifolia*. *J. Nat. Prod.* 61, 152–5.

Kanegawa, K., Harada, H., Myouga, H., Katakura, Y., Shirahata, S. and Kamei, Y. (2000) *Cytotechnology* 33, 221–7.

Kashiwagi, M., Mynderse, J.S., Moore, R.E. and Norton, T.R. (1980) Antineoplastic evaluation of Pacific basin marine algae. *J. Pharm. Sci.* 69 735–8.

Kaya, K., Sano, T. and Shiraishi, F. (1995) Astasin, a novel cytotoxic carbohydrate-conjugated ergosterol from the colorless euglenoid, *Astasia longa*. *Biochim. Biophys. Acta* 1255, 201–4.

Kerr, R.G. and Kerr, S.S. (1999) Marine natural products as therapeutic agents. *Exp. Opin. Ther. Patents* 9, 1207–22.

Ketchum, R.E.B., Tandom, M., Gibson, D.M., Begley, T. and Shuler, M.L. (1999) Isolation of labeled 9-dihydrobaccatin III and related taxoids from cell cultures of *Taxus canadiensis* elicited with methyl jasmonate. *J. Nat. Prod.* 62, 1395–8.

Kintzios, S. and Barberaki, M. (2000) The biotechnology of *Viscum album*: tissue culture, somatic embryogenesis and protoplast isolation. In: *Medicinal and Aromatic Plants – Industrial Approaches: The Genus Viscum* (Ed. A. Bussing) Harwood Publishers, Amsterdam, pp. 95–100.

Kintzios, S., Barberaki, M., Tourgelis, P., Aivalakis, G. and Volioti, An. (2002) Preliminary evaluation of somaclonal variation for the *in vitro* production of new toxic proteins from *Viscum album* L. *J. Herbs, Spices & Medicinal Plants* 9, 217–22.

Kobayashi, J. (1989) Pharmacologically active metabolites from symbiotic microalgae in Okinawan marine invertebrates. *J. Nat. Prod.* 52, 225–38.

Kobayashi, J., Ishibashi, M., Nakamura, H., Ohizumi, Y., Yamasu, T., Sasaki, T. and Hirata, Y. (1986) Amphidinolide-A, a novel antineoplastic macrolide from the marine dinoflagellate *Amphidinium* sp. *Tetrahedron Lett.* 27, 5755–8.

Kobayashi, J., Ishibashi, M., Nakamura, H., Ohizumi, Y., Yamasu, T., Hirata, Y., Sasaki, T., Ohta, T. and Nozoe, S. (1989) Cytotoxic macrolides from a cultured marine dinoflagellate of the genus *Amphidinium*. *J. Nat. Prod.* 52, 1036–41.

Kobayashi, J., Ishibashi, M., Wälchli, M.R., Nakamura, H., Hirata, Y., Sasaki, T. and Ohizumi, Y. (1988a) Amphidinolide C: the first 25-membered macrocyclic lactone with potent antineoplastic activity from the cultured dinoflagellate *Amphidinium* sp. *J. Am. Chem. Soc.* 110, 490–4.

Kobayashi, J., Ishibashi, M., Nakamura, N., Hirata, Y., Yamasu, Y., Sasaki, T. and Ohizumi, Y. (1988b) Symbioramide a novel Ca^{2+}-ATPase activator from the cultured dinoflagellate *Symbiodinium* sp. *Experientia* 44, 800–2.

Koehnlechner, M. (1987) *Leben ohne Krebs*. Knaur, Munich.

Koenig, G.M., Wright, A.D. and Linden, A. (1999) *Plocamium hamatum* and ist monoterpenes: chemical and biological investigations of the tropical marine red alga. *Phytochemistry* 52, 1047–53.

Konishi, F., Tanaka, K., Himeno, K., Taniguchi, K. and Nomoto, K. (1985) Antitumor effect induced by a hot water extract of *Chlorella vulgaris*: resistance to Meth-A tumor growth mediated by CE-induced polymorphonuclear leukocytes. *Cancer Immunol. Immunother.* 19, 73–8.

Konishi, F., Mitsuyama, M., Okuda, M., Tanaka, K., Hasegawa, T., Nomoto, K. (1996) Protective effect of an acidic glycoprotein obtained from culture of *Chlorella vulgaris* against myelosuppression by 5-fluorouracil. *Cancer Immunol. Immunother.* 42(5), 268–74.

Ktari, L. and Guyot, M. (1999) A cytotoxic oxysterol from the marine alga *Padina pavonica* (L.) Thivy. *J. Appl. Phycol.* **11**, 511–13.

Ktari, L., Blond, A. and Guyot, M. (2000) 16β-Hydroxy-5α-cholestane-3,6-dione, a novel cytotoxic oxysterol from the red alga *Jania rubens*. *Bioorg. Med. Chem. Lett.* **10**, 2563–5.

Kubota, T., Tsuda, M. and Kobayashi, J. (2000) Amphidinolide V, novel 14-membered macrolide from marine dinoflagellate *Amphidinium* sp. *Tetrahedron Lett.* **41**, 713–16.

Langer, M., Zinke, H., Eck, J., Mockel, B. and Lentzen, H. (1997) Cloning of the active principle of mistletoe: the contributions of mistletoe lectin single chains to biological functions. *Eu. J. Cancer* **33**, 24.

Lemée, R., Pesando, D., Durand-Clément, M., Dubreuil, A., Meinesz, A., Guerriero, A. and Pietra, F. (1993a) Preliminary survey of toxicity of the green alga *Caulerpa taxifolia* introduced into the Mediterranean. *J. Appl. Phycol.* **5**, 485–93.

Lemée, R., Pesando, D., Durand-Clement, M. and Dubreuil, A. (1993b) Preliminary survey of toxicity of the green alga *Caulerpa taxifolia* introduced into the Mediterranean. *J. Appl. Phycol.* **5**, 485–93.

Lewin, R.A. and Cheng, L. (1989) *Prochloron*. Chapman and Hall, New York.

Lichtenberg-Kraag, B., Klinker, J.F., Muhlbauer, E. and Rommelspacher, H. (1997) The natural beta-carbolines facilitate inositol phosphate accumulation by activating small G-proteins in human neuroblastoma cells (SH-SY5Y). *Neuropharmacology* **36**(11–12), 1771–8.

Lincoln, R.A., Strupinski, K. and Walker, J.M. (1991) Bioactive compounds from algae. In: *Life Chemistry Reports* **8**, Harwood Academic Publishers GmbH, Singapore, pp. 97–183.

Lorch, E. and Troger, W. (2000) Pharmaceutical quality of mistletoe preparations. In: *Medicinal and Aromatic Plants – Industrial Approaches: The Genus Viscum* (Ed. A. Bussing) Harwood Publishers, Amsterdam, pp. 223–236.

Luesch, H., Yoshida, W., Moore, R.E., Paul, V.J. and Mooberry, S.L. (2000a) Isolation, structure determination and biological activity of Lyngbyabellin A from the marine cyanobacterium *Lyngbya majuscula*. *J. Nat. Prod.* **63**, 611–15.

Luesch, H., Yoshida, W., Moore, R.E. and Paul, V.J. (2000b) Isolation and structure of the cytotoxin Lyngbyabellin B and absolute configuration of Lyngbyapeptin A from the marine cyanobacterium *Lyngbya majuscula*. *J. Nat. Prod.* **63**, 1437–9.

Luesch, H., Yoshida, W., Moore, R.E., Paul, V.J. and Corbett, T.H. (2001) Total structure determination of Apratoxin A, a potent novel cytotoxin from the marine cyanobacterium *Lyngbya majuscula*. *J. Am. Chem. Soc.* **123**, 5418–23.

Márquez, B., Verdier-Pinard, P., Hamel, E. and Gerwick, W.H. (1998) Curacin D, an antimitotic agent from the marine cyanobacterium *Lyngbya majuscula*. *Phytochemistry* **49**, 2387–9.

McConnell, O.J., Longley, R.E. and Koehn, F.E. (1994) The discovery of marine natural products. *Biotechnol. Ser.* **26**, 109–74.

Manríquez, C.P., Souto, M.L., Gavín, J.A., Norte, M. and Fernández, J.J. (2001) Several new squalene-derived triterpenes from *Laurencia*. *Tetrahedron* **57**, 3117–23.

Margiolis, R.L. and Wilson, L. (1977) Addition of colahicine – tubulin complex to microtubule ends: the mechanism of sub-stoichiometric colchicine poisoning. *Proc. Natl. Cancer Inst., USA* 74, 3466.

Maruyama, H., Nakajima, J. and Yamamoto, I. (1987) A study on the anticoagulant and fibrinolytic activities of a crude fucoidan from the edible brown seaweed *Laminaria religiosa*, with special reference to its inhibitory effect on the growth of Sarcoma-180 ascites cells subcutaneously implanted into mice. *Kitasaro Arch. Exp. Med.* **60**, 105–21.

Matzusawa, S., Suzuki, T., Suzuki, M., Matsuda, A., Kawamura, T., Mizuno, Y. and Kikuchi, K. (1994) Thyrsiferyl 23-acetate is a novel specific inhibitor of protein phosphatase PP2A. *FEBS Lett.* **356**, 272–4.

Mayer, A.M.S. (1998) Marine pharmacology in 1998: Antitumor and Cytotoxic compounds. *The Pharmacologist* 41(4), pp. 159–64.

Mayer, A.M.S. and Lehmann, V.K.B. (2001) Marine pharmacology in 1999: antitumor and cytotoxic compounds. *Anticancer Res.* 21, 2489–500.

Meijer, L., Guidet, S. and Philippe, M. (eds) (1997) *Progress in Cell Cycle Research*, vol. 3. Plenum Press, New York, pp. 261–9.

Meiun, T. (1981) Mannugulunoglycans as neoplasm inhibitors from seaweeds (*Sargassum fulvellum*). *Jap. Kokai Tokkyo Koho* 81(79), 101–4.

Minehan, T.G., Cook-Blumberg, L., Kishi, Y., Prinsep, M.R. and Moore, R.E. (1999) Revised structure of Tolyporphin A. *Angew. Chem. Int. Ed.* 38, 926–8.

Miyairi, S., Shimada, H., Morita, H., Awata, N., Goto, J. and Nambara, T. (1991) Enzyme immunoassay for (3-amino-1-methyl-5H-pyrido[4,3-b]indole), a tryptophan pyrolysate. *Cancer Lett.* 56(3), 207–13.

Miyazawa, Y., Murayama, T., Ooya, N., Wang, L.F., Tung, Y.C. and Yamaguchi, N. (1988) Immunomodulation by a unicellular green algae (*Chlorella pyrenoidosa*) in tumor-bearing mice. *J. Ethnopharmacol.* 24, 135–46.

Mizuno, T., Usui, T., Kanemitsu, K., Ushiyama, M., Matsueda, S., Shinho, K., Hasegawa, T., Kobayashi, S., Shikai, K. and Arakawa, M. (1980) Studies on the carbohydrates of *Chlorella*. III. Fractionation and some biological activities of the *Chlorella* polysaccharides. *Shizuoka Daig. Nogakuku Kenkyu Hokoku* 30, 51–9.

Mockel, B., Schwarz, T., Eck, J., Langer, M., Zinke, H. and Lentzen, H. (1997) Apoptosis and cytokine release are biological responses mediated by recombinant mistletoe lectin *in vitro*. *Eur. J. Cancer* 33, 35.

Moore, R.E. (1981) In: *The Water Environment Algal Toxins and Health* (Ed. W.W. Carmichael) Plenum Press, New York, pp. 15–23.

Moore, R.E. (1982) Toxins, anticancer agents and tumor promoters from marine prokaryotes. *Pure Appl. Chem.* 54, 1919–34.

Mori, K. and Koga, Y. (1992) Synthesis and absolute configuration of (-)-Stypoldione. *Bioorg. Med. Chem. Lett.* 2, 391–4.

Morimoto, T., Nagatsu, A., Murakami, N., Sakakibara, J., Tokuda, H., Nishino, H. and Iwashima, A. (1995) Anti-tumour-promoting glyceroglycolipids from the green alga, *Chlorella vulgaris*. *Phytochemistry* 40, 1433–7.

Morlière, P., Mazière, J.-C., Santus, R., Smith, C.D., Prinsep, M.R., Stobbe, C.C., Fenning, M.C., Goldberg, J.L. and Chapman, J.D. (1988) Tolyporphin: a natural product from cyanobacteria with potent photosensitizing activity against tumor cells *in vitro* and *in vivo*. *Cancer Res.* 58, 3571–8.

Munro, M.H., Bunt, J.W., Dumdei, E.J., Hickford, S.J., Lill, R.E., Li, S., Battershill, C.N. and Duckworth, A.R. (1999) The discovery and development of marine compounds with pharmaceutical potential. *J. Biotechnol.* 70, 15–25.

Mynderse, J.S. and Faulkner, D.J. (1978) Variations in the halogenated monoterpene metabolites of *Plocamium cartilagineum* and *P. violaceum*. *Phytochemistry* 1(2), 237–40.

Mynderse, J.S. and Moore, R.E. (1978) Toxins from blue green algae: structures of Oscillatoxin A and three related bromine-containing toxins. *J. Org. Chem.* 43, 2301–3.

Mynderse, J.S., Moore, R.E., Kashiwagi, M. and Norton, T.R. (1977) Antileukemia activity in the oscillatoriaceae: isolation of debromoaplysiatoxin from *Lyngbya*. *Science* 196, 538–40.

Nagao, M., Wakabayashi, K., Fujita, Y., Tahira, T., Ochiai, M., Takayama, S. and Sugimura, T. (1985) Nitrosatable precursors of mutagens in vegetables and soy sauce. *Princess Takamatsu Symp.* 16, 77–86.

Nagasawa, H., Konishi, R., Sensui, N., Yamamoto, K. and Ben-Amotz A. (1989) Inhibition by beta-carotene-rich algae *Dunaliella* of spontaneous mammary tumourigenesis in mice. *Anticancer Res.* 9(1), 71–5.

Nagasawa, H., Fujii, Y., Kageyama, Y., Segawa, T. and Ben-Amotz, A. (1991) Suppression by beta-carotene-rich algae *Dunaliella bardawil* of the progression, but not the development, of spontaneous mammary tumours in SHN virgin mice. *Anticancer Res.* 11(2), 713–17.

Nagle, D.G., Geralds, R.S., Yoo, H.-D., Gerwick, W.H., Kim, T.-S., Nambu, M. and. White, J.D. (1995) Absolute configuration of Curacin A, a novel antimitotic agent from the tropical marine cyanobacterium *Lyngbya majuscula*. *Tetrahedron Lett.* 36(8), 1189–92.

Nakano, T., Noro, T. and Kamei, Y. (1997) *In vitro* promoting activity of human interferon β production by extracts of marine algae from Japan. *Cytotechnology* **25**, 239–41.

Nagumo, T. (1983) Antitumor polysaccharides from seaweeds *Jap. Kokai Tokkyo Koho JP* **58**, 174, 329.

Nakano, H., Moriwaki, M., Washino, T., Kino, T., Yoshizumi, H. and Kitahata, S. (1994) Purification and some properties of a trehalase from a green alga, *Lobosphaera* sp. *Biosci. Biotech. Biochem.* **58**(8), 1430–4.

Nakazawa, S. and Ikeda, O. (1972) Anticarcinogenic substance from marine algae. *Jap. Patent* 72 33, 122.

Nakazawa, S., Kuroda, H., Abe, F., Nishino, T., Ohtsuki, M. and Umezaki, I. (1974) Antitumor effect of water extracts from marine algae (I). *Chemotherapy* (Japan) **22**, 1435–44.

Nakazawa, S., Abe, F., Kuroda, H., Kohno, K., Higashi T. and Umezaki, I. (1976a) Antitumor effect of water extracts from marine algae (III) *Codium pugniformis* Okamura. *Chemotherapy* (Japan) **24**, 448–50.

Nakazawa, S., Abe, F., Kuroda, H., Kohno, K., Higashi, T. and Umezaki, I. (1976b) Antitumor effect of water extracts from marine algae (II) *Sargassum horneri* (Turner) C. Agarth. *Chemotherapy* (Japan) **24**, 433–47.

Noda, H., Amano, H., Arashima, K., Hashimoto, S. and Nisizawa, K.(1989) Studies on the antitumor activity of marine algae. *Nippon Suisan Gakkaishi* **55**(7), 1259–64.

Noda, H., Amano, H., Arashima, K. and Nisizawa, K. (1990) Antitumor activity of marine algae. *Hydrobiologia* **204/205**, 577–84.

Noda, H., Amano, H., Arashima, S., Hashimoto, S. and Nisizawa, K. (1982) Studies on the antitumor activity of marine algae. *Nippon Suisan Gakkaishi* **55**, 1259–64.

Noda, K., Ohno, N., Tanaka, K., Kamiya, N., Okuda, M., Yadomae, T., Nomoto, K. and Shoyama, Y. (1969) A water-soluble antitumor glycoprotein from *Chlorella vulgaris. Planta Med.* **62**, 423–6.

Nomoto, K., Yokokura, T., Satoh, H. and Mutai, M. (1983) Antitumor activity of *Chlorella* extract, PCM-4, by oral administration. *Gan To Kagaku Ryoho* (Japanese) **10**(3), 781–5.

Norte, M., Fernandez, J., J., Souto, M., L. and Garcia-Gravalos, M.D. (1996) Two new antitumoral polyether squalene derivatives. *Tetrahedron Lett.* **37**, 2671–4.

Norte, M., Fernandez, J.J., Souto, M.L., Gavín, J.A. and Garcia-Gravalos, M.D. (1997) Thyrsenols A and B, two unusual polyether squalene derivatives. *Tetrahedron* **53**, 3173–8.

Nossal, G.J.V. (1993) Life, death and the immune system. *Sci. American* **269**, 20–31.

Numata, A., Kanbara, S., Takahashi, C., Fujiki, R., Yoneda, M., Usami, Y. and Fujita, E. (1992) A cytotoxic principle of the brown alga *Sargassum tortile* and structures of chromenes. *Phytochemistry* **31**, 1209–13.

O'Brien, E.T., White, S., Robert, J.S., Boder, B.G. and Wilson, L. (1984) Pharmacological properties of a marine natural product, stypoldione, obtained from the brown alga *Stypopodium zonale. Cuimr. Hydrobiologia*, **116/117**, 141–5.

Ohta, K., Mizushina, Y., Hirata, N., Takemura, M., Sugawara, F., Matsukage, A., Yoshida, S. and Sakaguchi, K. (1999) Action of a new mammalian DNA polymerase inhibitor sulfoquinovosyldiacyl glycerol. *Biol. Pharm. Bull.* **22**, 111–16.

Okai, Y., Higashi-Okai, K., Nakamura, S., Yano, Y. and Otani, S. (1994) Suppressive effects of the extracts of Japanese edible seaweeds on mutagen-induced umu C gene expression in *Salmonella typhimurium* (TA 1535/pSK 1002) and tumor promotor-dependent ornithine decarboxylase induction in BALB/c 3T3 fibroblast cells. *Cancer Lett.* **87**(1), 25–32.

Okai, Y., Higashi-Okai, K., Yano, Y. and Otani, S. (1996) Identification of antimutagenic substances in an extract of edible red alga, *Porphyra tenera* (Asakusa-nori). *Cancer Lett.* **100**(1–2), 235–40.

Old, L.J. (1996) Immunotherapy for cancer. *Sci. American* **275**, 102–9.

Orjala, J., Nagle, D. and Gerwick, W.H. (1995) Malyngamide H, an Ichthyotoxic amide possessing a new carbon skeleton from the Caribbean cyanobacterium *Lyngbya majuscula*. *J. Nat. Prod.* **58**, 764–8.

Patterson, G.M.L. and Bolis, C.M. (1993) Determination of scytophycins in cyanophyte cultures by high performance liquid chromatography. *Hawau, J. Liquid Chrom.* **16**(2), 475–86.

Paul, V.J. and Fenical, W. (1983) Isolation of Halimedatrial: chemical defense adaptation in the calcareous reef-building alga *Halimeda*. *Science* **221**, 747–8.

Payne, G., Bringi, V., Prince, C. and Shuler, M. (1991) *Plant Cell and Tissue Culture in Liquid Systems*. Hanser Publishers, Munich, pp. 3–9.

Pec, M.K., Hellan, M., Moser-Thier, K., Fernández, J.J., Souto, M.L. and Kubista, E. (1998) Inhibitory effects of a novel marine terpenoid on sensitive and multidrug resistant KB cell lines. *Anticancer Res.* **18**, 3027–32.

Pec, M.K., Moser-Thier, K., Fernández, J.J., Souto, M.L. and Kubista, E. (1999) Growth inhibition by dehydrothyrsiferol – a non-Pgp modulator, derived from a marine red alga – in human breast cancer cell lines. *Inter. J. Oncology* 14, 739–43.

Pesando, D., Lemée, R., Ferrua, C., Amade, P. and Girard, J.-P. (1996) Effects of caulerpenyne, the major toxin from *Caulerpa taxifolia* on mechanisms related to sea-urchin egg cleavage. *Aquatic Toxicol.* **35**, 139–55.

Pesando, D., Huitorel, P., Dolcini, V., Amade, P. and Girard, J.-P. (1998) Caulerpenyne interferes with microtubule-dependent events during the first mitotic cycle of sea urchin eggs. *Eur. J. Cell Biol.* 77, 19–26.

Petit, G.R., Pierson, F.H. and Herald, C.L. (1994) *Anticancer Drugs from Animals, Plants and Microorganisms.* Wiley, New York, pp. 49–158.

Pettit, G.R., Kamano, Y., Holzapfel, C.W., van Zyl, W.J., Tuinman, A.A., Herald, C.L., Baczynskyj, L. and Schmidt, J.M. (1987) The structure and synthesis of Dolastatin 3. *J. Am. Chem. Soc.* **109**, 7581–2.

Pfuller, U. (2000) Chemical constituents of European mistletoe (*Viscum album* L.) In: *Medicinal and Aromatic Plants – Industrial Approaches: The Genus Viscum* (Ed. A. Bussing) Harwood Publishers, Amsterdam, pp. 101–22.

Piattelli, M., Tringali, C., Neri, P. and Rocco, C. (1995) Stereochemistry and conformation of dolabellane diterpenes: a NMR and molecular mechanics study. *J. Nat. Prod.* **58**, 697–704.

Pietra, F. (1990) *A Secret World.* Birkhauser Verlag, Basel, Switzerland.

Podlech, D. (1996) *Herbs and Medicinal Plants of Britain and Europe.* HarperCollinsPublishers, London. pp. 240–6.

Poncet, J. (1999) The Dolastatins, a family of promising antineoplastic agents. *Curr. Phar. Des.* **5**, 139–62.

Prinsep, M.R., Caplan, F.R., Moore, R.E., Patterson, G.M.L. and Smith, C.D. (1992) Tolyporphin, a novel multidrug resistance reversing agent from the blue-green alga *Tolypothrix nodosa. J. Am. Chem. Soc.* 114, 385–6.

Prinsep, M.R., Patterson, G.M.L., Larsen, L.K. and Smith, C.D. (1995) Further tolyporphins from the blue-green alga *Tolypothrix nodosa. Tetrahedron* **51**, 10523–30.

Prinsep, M.R., Patterson G.M.L., Larsen, L.K. and Smith, C.D. (1998) Tolyporphin J and K, two further porphinoids from the cyanobacterium *Tolypothrix nodosa. J. Nat. Prod.* **61**, 1133–6.

Reddy, B.S., Sharma, C. and Mathews, L. (1984) Effects of Japanese seaweed (*Laminaria angustata*) extracts on the mutagenicity of 7,12 dimethylbenz(a) anthracene, a breast carcinogen and of 3,2′ dimethyl-4-aminobiphenyl, a colon and breast carcinogen. *Mutat. Res.* **127**, 113–58.

Ren, D.L., Wang, J.Z., Noda, H., Amano, H. and Ogawa, S. (1995) The effects of an algal polysaccharide from *Gloiopeltis tenax* on transplantable tumors and immune activities in mice. *Planta Med.* **61**, 120–5.

Renau, T.E., Lee, J.S., Kim, H., Young, C.G., Wotring, L.L., Townsend, L.B. and Drach, J.C. (1994) Relationship between cytotoxicity and conversion of thiosangivamycin and related analogs to toyocamycin and toyocamycin analogs in cell culture medium. *Biochem. Pharmacol.* **48**, 801–7.

Rickards, R.W., Rothschild, J.M., Willis, A.C., de Chazal, N.M., Kirk, J., Kirk, K., Saliba, K.J. and Smith, G.D. (1999) Calothrixins A and B, novel pentacyclic metabolites from *Calothrix* cyanobacteria with potent activity against malaria parasites and human cancer cells. *Tetrahedron* **55**, 13513–20.

Riguera, R. (1997) Isolating bioactive compounds from marine organisms. *J. Mar. Biotechnol.* 5, 187–93.

Riou, D., Colliec-Jouault, S., Pinczon du Sel, D., Bosch, S., Siavoshian, S., Le Bert, V., Tomasoni, C., Sinquin, C., Durand, P. and Roussakis, C. (1996) Antitumor and antiproliferative effects of a fucan extracted from *Ascophyllum nodosum* against a non-small-cell bronchopulmonary carcinoma line. *Anticancer Res.* 16, 1213–18.

Robinson, T. (1964) *The Organic Constituents of Higher Plants.* Burgess Publishing Company, Minneapolis.

Sakai, R., Rinehart, K.L., Guan, Y. and Wang, A.H. (1992) Additional antitumor ecteinascidins from a Caribbean tunicate: crystal structures and activities *in vivo. Proc. Natl. Acad. Sci. USA* **89**(23), 11456–60.

Schwartz, J.L. and Shklar, G. (1989) A cyanobacteria extract and β-carotene stimulate an antitumor host response against an oral cancer line. *Phytotherapy Res.* 3, 243–348.

Schwartz, J.L., Suda, D. and Light, G. (1986) Beta carotene is associated with the regression of tumour necrosis macrophages. *Biochem. Biophys. Res. Commun.* 136, 1130.

Schweppe, K.W. (1982) The attempts at drug therapy of cancer by Anton Storck (1731–1803). History of experimental pharmacology in the old Vienna Medical School Probst C *Wien Med. Wochenschr.* 15132, (5), 107–17.

Shan, B.E., Yoshida, Y., Kuroda, E. and Yamashita, U. (1999) Immunomodulating activity of seaweed extract on human lymphocytes *in vitro*. *Int. J. Immunopharmacol.* 21, 59–70.

Sheu, J.-H., Huang, S.-Y. and Duh, C.-Y. (1996) Cytotoxic oxygenated desmosterols of the red alga *Galaxaura marginata*. *J. Nat. Prod.* 59, 23–6.

Sheu, J.-H., Huang, S.-Y., Wang, G.-H. and Duh, C.-Y. (1997a) Study on cytotoxic oxygenated desmosterols isolated from the red alga *Galaxaura marginata*. *J. Nat. Prod.* 60, 900–3.

Sheu, J.H., Liaw, C.-C. and Duh, C-Y. (1995) Oxygenated clerosterols isolated from the marine alga *Codium arabieum*. *J. Nat. Prod.* 58, 1521–6.

Sheu, J.H., Wang, G.H., Sung, P.J. and Duh, C.Y. (1999) New cytotoxic oxygenated fucosterols from the brown alga *Turbinaria conoides*. *J. Nat. Prod.* 62, 224–7.

Sheu, J.H., Wang, G.H., Sung, P.J., Chiu, Y.H. and Duh, C.Y. (1997b) Cytotoxic sterols from the formosan brown alga *Turbinaria ornata*. *Planta Med.* 63, 571–2.

Shimizu, Y. (1993) Microalgal metabolites. *Chem Rev.* 93, 1685–98.

Shimizu, Y. (1996) Patent. US5494930.

Shinho, K. (1986) Antitumor glycoproteins from *Chlorella* and other species. *Jap. Kokai Tokkyo Koho JP* 61, 728.

Shinho, K. (1987) Extraction of an anticancer agent from *Chlorella*. *Eur. Pat. Appl. EP* 209, 78.

Shklar, G. and Schwartz, J.L. (1988) Tumour necrosis factor in experimental cancer regression with alpha-tocopherol, beta-carotene, canthaxanthin and algae extract. *Eur. J. Cancer. Clin. Oncol.* 244, 839–50.

Shklar, G., Schwartz, J., Trickler, D. and Cheverie, S.R. (1993) The effectiveness of a mixture of beta-carotene, alpha-tocopherol, glutathione, and ascorbic acid for cancer prevention. *Nutr. Cancer* 20(2), 145–51.

Sidransky, D. (1996) Advances in cancer detection. *Sci. American* 275, 70–2.

Sieburth, J.M.N. (1979) *Sea Microbes*. Oxford University Press, New York.

Sitachitta, N. and Gerwick, W.H. (1998) Grenadadiene and grenadamide, cyclopropyl containing fatty acid metabolites from the marine cyanobacterium *Lyngbya majuscula*. *J. Nat. Prod.* 61, 681–4.

Sitachitta, N., Williamson, R.T. and Gerwick, W.H. (2000) Yanucamides A and B, two new depsipeptides from an assemblage of the marine cyanobacteria *Lyngbya majuscula* and *Schizothrix* species. *J. Nat. Prod.* 63, 197–200.

Smith, C.D., Carmeli, S., Moore, R.E. and Patterson, G.M.L. (1993) Scytophycins: novel microfilament-depolymerizing agents which circumvent P-glycoprotein mediated multidrug resistance. *Cancer Res.* 53, 1343–7.

Smith, C.D., Zhang, X., Mooberry, S.L., Patterson, G.M.L. and Moore, R.E. (1994a) Cryptophycin: a new antimicrotubule agent active against drug-resistant cells. *Cancer Res.* 54(14), 3779–84.

Smith, C.D., Prinsep, M.R., Caplan, F.R., Moore, R.E. and Patterson, G.M.L. (1994b) Reversal of multiple drug resistance by tolyporphin, a novel cyanobacterial natural product. *Oncol. Res.* 6, 211–18.

Smollny, T., Wichers, H., Kalenberg, S., Shahsavari, A., Petersen, M. and Alfermann, A.W. (1998) Accumulation of podophyllotoxin and related lignans cell suspension cultures of *Linum album*. *Phytochemistry* 48(6), 975–9.

Soeder, C.J. (1976) Use of microalgae for nutritional purpose. Zur Verwendung von Mikroalgen fur Ernahrungszwecke. *Naturwissenschaften* 63, 131–8.

Stewart, J.B., Bonremann, V., Chen, J.L., Moore R.E., Chaplan, F.R., Karuso, H., Larsen, L.K. and Patterson, G.M.L. (1988) Cytotoxic, fungicidal nucleosides from the blue-green algae belonging to the Scytonemataceae. *J. Antibiotics* XLI(8), 1048–56.

Sugimura, T. (1986) Past, present, and future of mutagens in cooked foods. *Environ. Health Perspect.* 67, 5–10.

Sukumaran, K. and Kuttan, R. (1991) Screening of 11 ferns for cytotoxic and antitumor potential with special reference to *Pityrogramma calomelanos*. *J. Ethnopharmacol.* 34(1), 93–6.

Suzuki, Y., Yamamoto, I. and Umezawa, I. (1980) Antitumor effect of seaweed: partial purification and the antitumor effect of polysaccharides from *Laminaria angustata* Kjellman var. longissima Miyabe. *Chemotherapy* 28, 165–70.

Suzuki, M., Matsuo, Y., Takahashi, Y. and Masuda, M. (1995) Callicladol, a novel bromotriterpene polyether from a Vietnamese species of the red algal genus *Laurencia*. *Chem. Lett.* 1045–6.

Takahashi, M. (1983) Studies on the mechanisms of hostmediated antitumor action of crude fucoidan from a brown marine alga *Eisenia bicyclis*. *J. Jap. Soc. Reticuloendothel. Syst.* 22, 269–83.

Tan, L.-T., Williamson, R.T., Gerwick, W.H., Watts, K.S., McGough, K. and Jacobson, R. (2000a) *cis, cis*-and *trans, trans*-Ceratospongamide, new bioactive cyclic heptapeptides from the Indonisian red alga *Ceratodictyon spongiosum* and symbiotic sponge *Sigmadocia symbiotica*. *J. Org. Chem.* 65, 419–25.

Tan, L.T., Okino, T. and Gerwick, W.H. (2000b) Hermitamides A and B, toxic Malyngamide-type natural products from the marine cyanobacterium *Lyngbya majuscula*. *J. Nat. Prod.* 63, 952–5.

Tanaka, K., Konishi, F., Himeno, K., Taniguchi, K. and Nomoto, K. (1984) Augmentation of antitumor resistance by a strain of unicellular green algae, *Chlorella vulgaris*. *Cancer Immunol. Immunother.* 17, 90–4.

Tanaka, K., Tomita, Y., Tsuruta, M., Konishi, F., Okuda, M., Himeno, K. and Nomoto, K. (1990a) Oral administration of *Chlorella vulgaris* augments concomitant antitumor immunity. *Immunopharmacol. Immunotoxicol.* 12(2), 277–91.

Tanaka, K., Yamada, A., Noda, K., Shoyama, Y., Kubo, C. and Nomoto, K. (1997) Oral administration of a unicellular green algae, *Chlorella vulgaris*, prevents stress-induced ulcer. *Planta Med.* 63, 465–6.

Tanaka, K., Tomita, Y., Tsuruta, M., Konishi, F., Okuda, M., Himeno, K. and Nomoto, K. (1990b) Oral administration of *Chlorella vulgaris* augments concomitant antitumor immunity. *Immunopharmacol. Immunotoxicol.* 12, 277–91.

Tanaka, K., Yamada, A., Noda, K., Hasegawa, T., Okuda, M., Shoyama, Y. and Nomoto, K. (1998) A novel glycoprotein obtained from *Chlorella vulgaris* strain CK22 shows antimetastatic immunopotentiation. *Cancer Immunol. Immunother.* 45, 313–20.

Teas, J. (1983) The dietary intake of *Laminaria*, a brown seaweed, and breast cancer prevention. *Nutr. Cancer* 4(3), 217–22.

Teas, J., Harbison, M.L. and Gelman, R.S. (1984) Dietary seaweed (*Laminaria*) and mammary cardiogenesis in rats. *Cancer Res.* 44, 2758–61.

Thorndike, J. and Beck, W.S. (1977) Production of formaldehyde from N5-ethyltetrahydrofolate by normal and leukemic leukocytes. *Cancer Res.* 37(4), 1125–32.

Tomita, Y., Himeno, K., Nomoto, K., Endo, H. and Hirohata, T. (1987) Augmentation of tumourimmunity against syngeneic tumours in mice by beta carotene. *J. Natl. Cancer. Inst.* 78, 679.

Torigoe, K., Murata, M. and Yasumoto, T. (1988) Prorocentrolide, a toxic nitrogenous macrocycle from a marine dinoflagellate, *Prorocentrum lima*. *J. Am. Chem. Soc.* 110, 7876–7.

Trichopoulos, D., Li, F.P. and Hunter, D.J. (1996) What causes cancer? *Sci. American* 275, 50–5.

Tringali, C. (1997) Bioactive metabolites from marine algae: recent results. *Current Organic Chem.* 1, 375–94.

Tsukamoto, S., Painuly, P., Young, K.A., Yang, X. and Shimizu, Y. (1993) Microcystilide A: a novel cell-differentiation-promoting depsipeptide from *Microcystis aeruginosa* NO-15-1840. *J. Am. Chem. Soc.* 115, 11046–7.

Urones, J.G., Basabe, P., Marcos, I.S., Pineda, J., Lithgow, A.M., Moro, R.F., Brito Palma, F.M.S., Araújo, M.E.M. and Gravalos, M.D.G. (1992a) Meroterpenes from *Cystoceira usneoides*. *Phytochemistry* 31, 179–82.

Urones, J.G., Araújo, M.E.M., Brito Palma, F.M.S., Basabe, P., Marcos, I.S., Moro, R.F., Lithgow, A.M. and Pineda, J. (1992b) Meroterpenes from *Cystoceira usneoides* II. *Phytochemistry* 31, 2105–9.

Usui, T., Asari, K. and Mizuno, T. (1980) Isolation of highly purified fucoidan from *Eisenia bicyclis* and its anticoagulant and antitumor activities. *Agric. Biol. Chem.* 44, 1965–6.

Valeriote, F.A., Demchik, L., Corbet, T.H., Golakoti, T., Heltzel, C.E., Ogino, J., Patterson, G.M.L. and Moore, R.E. (1995) Anticancer Activity of Cryptophycin Analogs, *Proc. Amer. Assoc. Cancer Res.* 36, 303.

Valls, R., Banaigs, B., Piovetti, L., Archavlis, A. and Artaud, J. (1993) Linear diterpene with antimitotic activity from the brown alga *Bifurcaria bifurcata*. *Phytochemistry* 34, 1585–8.

Valls, R., Piovetti, L., Banaigs, B., Archavlis, A. and Pellegrini, M. (1995) (S)-13-Hydroxygeranylgeraniol-derived furanoditerpenes from *Bifurcaria bifurcata*. *Phytochemistry* 39, 145–9.

Verdier-Pinard, P., Lai, J.-Y., Yoo, H.D., Yu, J., Marquez, B., Nagle, D.G., Nambu, M., White, J.D., Falck, J.R., Gerwick, W.H., Day, B.W. and Hamel, E. (1998) Structure–activity analysis of the interaction of curacin A, the potent colchicine site antimitotic agent, with tubulin and effects of analogs on the growth of MCF-7 breast cancer cells. *Mol. Pharmacol.* 53, 62–76.

Wakabayashi, K., Nagao, M., Ochiai, M., Fujita, Y., Tahira, T., Nakayasu, M., Ohgaki, H., Takayama, S. and Sugimura, T. (1987) Recently identified nitrite-reactive compounds in food: occurrence and biological properties of the nitrosated products. *IARC Sci. Publ.* 84, 287–91.

Wakabayashi, K., Totsuka, Y., Fukutome, K., Oguri, A., Ushiyama, H. and Sugimura, T. (1997) Human exposure to mutagenic/carcinogenic heterocyclic amines and comutagenic beta-carbolines. *Mutat. Res.* 376(1–2), 253–9.

Wall, M.E., Wani, M.C., Manikumar, G., Taylor, H., Hughes, T.J., Gaetano, K., Gerwick, W.H., McPhail, A.T. and McPhail, D.R. (1989) Plant antimutagenic agents, 7. Structure and antimutagenic properties of cymobarbatol and 4-isocymobarbatol, new cymopols from green alga (*Cymopolia barbata*). *J. Nat. Prod.* 52, 1092–9.

Yamamoto, I. and Maruyama, H. (1985) Effects of dietary seaweed preparations on 1,2 dimethylhydrazine-induced intestinal carcinogenesis in rats. *Cancer Lett.* 26, 241–51.

Yamamoto, I., Maruyama, H. and Moriguchi, M. (1987) The effect of dietary seaweeds on 7,12 dimethylbenz(a) anthracene-induced mammary tumorigenesis in rats. *Cancer Lett.* 35, 109–18.

Yamamoto, I., Maruyama, H., Takahashi, M. and Komiyama, K. (1986) The effect of dietary of intraperitoneally injected seaweed preparations on the growth of Sarcoma-180 cells subcutaneously implanted into mice. *Cancer Lett.* 30, 125–31.

Yamamoto, I., Nagumo, T., Yagi, K., Tominaga, H. and Aoki, M. (1974) Antitumor effects of seaweeds, I. Antitumor effect of extracts from *Sargassum* and *Laminaria*. *Jap. J. Exp. Med.* 44, 543–6.

Yamamoto, I., Nagumo, T., Fugihara, M., Takahashi, M., Ando, Y., Okada, M. and Kawai, K. (1977) Antitumor effect of seaweeds. II. Fractionation partial characterization of the polysaccharide with antitumor activity from *Sargassum fulvellum*. *Jap. J. Exp. Med.* 47, 133–40.

Yamamoto, I., Nagumo, T., Takahashi, M., Fugihara, M., Suzuki, Y. and Iizima, N. (1981) Antitumor effects of seaweeds III. Antitumor effect of an extract from *Sargassum kjellmanianum*. *Jap. J. Exp. Med.* 51, 187–9.

Yamamoto, I., Takahashi, M., Tamura, E. and Maruyama, H. (1982) Antitumor activity of crude extract from edible marine algae against L-1210 leukemia. *Bot. Mar.* 25, 455–7.

Yamamoto, I., Takahashi, M., Tamura, E., Maruyama, H. and Mori, H. (1984a) Antitumor activity of edible marine algae: effects of crude fucoidan fractions prepared from edible brown seaweeds against L1210 leukemia. *Hydrobiologia* 116/117, 145–8.

Yamamoto, I., Takahashi, M., Suzuki, T., Seino, H. and Mori, H. (1984b) Antitumor effects of seaweeds IV. Enhancement of antitumor activity by sulfation of a crude fucoidan fraction from *Sargassum kjellmanianum*. *Jap. Exp. Med.* 54, 143–51.

Yoo, H.-D. and Gerwick, W.H. (1995) Curacins B and C, new antimitotic natural products from the marine cyanobacterium *Lyngbya majuscula*. *J. Nat. Prod.* 58, 1961–5.

Zhang, L.-H., Longley, R.E. and Koehn, F.E. (1997) Antiproliferative and immunosuppressive properties of Microcolin A, a marine derived lipopeptide. *Life Sci.* 60, 751–62.

Zhang, X. and Smith, C.D. (1996) Microtubule effects of welwistatin, a cyanobacterial indolinone that circumvents multiple drug resistance. *Mol. Pharmacol.* 49, 288–94.

Zhuang, C., Itoh, H., Mizuno, T. and Ito, H. (1995) Antitumor active fucoidan from the brown seaweed, umitoranoo (*Sargassum thunbergii*). *Biosci. Biotechnol. Biochem.* 59(4), 563–7.

World Wide Web sites

http://www.alohatropicals.com Aloha Tropicals
http://www.bali-travelnews.com Bali Travel News
http://biotech.icmb.utexas.edu/botany BioTech
http://www.botany.com Botany.com
http://www.chemfinder.com ChemFinder.com
http://www.colciencias.gov.co/simbiosis/ingles/projects.html Colciencias
http://www.desert-tropicals.com Desert-Tropicals.com
http://django.harvard.edu Django
http://.epnws1.ncifcrf.gov.2345
http://ericir.syr.edu
http://www.floridata.com Floridata
http://www.gi.alaska.edu Geophysical Institute
http://healthcentral.com Health Central
http://www.hortpix.com Horticopia® Plant Information
http://www.hort.purdue.edu Horticulture and Landscape Architecture
http://www.ird.nc IRD
http://lamington.nrsm.uq.edu.au Lamington National Park Website
http://www.ncbi.nlmnih..../query.fcgi?cmd National Centre for Biotechnology Information
http://www.npwrc.usgs.gov Northern Prairie Wildlife Research Centre
http://www.pasificrimginseng.com/world.html
http://www.pfc.cfs.nrcan.gc.ca/ecosystem/yew/taxol.html Pacific Forestry Centre
http://www.pubmed Pubmed
http://www.rsnz.govt.nz Royal Society of New Zealand
http://www.sare.org Sustainable Agriculture Network
http://scisun.nybg.org
http://www.streetside.com Streetside.com
http://stuartxchange.org Stuart Xchange
http://www.Taxol.com/timeli.html Taxol
http://www.tropilab.com Tropilab® Inc
http://uvalde.tamu.edu Agriculture Research & Extension Center-Uvalde
http://www.uog.edu University of Guam
http://www.vedamsbooks.com/no13147.htm Vedams
http://wiscinfo.doit.wisc.edu/herbarium University of Wisconsin

Index